AN INTRODUCTION TO METAMORPHIC PETROLOGY

LONGMAN EARTH SCIENCE SERIES
Edited by Professor J. Zussman and Professor W. S. MacKenzie,
University of Manchester

B. C. M. Butler and J. D. Bell: Interpretation of Geological Maps
P. R. Ineson: Introduction to Practical Ore Microscopy
Bruce W. D. Yardley: An Introduction to Metamorphic Petrology

Bruce W. D. Yardley is co-author of *Atlas of Metamorphic Rocks and their Textures* also published by Longman.

AN INTRODUCTION TO METAMORPHIC PETROLOGY

Bruce W. D. Yardley

 LONGMAN

Addison Wesley Longman Limited
Edinburgh Gate
Harlow
Essex CM20 2JE, England
and Associated Companies throughout the world

First published 1989
Reprinted 1990, 1991, 1993, 1994, 1995, 1996, 1997

British Library Cataloging in Publication Data
Yardley, B. W. D. (Bruce W. D.)
    An introduction to metamorphic petroloy
    1. Metamorphic rocks. Petrology
    I. Title
    552'.4
    ISBN 0–582–30096–7

Library of Congress Cataloging-in-Publication Data
A catalog entry for the title is available from the Library of Congress.

Set in Linotron Ehrhardt

Produced by Longman Singapore Publishers (Pte) Ltd.
Printed in Singapore

To my wife, Nick

CONTENTS

Preface   xi

CHAPTER 1   THE CONCEPT OF METAMORPHISM   1
The development of modern ideas of metamorphism   1
Types of metamorphic change   5
Metamorphic studies in geology   6
   Some examples of metamorphism   7
The settings of metamorphism   12
The controlling factors of metamorphism   13
Terminology of metamorphic rocks   21
Further reading   27

CHAPTER 2   CHEMICAL EQUILIBRIUM IN METAMORPHISM   29
Equilibrium – an introduction   29
   The phase rule   30
Metamorphic phase diagrams   32
Application of the phase rule to natural rocks   33
Metamorphic reactions – some first principles   34
Metamorphic reactions – some second thoughts   38
The influence of fluids on metamorphic phase equilibria   41
Application of chemical equilibrium to natural rocks: an example   42
Evidence for equilibrium in metamorphism: summary and critique   46
Metamorphic facies   49
Determination of pressure–temperature conditions of metamorphism   51
Summary   58
Further reading   58

CHAPTER 3   METAMORPHISM OF PELITIC ROCKS   60
Representation of pelite assemblages on phase diagrams   60
Pelitic rocks at low grades   63
Metamorphism of pelite in the Barrovian zonal scheme   64
Variations on the Barrovian zonal pattern   74
   High temperature metamorphism of pelites   74
   Metamorphism of pelites at low pressures   78
   Metamorphism of pelites at high pressures   82
Pressures and temperatures of metamorphism of pelitic rocks   85

CHAPTER 4   METAMORPHISM OF BASIC IGNEOUS ROCKS   91
The facies classification   92
Metamorphism of basic rocks at low grades: zeolite and prehnite–pumpellyite
    facies   95
Metabasites from the Barrovian zones: greenschist and amphibolite facies   99
Effects of lowered pressure: hornfels facies   101
Basic igneous rocks metamorphosed at high pressures: blueschist and eclogite
    facies   102
High temperature metamorphism: granulite facies   109
The *P–T* conditions of formation of metabasic rock types   113
Hydrothermal metamorphism of basaltic rocks   120

CHAPTER 5   METAMORPHISM OF MARBLES AND CALC-SILICATE
              ROCKS   126
Marbles   127
    Calcite marbles   127
    Dolomitic marbles   129
Controls on the fluid composition in marbles   133
A petrogenetic grid for reactions in marbles   139
Metamorphism of calc-silicates   141
Summary and discussion   145

CHAPTER 6   METAMORPHIC TEXTURES AND PROCESSES   147
Metamorphic textures – the underlying principles   147
    Diffusion in solids   150
    Nucleation and growth of mineral grains   151
The textures of metamorphic rocks   154
    Textures of recrystallisation   154
    Textures of crystallisation   158
Disequilibrium textures   161
Metamorphic textures as a guide to the mechanisms of metamorphic reactions   164
The influence of rock deformation on metamorphic textures and processes   167
Relationships between metamorphism and deformation   170
    Metamorphic textures and the relative timing of metamorphism and
        deformation   170
    Interactive relationships between metamorphism and deformation   175
Rates of metamorphic processes   177
The duration of a metamorphic cycle   178
Rates of metamorphic reactions   181
Further reading   186

CHAPTER 7   THE RELATIONSHIPS BETWEEN REGIONAL
              METAMORPHISM AND TECTONIC PROCESSES   187
Metamorphism, geothermal gradient and paired metamorphic belts   188
Plate tectonic interpetation of paired metamorphic belts   191
Modern convergent margins: implications for metamorphism   192
Time as a variable in metamorphism   197
Preservation of high pressure rocks after metamorphism   202
Tectonic setting of low pressure metamorphism   204
Metamorphism and continental collision   207

Contents   ix

Metamorphism related to ophiolites   213
Variation in metamorphism through geological time   214
Summary and conclusions   215

APPENDIX   SCHREINEMAKERS METHODS FOR THE
            CONSTRUCTION OF PHASE DIAGRAMS   217

References   223

Glossary of mineral names and abbreviations used in the text   238

Index   242

PREFACE

Metamorphic processes have been taking place on a massive scale throughout the Earth's history, and have affected the bulk of the rocks now present in the crust. Despite this, they are not as well understood as sedimentary or volcanic processes, because metamorphism can scarcely ever be observed directly, and the study of metamorphic rocks is instead based on observation, inference and logic, founded in relatively simplistic experimental studies and the basic principles of chemistry and physics.

In writing this book I have attempted to give an idea of the deductive methods of the metamorphic geologist, as well as introducing some of the known facts and currently fashionable hypotheses about metamorphic rocks. My aim has been to provide a broad overview of the subject for the student who will only take a single course in metamorphic petrology and needs to know how it is relevant to other areas of geology, while at the same time providing an adequate introduction for those who may go on to make their own contributions to the field. The largest part of the book is devoted to metamorphic mineral assemblages that develop in particular rock types under particular conditions, and is essentially based on the premise that rocks react until they are in chemical equilibrium during metamorphism. In reality this is not always the case, but even so the equilibrium approach serves to define the goalposts towards which natural processes are moving, and hence it is an essential pre-requisite for the study of metamorphic processes which is introduced next. Finally, metamorphic petrology is not just of interest for its own sake; it also plays an important role in attempts to study the past tectonic behaviour of the crust, and so the last part of the book deals with the very rapidly developing subject of tectonic causes of metamorphism.

The most difficult problem to be faced in writing a book on metamorphic petrology is deciding to what extent to introduce the theoretical chemistry background of much modern work. On the one hand a good knowledge of chemical thermodynamics is essential for anyone who plans to do research in metamorphic petrology, but on the other hand it is perfectly possible to understand the general methodology, aims and achievements of metamorphic petrology without taking a course in thermodynamics, and many geologists who require only a general knowledge of the subject will legitimately expect a book to treat the subject at this level in the first instance. My aim has been to write a text that will be comprehensible to the student who has little knowledge of thermodynamics, while at the same time introducing some of the key thermodynamic variables and showing their relevance to metamorphic petrology.

Metamorphic petrology research is likely to remain a relatively small field of endeavour within the geological sciences for the foreseeable future, but I hope that I will be able to persuade my readers that metamorphism is not such a complex, difficult or illogical subject as is often supposed by students, and instead communicate some of the excitement of modern research, and show that it can help answer some of the key questions posed in other areas of earth sciences.

Bruce Yardley
Leeds, February 1988

ACKNOWLEDGEMENTS

Work on this book was started while I was on the staff of the University of East Anglia, and I am indebted to my students there for their patience and critical comments as the approach I have adopted was being developed. I would also like to acknowledge Brian Chadwick, Bernard Leake and especially Bernard Evans for inspiring my interest and helping me to understand various aspects of metamorphic rocks.

Thanks to Sue Winston, Pauline Blanch and Leslie Enoch for typing, and David Mew and Richard Hartley for assistance with some of the figures. In addition to the support of the Universities of East Anglia and Leeds, I also thank the Institut für Mineralogie und Petrographie, E.T.H. Zurich, where Chapter 6 was largely written.

The manuscript was greatly improved by the perceptive comments of series editor Professor W. S. MacKenzie and of Giles Droop, and in addition Bob Cliff, John Ridley, Rob Knipe and Casey Moore made invaluable comments on individual chapters. The remaining mistakes and obscurities are my own.

Finally, I would like to thank my wife, Nick, without whose promptings this book would never have been completed, and the editorial and production staff at Longman for their patience and faith that a manuscript would eventually appear, and for their help and efficiency in turning it into a book.

We are grateful to the following for permission to reproduce copyright material:

Academic Press Inc., Florida and the author for fig. 7.2 from fig. 3 (Karig 1983); American Journal of Science and the authors for figs. 2.4 from fig. 2 (Trommsdorff & Evans 1972), 3.13 from fig. 2 (Carmichael 1978) and 4.9a & b from figs. 2A & 2B (Laird & Albee 1981); The Geological Society and the authors for figs. 5.10 from fig. 1 (Ferry 1983b) and 7.5 from fig. 1 (England & Richardson 1977); The Royal Society and the authors for figs. 7.3 from figs. 3b & 9 (Walcott 1987), 7.8 from fig. 6b (Platt 1987) and 7.9 from figs. 7 & 10 (Yardley, Barber & Gray 1987); Springer-Verlag, Heidelberg and the authors for figs. 2.11a from fig. 3 (Ferry & Spear 1978), 6.6b from fig. 1c (Yardley 1977b) and 7.6 from fig. 6 (Yardley 1982); Springer-Verlag, New York for fig. 6.14 from fig. 3d (Yardley 1986); the author, Prof. V. Trommsdorff for fig. 5.7b from fig. 2 (Trommsdorff 1972); University of Chicago Press and the author for figs. 6.16a & b from fig. 9 (Carlson & Rosenfeld 1981) © 1981 University of Chicago Press.

1 THE CONCEPT OF METAMORPHISM

Metamorphism means the processes of change, whereby a rock that formed originally in an igneous or sedimentary environment recrystallises in response to new conditions, to produce a **metamorphic rock**. Most metamorphic rocks retain some of the characteristics of the parent material, such as bulk chemical composition or gross features such as bedding, while developing new textures and often new minerals. Examples of rocks that retain some obvious parental features but have also developed new metamorphic minerals and textures are shown in Fig. 1.1.

There is a wide variety of processes that can cause metamorphism, ranging from progressive burial and consequent heating of thick sedimentary sequences, through igneous activity to the rare impacts of large meteorites with the earth's surface. However, most metamorphism probably occurs in the vicinity of active plate margins, and the nature of this relationship is explored in the last part of this book.

Although it is not normally possible to see metamorphism taking place in the same way that we can watch some igneous or sedimentary rocks being formed, there are certain types of metamorphism occurring near the surface today. For example in high temperature geothermal fields, such as those exploited for power in Iceland, Italy, New Zealand and elsewhere, volcanic glass and high temperature basalt minerals are being actively converted to clays, chlorite, zeolites, epidote and other minerals that are more stable in the cooler, but wet, environment of the geothermal system. This metamorphism is taking place at depths of only a few hundred metres.

THE DEVELOPMENT OF MODERN IDEAS OF METAMORPHISM

Modern ideas about the origins of metamorphic rocks can be traced back at least as far as Hutton, whose *Theory of the Earth* published in 1795 had a profound influence on later geological thought. Hutton recognised that some of the 'Primitive Strata' of the Scottish Highlands had originally been sediments and had been changed by the action of heat deep in the earth. In the early nineteenth century a distinction was recognised between **contact metamorphism**, which constitutes the changes caused by heating in the immediate vicinity of an igneous intrusion, and **regional metamorphism**, which takes place over a large area, has no distinct focus, and is usually accompanied by intense deformation of the rocks concerned. The study of metamorphic rocks was brought to a wider audience by the publication of Lyell's *Principles of Geology* in 1833, and by the middle of the century some geologists mapping metamorphic rocks in the field had clearly reached a sophisticated

Fig. 1.1 Examples of metamorphic rocks retaining primary sedimentary or igneous features. *a*) Metamorphosed thinly bedded sediments with cross-laminated siltstones preserving original sedimentary structures alternating with original clay beds, now pelitic schist. The dark spots in the schist layers (some examples are arrowed) are crystals of staurolite, indicating that metamorphic temperatures reached at least 550 °C. Rangeley district, Maine, USA. *b*) Metamorphosed pillow-lavas retaining their original pillow shape undeformed. These rocks contain abundant pumpellyite, which gives the outcrop a distinctive blue–green colour. Riverton, South Island, New Zealand.

1.1c) Porphyritic diabase (or dolerite) displaying intense hydrothermal metamorphism in the vicinity of fractures. The original fractures are infilled with pale carbonate, and the adjacent rock is distinctly darker in colour due to growth of chlorite and breakdown of feldspar. Lake Cushman, Olympic Peninsula, Washington, USA.

level of understanding of their problems. For example the pioneer Irish geologist Patrick Ganly described the use of sedimentary cross-bedding structures to determine the way-up of deformed beds in 1856.

In the latter part of the nineteenth century several schools of thought arose as to the main causes of metamorphism, and the influence of most of them can still be seen today.

Hutton had emphasised the role of heat, and was probably the first geologist to appreciate that when rocks are heated under the pressures that pertain inside the earth, vapours are not as readily given off as under surface conditions because of the great pressure. The conclusion was tested experimentally by a friend of Hutton, Sir James Hall, who had powdered chalk sealed into a cannon barrel that was then heated in the furnace of his iron foundry. On opening, the barrel was found to contain coarsely crystalline marble rather than the calcium oxide that would have formed on heating at atmospheric pressure. Hutton's influence can be seen a century later in the work of Barrow who made one of the first systematic studies of the variation in degree of metamorphism, across a region in the Scottish Highlands, and ascribed the metamorphism to the heat from granite intrusions.

Many German and Swiss geologists followed von Rosenbusch in emphasising the role of pressure and deformation in causing regional metamorphism. Grubenmann popularised a classification scheme in which different types of metamorphism were assigned to three depth zones. The upper **epizone** was characterised by low temperature but intense deformation; the intermediate **mesozone** by moderate temperature and deformation; and the lower **catazone** by high temperatures but little deformation. The term **anchizone** is sometimes used for the region transitional between diagenesis and metamorphism. The role of deformation was also emphasised in the 1920s by the British petrologist Harker, who believed that some minerals which he termed 'stress minerals' could only form in rocks that were undergoing deformation, and hence were restricted to regionally metamorphosed rocks. This concept has been shown to be almost entirely fallacious by modern experimental studies.

A third, predominantly French, school emphasised the role of fluids and emanations, and believed that they often caused considerable changes in rock chemistry during metamorphism. This approach has sometimes been very popular and is seen in the debates about the origin of granite in the 1940s and 1950s. Perhaps it is not surprising that today we are indebted to French scientists for pioneering the study of trapped bubbles of metamorphic fluids (**fluid inclusions**) in the minerals of metamorphic rocks.

Our present understanding of the conditions under which metamorphism takes place is very largely due to experimental studies carried out at high pressures and temperatures since the early 1950s. However, the origins can be traced back to the beginning of the century, and in particular to the work of Goldschmidt in the Oslo region of Norway (Goldschmidt, 1911). In 1912 Goldschmidt used the new methods of thermodynamics to calculate the conditions at which wollastonite would form at the expense of calcite and quartz, and hence the temperature of formation of wollastonite-bearing rocks. This pre-empted by over 50 years the widespread application of such methods in metamorphic petrology.

Shortly afterwards, the Finnish geologist Eskola made a study of metamorphism in south-west Finland in which he also applied the principles of chemical equilibrium to the interpretation of mineral assemblages. Comparison of his results with those of Goldschmidt led him to develop a scheme of **metamorphic facies** (Chs 2 and 4) whereby different metamorphic mineral assemblages were recognised as characterising different pressure and temperature regimes of metamorphism. He established the relative temperatures and pressures of the different facies and was able to show that the glaucophane schists, assigned by Grubenmann to the epizone, had in fact formed at considerable depths. Eskola's ideas put forward in the 1920s were only slowly accepted but are seen now as central to the understanding of metamorphic rocks.

The major developments of the latter part of the twentieth century are based on new technology. Many metamorphic minerals and **mineral assemblages** (i.e. associations of coexisting minerals) have been synthesised under carefully controlled experimental conditions and the thermodynamic properties of a number of minerals are now quite well known, providing additional insights into the conditions of their formation. New, rapid and sophisticated techniques for the analysis of natural rocks and minerals are available, while widespread age dating has permitted much better understanding of the evolution of regionally metamorphosed rock masses. Developments in tectonics have provided a framework in which to begin to investigate the underlying causes for the formation of particular kinds of metamorphic rocks. In recent years the major emphasis in metamorphic studies has been on determining the pressure and temperature of formation of metamorphic rocks, and many other important problems are only just beginning to be tackled. For example the rates of metamorphism, the way in which metamorphic conditions evolve with time, the relation of deformation and tectonics to metamorphism and the behaviour of fluids during metamorphism.

TYPES OF METAMORPHIC CHANGE

Rocks undergo two principal types of readily observable change during metamorphism: replacement of original minerals by new **metamorphic minerals** due to chemical reaction (i.e. **phase changes**) and recrystallisation of minerals to produce new textures such as the alignment of platy minerals in slate or the progressive coarsening of limestone to marble (**textural changes**). These sorts of change may occur together or take place more or less independently, according to the cause of metamorphism and the type of rock involved.

One important feature of all metamorphism is that it takes place essentially in the **solid state**. In other words the rock is never completely disaggregated (as would be the case if extensive melting occurred) and so features such as original compositional layering of sediments are retained, though they are often distorted and disrupted by deformation. The description of metamorphism as comprising solid state changes should not be taken too literally: small amounts of fluid may be present in the pore spaces of rocks during metamorphism and these fluids are believed to play a very important role in facilitating metamorphic changes.

It has been found in many studies that metamorphic rocks have the same chemical composition as their sedimentary or igneous parents, except for removal or addition of volatile species (usually H_2O). Metamorphism of this type is known as **isochemical metamorphism**, and is illustrated by the analyses in Table 1.1. Sometimes, however, the chemical composition of a rock is changed more extensively as it recrystallises. This usually happens where hot solutions are able to circulate freely through rocks, dissolving some substances and precipitating others, or where two adjacent original layers are of contrasting chemical compositions and react together at their interface. The process of chemical change during metamorphism is known as **metasomatism** and the rocks so produced are **metasomatic rocks**.

Table 1.1   Comparison of the chemical compositions of metamorphic rocks and their sedimentary or igneous precursors

| | PELITIC SEDIMENTS | | | | BASIC ROCKS | | | |
| | Low grade slates | | High grade schists | | Representative analyses | | | |
	Mean	σ	Mean	σ	1	2	3	4
SiO_2	59.93	6.33	63.51	8.94	43.01	51.98	50.92	49.17
TiO_2	0.85	0.57	0.79	0.67	2.99	1.21	0.60	1.04
Al_2O_3	16.62	3.33	17.35	5.08	14.55	14.48	16.83	18.02
Fe_2O_3	3.03	2.08	2.00	1.66	5.60	1.37	1.11	1.64
FeO	3.18	1.84	4.71	2.44	8.55	8.92	9.78	7.09
MgO	2.63	1.98	2.31	1.82	8.45	7.59	7.99	10.67
CaO	2.18	2.54	1.24	0.92	10.21	10.33	9.87	10.64
Na_2O	1.73	1.27	1.96	1.06	1.82	2.04	1.15	2.74
K_2O	3.54	1.33	3.35	1.31	1.02	0.84	1.12	0.63
H_2O	4.34	2.38	2.42	1.53	2.87	0.88	0.96	—
CO_2	2.31	2.60	0.22	0.22	—	—	—	—

Sources   Pelitic sediments: mean analyses from Shaw (1956), with standard deviations.
Basic Rocks: 1. Olivine tholeiite, Principe Island, Gulf of Guinea (Barros, 1960).
2. Chilled diabase (dolerite), base of Palisades Sill, New Jersey (Walker, 1969).
3. Eclogite, Lyell Highway, Tasmania (Spry, 1963a).
4. Ophitic metadolerite, greenschist facies, south-west Highlands of Scotland (Graham, 1976). This analysis is totalled to 100 per cent on an anhydrous basis.

METAMORPHIC STUDIES IN GEOLOGY

Many geologists tend to think that metamorphism begins with the growth of large and attractive crystals in rocks! However, logical definition of the *scope of metamorphism* must include *any rock that has recrystallised without totally melting so that it now either contains minerals not stable in the original sedimentary or igneous environments in which it first formed, or has developed new textures*. In practice, changes that take place during deep burial and diagenesis of sediments are conventionally excluded from the scope of metamorphism, while processes that lead to the formation of ore deposits are also usually treated separately. The boundaries of the subject of metamorphism are therefore to some extent arbitrary, and certain sorts of metamorphic processes can be treated scientifically in exactly the same way as analogous processes in ore formation or diagenesis.

The understanding of metamorphism has advanced more slowly than that of igneous and sedimentary processes, perhaps because we cannot usually see metamorphism taking place and are reduced to inferring what has probably happened from the rocks produced. On the other hand metamorphism is extremely important to our understanding of the way the earth behaves, because the majority of the rocks of the earth's crust have undergone metamorphism of one sort or another. Thanks to the technological developments of the past 30 years we can often obtain a good idea of the sorts of physical conditions under which particular metamorphic rocks have formed, and this is proving an invaluable source of information about past conditions in the earth's crust.

SOME EXAMPLES OF METAMORPHISM

Before tackling some of the rather abstract concepts of how and why metamorphism takes place, it is worth having some sort of appreciation of how metamorphic rocks occur in the field, and what they actually look like. In this section some contrasting areas of metamorphic rocks are described to illustrate some of the possible modes of occurrence. Although much modern metamorphic petrology is based on theoretical or experimental studies, the laboratory work is of little value unless it relates to the compositions and physical conditions of formation of natural rocks, and so field studies will always be the cornerstone of the subject.

The southern Highlands of Scotland

One of the classic areas for metamorphic geology is in the south-east Highlands of Scotland, shown in Fig. 1.2. It was here that G.M. Barrow (1893, 1912) made one of the first studies to show a systematic variation in the mineralogy of metamorphic rocks, which could be clearly related to changing conditions (of temperature in particular) to which the rocks had been subjected. Subsequently, Barrow's work was refined by Tilley (1925) and extended to the south-west.

The rocks of the area are metasediments (with some basic igneous rocks) of the Eocambrian–Cambrian Dalradian Supergroup. Original quartz-rich sandstones show relatively little change across the area, but Barrow found that the **pelitic rocks** (metamorphosed argillaceous sediments) could be divided into a series of **metamorphic zones**, each characterised by the appearance of a new metamorphic mineral produced as the metamorphism became more intense. The mineralogical changes are accompanied by a general coarsening of the grain size as the pelites progress from fine grained slaty rocks to the coarse schists. The slates are the least changed from the original sediments and are

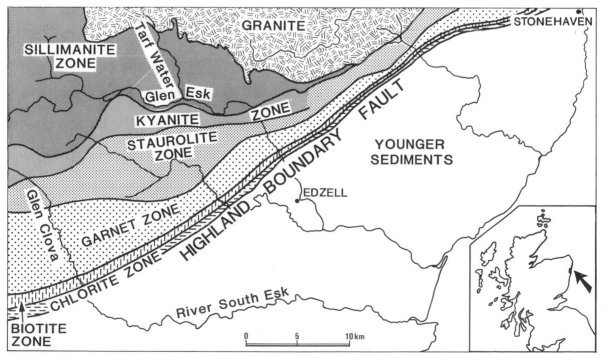

Fig. 1.2 Metamorphic zones defined by the mineralogy of pelitic schists in the eastern Highlands of Scotland. Based on Tilley (1925).

referred to as **low grade** rocks. (**Grade** is a loosely defined term used in a relative sense to refer to the extremity of conditions of temperature, or more rarely pressure, to which a rock has been subjected. A high grade rock has been metamorphosed at temperatures approaching those of some magmas.) Barrow referred the lowest grade slaty rocks to a 'zone of undigested clastic micas', but this was renamed by Tilley as the 'chlorite zone'. Successive metamorphic zones are recognised by the appearance of new minerals not present at lower grades. The sequence of zones now recognised is as follows:

Chlorite zone: pelites typically contain chlorite with muscovite, quartz and albite.
Biotite zone: marked by the appearance of red–brown or green–brown biotite; chlorite, muscovite, albite and quartz usually persist.
Garnet zone: red almandine garnet is typical and is often conspicuous in hand specimen. At this grade pelitic rocks are schists and usually contain biotite, chlorite, muscovite, quartz and albite or oligoclase feldspar in addition to garnet. However, chlorite may be absent.
Staurolite zone: staurolite occurs with biotite, muscovite, quartz, a sodic plagioclase feldspar and usually garnet. A little chlorite may persist.
Kyanite zone: kyanite schists contain biotite, quartz, plagioclase, muscovite and garnet in most cases. Staurolite is also often present and may form complex intergrowths with kyanite.
Sillimanite zone: sillimanite occurs with biotite, muscovite, quartz, plagioclase, garnet and sometimes staurolite. Kyanite may also be present even though kyanite and sillimanite both have the same chemical composition (Al_2SiO_5).

Under the microscope it is readily apparent that in addition to the major silicate phases listed above, metamorphosed pelitic rocks contain accessory minerals such as tourmaline, apatite and zircon, and often significant amounts of opaque minerals of which magnetite, ilmenite, pyrite, pyrrhotite and graphite are the most common.

The minerals that characterise each zone (e.g. chlorite, biotite, etc.) are termed **index minerals**, but note that each index mineral persists to higher grades than the zone which it characterises. The way in which the zones were mapped was to plot the location of the different index minerals on a map and to draw lines through the *first appearance* of each index mineral in turn with increasing grade. Some index minerals will appear at somewhat different temperatures in layers of slightly different composition, and so this method of drawing the boundaries was intended to smooth out the effects of random variation in rock composition. It means that, for example, some pelitic rocks from the garnet zone will not in fact contain garnet, and will be indistinguishable from biotite zone rocks (Fig. 1.3). Tilley termed the zone boundaries **isograds**, meaning lines of constant metamorphic grade. Isograds drawn in this way are known as **mineral appearance isograds**. Because this type of isograd cannot be drawn accurately in areas where rock compositions vary, it is now customary to try to define isograds more rigorously by basing them on the full mineral assemblage* of the rocks, rather than on the appearance of individual index materials.

This study of metamorphic zones illustrates a very important concept in metamorphism, that of **progressive metamorphism**. It is generally assumed that in most sorts of metamorphism the higher grade rocks formerly had mineral assemblages typical of lower grade zones, and have progressively recrystallised and changed their mineralogy as metamorphism proceeded, rather than being converted directly from, say, unmetamorphosed sediment to high grade schist. This assumption has been periodically challenged, but there is a lot of evidence to support it. For example, minerals typical of lower grade zones may be preserved as **inclusions** inside larger grains of other minerals, and so persist into

* A particular assemblage of coexisting minerals is sometimes also known as a **mineral paragenesis**.

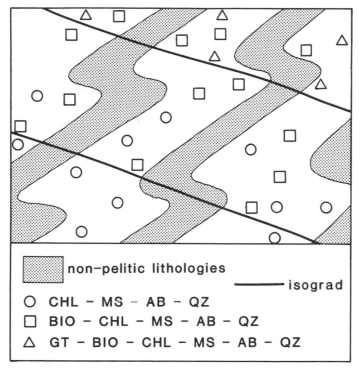

Fig. 1.3 Schematic map illustrating the distribution of mineral assemblages (shown by open symbols) of pelitic lithologies in a regional metamorphosed sequence, and the mineral-appearance isograds drawn on the basis of the assemblage information. (See Glossary for mineral abbreviations.)

higher grade zones where they would not otherwise be present. This is a topic that will be returned to in Chapter 6.

The metamorphism in the south-east Highlands is of a type known as **regional metamorphism** because it affects a large area and, contrary to the original opinion of Barrow, has no specific focus such as an igneous intrusion. The fact that granites occur in the area of Fig. 1.2 is now believed to be coincidental. The zonal scheme found in these Scottish pelites is often referred to as **Barrovian zoning** and has been reported from many parts of the world.

The eastern Ligurian Apennines of north-west Italy   The rocks of eastern Liguria include an **ophiolite suite** of altered ultrabasic rocks, metamorphosed gabbros locally cut by basic dykes, pillow lavas and a sedimentary sequence including cherts, turbidites and shales. This suite has been interpreted as a slice of oceanic crust that has become faulted into the continental crust during orogeny. The metamorphism of a 225 m thick sequence of lava flows and feeder dykes was described by Spooner & Fyfe (1973) who reported the following zonal sequence from the stratigraphic top of the lavas downwards.

Red pillow lavas: characterised by abundant red hematite. Albite, smectite clays, calcite and sphene are also present and chlorite occurs in the middle and lower part of the zone. In the lowermost part pumpellyite appears.

Blue–green pillow lavas: in this zone smectite is absent and calcite and hematite only occur in minor amounts. The rock's colour reflects the abundant chlorite and pumpellyite.

Yellow–green pillow lavas: here epidote and actinolite have appeared, although chlorite, pumpellyite, albite and sphene persist.

Feeder dykes; at the base of the series hornblende appears with actinolite in the dykes, and magnetite replaces hematite.

This sequence of metamorphic zones is very different from the previous example and part of the reason is the different types of parent-rock involved. In this example Ca-bearing minerals such as calcite, pumpellyite and amphiboles are important because basic lavas are rich in Ca, whereas in the pelitic rocks K- or Al-rich minerals such as micas, staurolite and kyanite are characteristic.

However, although the area forms part of a complex regional belt of deformed rocks, the metamorphism probably has a different origin from that in the Scottish Highlands. Also affected by the deformation are breccias which overlie, and are intercalated with the upper part of this pillow lava sequence. These breccias contain fragments of the metamorphosed basalts *even though the breccia matrix is not metamorphosed*. This means that the lavas were metamorphosed very shortly after they were formed and before the deposition of the breccias. For this reason the metamorphic sequence in eastern Liguria is believed to be an example of **sea-floor metamorphism** – metamorphism caused by hot hydrothermal fluids circulating in newly formed oceanic crust. It is also worth noting that in at least the upper part of the sequence the rocks have been subjected to some metasomatism as well as metamorphism, because fresh lavas are far too reduced to develop hematite. Abundant oxygen, together with CO_2 and H_2O has been added to the rock from the circulating fluid, and it seems probable that other chemical changes may have occurred also.

Glinsk,
Co.Donegal,
Ireland

A third type of metamorphism, again affecting pelitic rocks of the Dalradian Supergroup as in the first example, is present near Glinsk, Ireland, where it occurs in a narrow region around a body of granite – the Fanad Pluton (Fig. 1.4). The metamorphism of this area was studied in detail by W.M. Edmunds and is described by Pitcher & Berger (1972).

The Dalradian rocks of Donegal underwent regional metamorphism similar to that of Scotland, and at Glinsk garnet zone conditions were attained. Subsequently, the garnets were partially replaced by chlorite. (The process of converting minerals formed at high temperature back to minerals characteristic of lower grades is known as **retrograde metamorphism**.)

The Fanad Pluton was emplaced after this episode of retrograde metamorphism. Its exact size is not known but it outcrops for 15 km along the coast. Distinctive metamorphic effects occur for a distance of up to 1750 m away from the granite.

The sequence of metamorphic zones is best seen along the east shore of Mulroy Bay where a single band of pelite runs northwards into the granite, and is illustrated in Fig. 1.4. The first change (visible in thin section) on approaching the granite is the growth of new biotite, some of which grows around the old garnet grains. This is followed by the appearance of small new grains of garnet around the old grains. Very close to this isograd andalusite appears while chlorite ceases to be present. Successive isograds mark the appearance of cordierite and of K-feldspar with sillimanite (initially in a fine-grained, fibrous form). Muscovite is no longer present in the sillimanite zone.

It is immediately apparent from the field relationships at Glinsk that the metamorphic zones are directly related to the intrusion of the granite; these rocks constitute a **meta-morphic aureole** around the granite which in this case was clearly the source of heat that caused the metamorphism. Hence this type of metamorphism is known as **contact metamorphism**.

Despite the similarity in the chemical composition of the rocks involved, the minerals produced in the Fanad aureole are not all the same as those in the south-east Highlands of Scotland. Cordierite and andalusite occur in the aureole, whereas staurolite is rare and

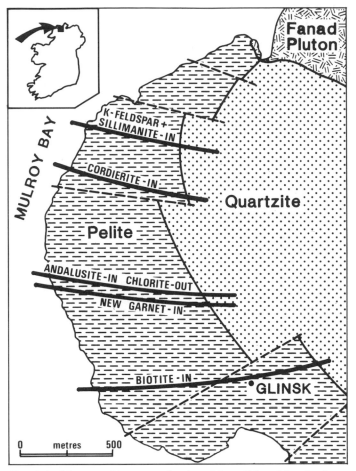

Fig. 1.4 Metamorphic zones developed in pelitic schists in the aureole of the Fanad pluton, Donegal, Ireland. Simplified from the work of W.M. Edmunds in Pitcher & Berger (1972).

kyanite absent. This suggests that the conditions of metamorphism were different in the two areas. In both examples there is a broad range of metamorphic grades present, with sillimanite at the highest grades, and it is therefore reasonable to assume that the temperature ranges of metamorphism of the two areas overlap. The mineralogical differences more probably reflect different depths, and hence pressures, of metamorphism. We can confirm this by looking at the different Al_2SiO_5 polymorphs that are found. In the southern Highlands of Scotland, kyanite is succeeded at higher grade by sillimanite, whereas at Fanad it is andalusite that is the first Al_2SiO_5 mineral to appear. Now the density of kyanite is $3.6 \, g/cm^3$ whereas that of andalusite is only $3.15 \, g/cm^3$. The effect of increased pressure is to favour the formation of relatively dense minerals, all else being equal. In this case, both andalusite and kyanite have the same chemical composition and appear to form over a similar temperature range, and so we can infer that the contact metamorphism at Glinsk took place at lower pressure than the regional metamorphism in the south-east Highlands.

THE SETTINGS OF METAMORPHISM

The examples of metamorphic rocks just described demonstrate that metamorphism can take place in a variety of different settings in the crust, and so it is convenient to have some sort of a genetic classification to distinguish them. There is no universally adopted classification scheme because ideas about metamorphism have changed considerably in the past 20 years. For example it is only recently that the process of sea-floor metamorphism has become well known.

Regional metamorphism gives rise to the large areas of metamorphic rocks characteristic of many mountain chains. Heating occurs without a close association with particular igneous bodies, although intrusions may be present and contribute to the overall rise in temperature. Regional metamorphism is almost invariably accompanied by deformation and folding, and regional metamorphic rocks often have a planar fabric (*schistosity* or *cleavage* according to whether the individual grains are coarse and readily visible, or very fine) which results from the deformation. **Burial metamorphism** is a form of regional metamorphism that may take place when a very thick sedimentary or volcano-sedimentary succession develops in a subsiding basin, so that low grade metamorphic conditions are attained at the base of the pile even though there has been none of the accompanying deformation and folding typical of regional metamorphism.

Contact or Thermal metamorphism is the metamorphism that results from rise in temperature in the surrounding **country rocks** near to igneous intrusions. Typically, aureole rocks are not deformed during metamorphic recrystallisation and so the grains grow together in an interlocking manner to form a tough rock known as **hornfels**. However, many hornfelses retain some vestiges of planar fabrics formed during earlier regional metamorphism, while in some metamorphic aureoles deformation accompanied igneous intrusion and has produced metamorphic rocks that in hand specimen are more akin to the products of regional metamorphism. The products of contact metamorphism are very varied; those most commonly encountered have formed around granite plutons in the upper or middle crust and the Fanad aureole is representative. However, in certain local settings, such as where volcanic rocks bake sediment very close to the surface, distinctive and unusual rocks and minerals may result.

Dynamic or cataclastic metamorphism is usually even more local in its occurrence than contact metamorphism, and is the metamorphism that takes place along fault planes or shear zones as a result of the intense deformation of rock in the immediate zone of movement. Often the mechanical comminution is accompanied by recrystallisation or by growth of hydrous minerals due to movement of fluid into the zone of deformation. **Mylonites** are the typical product of this type of metamorphism, and are fine grained rocks which are normally foliated (below, page 22).

Hydrothermal metamorphism involves chemical change (metasomatism) as an integral part of the process and is the result of the circulation of hot water through a body of rock along fissures and cracks (as seen, for example, in Fig. 1(*c*)). This sort of metamorphism is often associated with igneous activity because steep temperature gradients such as those present around high level intrusions are required to drive fluid convection. It is an important process in geothermal fields and is also responsible for the formation of many economic mineral deposits such as the porphyry–copper type. However, it is likely that the most widespread type of hydrothermal metamorphism is sea floor metamorphism taking place at mid-ocean ridges. The rocks of the eastern Ligurian Alps, described above, are believed to have formed in this way, and contemporary sea-floor metamorphism is described in Chapter 4.

Impact metamorphism has no genetic relation to other sorts of metamorphism and is produced by the impact of large, high velocity meteorites on a planetary surface. On some bodies in the solar system (e.g. Mercury or the Moon), impact metamorphism is perhaps the major geological process, but while it may be no less frequent on earth, other processes are so much more important that examples of impact metamorphism are very rare and have probably not played a significant part in shaping the earth's history since the Archaean. The shock wave generated on meteorite impact passes out through the surrounding rocks, subjecting them to extremely high pressures for a fraction of a second as it passes. The relaxation of mineral lattices after the shock wave has passed, causes a rise in temperature that may lead to melting or even vaporisation. One of the best known examples of impact metamorphism is at Meteor Crater, Arizona, where among other effects some quartz grains in Cretaceous sandstones have been converted to the dense, high pressure polymorphs of silica, coesite and stishovite. Coesite is otherwise only found naturally in rocks that have recrystallised at mantle depths, e.g. as inclusions in diamond, and so the clear implication is that pressures at least as great as those found in the earth's mantle must have been temporarily attained at the surface when the impact occurred. Impact metamorphism will not be considered further in this book, but some further reading is suggested at the end of this chapter.

With the exception of impact metamorphism, the other types of metamorphism do show similarities with one another in terms of the rocks that they affect and the minerals produced. For example, despite the textural differences between contact and regionally metamorphosed rocks, many minerals may form in either environment. The regionally metamorphosed rocks of many orogenic belts are cut by zones of cataclasis and dynamic metamorphism, and may also include slices of sea-floor rocks in which regional metamorphic mineral assemblages are superimposed on the products of early sea-floor metamorphism. Thus these distinctions between ideal metamorphic processes do not define mutually exclusive categories and a single rock may have undergone more than one type of metamorphism.

THE CONTROLLING FACTORS OF METAMORPHISM

Metamorphism takes place when a rock is subjected to a new chemical or physical environment in which its existing mineral assemblage is no longer the most stable. A 'new chemical environment' may mean the infiltration of fluid which reacts with the rock. Changes in 'physical environment' could mean change in the temperature or pressure to which the rock is subjected, or application of unequal stresses leading to deformation and recrystallisation of the rock, producing new textures. This section is intended to clarify what some of these parameters mean.

TEMPERATURE (T)

Temperature is of course a measure of how hot substances are. Heat flows from bodies at higher temperatures to bodies at lower temperatures until the temperature differential is eliminated, but different substances require different amounts of heat to raise their temperatures by any given amount. For example, it takes about 45 **joules** (J) of heat to raise the temperature of 22 cm^3 of quartz from 25 °C to 26 °C but approximately 75 J to make the same volume of magnetite undergo the same temperature change. The amount of heat needed to raise the temperature of 1 mol of a substance by 1 °C (at constant pressure) is known as the **heat capacity**, Cp, and is given in units of J/mol/K.

It is conventional to use the absolute or Kelvin temperature scale (on which water boils at 373.15 K, and freezes at 273.15 K), when dealing with thermodynamic properties of minerals or performing thermodynamic calculations, but the common centigrade temperature scale, °C, for descriptive purposes. To all intents and purposes $T\mathrm{K} = T°\mathrm{C} + 273$.

Rocks are fairly good insulators because they are slow to conduct heat. Large volumes of rocks such as are involved in regional metamorphism usually need tens of millions of years to undergo large temperature changes.

Temperature almost invariably increases with depth in the earth, and the rate at which it changes with depth is known as the **geothermal gradient**. Because heat flows along temperature gradients the geothermal gradient is closely related to heat flow through the crust. Geothermal gradients are usually in the range 15–30 °C/km, but extremes from 5–60 °C/km do occur. Measurements of heat flow near the surface at the present day provide a guide to the way in which the geothermal gradient will differ between different tectonic settings. The picture provided by early results was reviewed by Oxburgh (1974) and is summarized in Fig. 1.5. A comprehensive review of subsequent data by Sclater, Jaupart & Galson (1980) shows that in general the heat flow in continental crust is higher than that in the older ocean basins, but young oceanic crust gives higher heat flow values because it is still cooling from its original formation at magmatic temperatures. The effect is most marked for oceanic crust younger than 40 Ma (millions of years before present), but can be detected in crust of at least 120 Ma. The reason for the regional variations in heat flow is that there are three major contributions to surface heat flow providing different inputs in different situations. They are: (a) heat flowing into the base of the crust

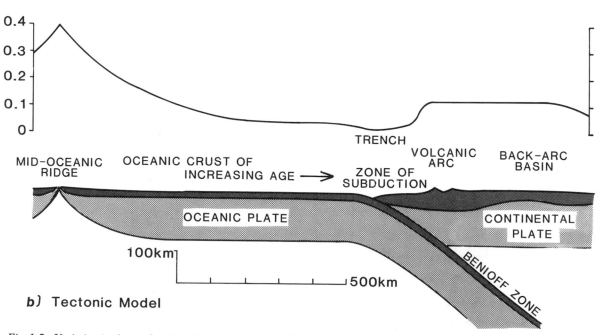

Fig. 1.5 Variation in the surface heat flow measured at different parts of the earth (*a*) shown in relation to plate tectonic setting (*b*). Compiled from Oxburgh (1974).

from the mantle: (b) heat generated by radioactive decay within the crust (greater for continental crust than oceanic crust); and (c) heat brought into the crust by rising bodies of magma. In addition, young mountain chains give the highest continental heat flow values, other than those associated with igneous activity, because of a fourth factor. The effect of rapid uplift and erosion is to bring up hot rocks to near the surface quickly, without much loss of heat. As a result, there is a steep temperature gradient, and hence high heat flow, near to the surface. Crustal stretching produces a similar effect.

In addition to heat inputs to the crust, the temperature gradient can also be affected by absorption of heat, for example by thick accumulations of cold sediment. The major sources and sinks of heat of relevance to metamorphism are summarised in Fig. 1.6.

Because of the way in which various factors contribute to heat flow in the crust, geothermal gradients vary somewhat with depth, and in general will tend to be steeper near the surface because of the greater amount of radiogenic heat. Below the lithosphere, mantle convection probably results in a relatively uniform temperature over considerable depths.

A **geotherm** is a curve on a graph of depth against temperature, showing the way in which temperature varies below a particular part of the earth's surface. It is made up of elements with different geothermal gradients. Two possible geotherms, for continental and oceanic regions, are shown in Fig. 1.7. In tectonically active areas where metamorphism is proceeding it is likely that temperature variations with depth will be much more complex and variable. As a matter of convenience, geotherms are usually shown on pressure–temperature (P–T) plots, rather than depth–temperature plots.

PRESSURE (P)

Pressure is a measure of the force per unit area to which a rock is subjected, and depends on the weight of overlying rock, and hence depth. The unit of pressure most usually used in geology is the bar or kilobar (kbar). 1 bar = 0.987 atmospheres = 14.5 pounds per square inch. The SI unit of pressure, the pascal, is however used increasingly; fortunately the conversion is simple: 1 bar = 10^5Pa, 1 kbar = 0.1 GPa (gigapascal).

The total pressure exerted at a point in the crust due to the weight of overlying rocks is known as the **lithostatic pressure** and is equal to ρgh where ρ is the mean density of overlying rock, h is depth and g is the acceleration due to gravity. In most metamorphic environments we assume that the pressure acting at a point is approximately uniform in all directions and is equal to the lithostatic pressure. The reason for this assumption is that if a rock is subject to a greater stress in one direction than in another it will yield provided the stress difference exceeds its strength. Hence rock strength proves an upper limit to stress differences in the crust. Experimental studies have shown that under most metamorphic conditions (low strain rates, moderate to high temperatures, water likely to be present) rocks are weak and can only sustain stress differences of a few tens, or at most hundreds, of bars (see Carter, 1976; Etheridge, 1983) which is very small compared to lithostatic pressures typically of several kilobars. Thus the pressure acting laterally on rocks at depths of a few kilometres must be within a few per cent of the pressure acting vertically. We therefore use the lithostatic pressure to approximate the **total confining pressure** to which a rock was subjected, and it is this value that is critical in evaluating the effect of pressure on the stability of metamorphic minerals. As a rough guide, the pressure exerted by a 10 km column of rock is in the range 2.6 to 3.2 kbar according to composition.

Lithostatic pressure does not of itself cause deformation, and indeed some rocks have been metamorphosed at very high pressures, equivalent to those obtaining at the base of the crust, without suffering significant distortion. Deformation is a result of unequal

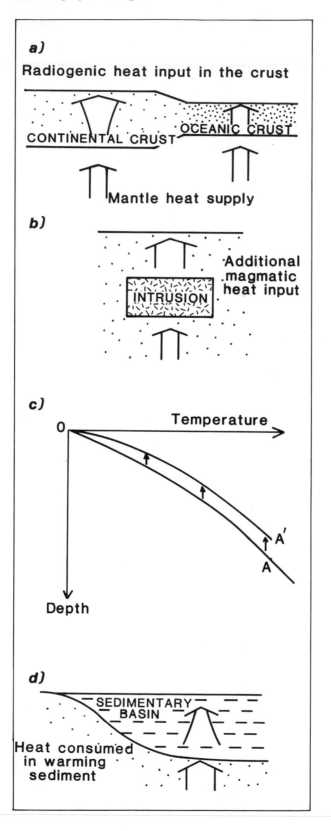

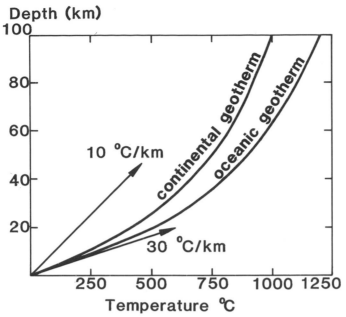

Fig. 1.7 Representative geotherms for stable continental and oceanic regions. Based on Clark & Ringwood (1964) and Brown & Mussett (1981).

stresses acting on a rock. For example, if a rectangular block of ice cream is briefly held at the bottom of a bucket of cold water, it will not be distorted, because the increased pressure due to the weight of overlying water acts equally in all directions. On the other hand if the ice cream block is placed on a bench and the bucket placed on top of it, the ice cream will be squashed flat. In this case the pressure acting on the sides of the block is still atmospheric pressure, while the pressure acting vertically is atmospheric pressure plus the pressure due to the weight of the bucket of water.

A rock experiencing different pressures in different directions is subject to **deviatoric stress**. The deformation resulting from deviatoric stresses plays a major role in determining the textural characteristics of rocks, but does not influence the mineral assemblage developed except in so far as deformation may catalyse reactions or permit movement of fluids. It was formerly thought by some geologists that certain minerals typical of regional metamorphism could only form in the presence of a deviatoric stress. Such minerals were termed 'stress minerals' by A. Harker in his classic text on metamorphism published in 1932. However, in the 1950s and 1960s virtually all the so-called stress minerals were synthesised experimentally in the absence of deviatoric stress. The main natural phase transition that is almost undoubtedly caused by deviatoric stress is the development of lamellae of clinoenstatite in enstatite host grains during faulting of certain peridotites (Coe, 1970).

Another role that has been suggested for deviatoric stresses is that they may lead to 'tectonic overpressures'. According to this hypothesis (discussed by Rutland, 1965) the

Fig. 1.6 opposite Causes of variation in surface heat flow. *a*) Illustration of the upwards increase in heat flow in continental crust due to a radiogenic input in addition to the mantle heat supply. *b*) The effect of intrusion on heat flow in the overlying region. *c*) The effect of rapid uplift and erosion, without significant cooling at depth, in enhancing thermal gradient. Rocks whose temperatures define geotherm A are displaced upwards, giving rise to geotherm A'. *d*) Reduction in heat flow due to absorption of heat by a blanket of cold sediment.

lateral pressures generated by tectonic squeezing may be so much greater than the vertical lithostatic pressure, that the mean pressure to which the rocks are subjected is significantly greater than that due to the overburden, causing anomalously high pressure mineral assemblages to develop at relatively shallow depths. The more recent work on rock strengths mentioned above shows that the possible overpressures that can be generated are far too small to be significant.

A very important pressure variable in metamorphism is **fluid pressure**, which is the pressure exerted by fluid present in pore spaces and along grain boundaries. Where the rock is dry, fluid pressure is effectively zero and the lithostatic pressure acts across grain boundaries, holding the grains together and making failure very difficult. However, if a fluid is present the fluid pressure tends to act in the opposite sense, reducing the **effective pressure** acting across the grain boundaries and so making cracking more likely. Even in deep diagenesis, sediments become buried sufficiently deep that fluid can no longer be squeezed out and pass directly to the surface along cracks. As a result high fluid pressures, approaching the lithostatic pressure, are sometimes generated. Compaction, coupled with release of mineralogically bound water and thermal expansion of existing fluid, is believed to result in the development of fluid pressures approximately equal to lithostatic pressure as metamorphic rocks are heated. If fluid pressure (P_f) exceeds lithostatic pressure (P_l) by more than the tensile strength of the rock (usually small), the rock is likely to burst due to **hydraulic fracturing** and in the process, fluid will escape along cracks (Norris and Henley, 1976). In marked contrast, as metamorphic rocks cool the fluid pressure is likely to drop to very low values because the small amount of pore fluid remaining is rapidly absorbed by mineralogical reactions (Yardley, 1981a).

A modern structural geology text should be consulted for a more complete and rigorous treatment of stresses. Those aspects described here are summarised in Fig. 1.8.

METAMORPHIC FLUIDS

It will be apparent already that pore fluids are generally believed to play an important role in some types of metamorphism (see Fyfe, Price & Thompson, 1979). The evidence for this is of two types. Firstly, we can argue from a theoretical standpoint that, since hydrous minerals such as micas, amphiboles, etc., are present in metamorphic rocks formed at elevated temperatures, water must have been present in the rock in some form or these minerals would have broken down and dehydrated. Furthermore, since volatiles such as H_2O or CO_2 are released from minerals as metamorphic reactions proceed, they must be present as fluid in the rock even if only for a short time before they are expelled.

Secondly, we can actually find samples of the metamorphic fluid preserved in many rocks as fluid inclusions, usually in milky vein quartz or in coarse matrix mineral grains. There is a growing body of evidence that some inclusions truly contain fluid trapped as metamorphism proceeded. (e.g. Roedder, 1972; Poty, Stalder & Weisbrod 1974; Touret, 1977). Although many fluid inclusions are filled with H_2O, other fluid species are also found, including CO_2, CH_4 and, more rarely, N_2. Some trapped fluids may be mixtures of more than one of these species, while others are relatively pure.

Given that fluids are very important during metamorphism, it is essential to understand something of their behaviour at metamorphic pressures and temperatures. It is well known that liquid water undergoes a phase change to steam by boiling at 100°C at atmospheric pressure, and that the boiling point of water is raised by increased pressure, as in a domestic pressure cooker. We can look on the process of boiling as involving a change in volume of a given mass of H_2O. It is convenient to refer to the **molar volume** of water or steam at any given pressure and temperature, which is the volume occupied by one gram

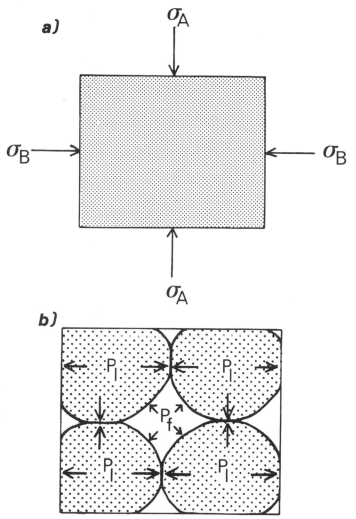

Fig. 1.8 Schematic summary of the stresses acting on rocks during metamorphism. *a*) Cross-section of a metamorphic rock showing σ_A, the normal stress due to the weight of overlying rocks and σ_B, a lateral normal stress. If $\sigma_A \neq \sigma_B$ the rock is subject to a deviatoric stress, but the difference $(\sigma_A - \sigma_B)$ is limited by rock strength. To a first approximation lithostatic pressure, $P_1 = \sigma_A = \sigma_B$. *b*) Interaction between fluid pressure (P_f) due to fluid in pores and lithostatic pressure (P_1) for a rock for which $\sigma_A = \sigma_B$. P_1 holds the grains together, and if the effective pressure ($P_e = P_1 - P_f$) becomes negative the fluid will tend to push the grains apart causing a crack.

formula weight (mol) of H_2O (18.015 g) at those conditions. The higher the pressure at which water is boiled, the smaller the volume change that accompanies the conversion of liquid to steam. This is illustrated in Fig. 1.9, which shows the molar volume of H_2O as a function of pressure and temperature. At 220 bar and 374 °C, there ceases to be any sharp discontinuity in volume between liquid water and steam; on the phase diagram the boiling curve separating these two phases simply comes to an end at this point, which is known as the **critical point**. If liquid water is at a pressure greater than 220 bar, it cannot be boiled no matter how hot it is made, although there may be a relatively rapid increase in volume over a fairly narrow temperature range. Water at a pressure greater than 220 bar or a temperature greater than 375 °C is referred to as a **supercritical fluid** rather than a liquid

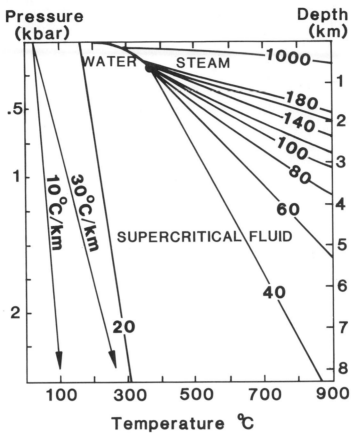

Fig. 1.9 Variation in the molar volume of H_2O with pressure and temperature. The right-hand vertical scale indicates approximate depth assuming that $P_f = P_l$. The boiling curve of water terminates at the critical point. Radial lines are of constant molar volume (numbered with values in cubic centimetres). Also shown are geothermal gradients of 10°/km and 30°/km, illustrating that only small variations in the volume of H_2O are likely over most metamorphic conditions (cf. Fig. 1.7). Data of Burnham, Holloway & Davis (1969).

or a gas. Note that the **critical pressure** (pressure of the critical point) of H_2O is low compared to most metamorphic pressures, it is even exceeded at the bottom of deep oceans. At pressures corresponding to depths of more than a few kilometres in the earth, the molar volume of supercritical water is comparable to that of liquid water in the laboratory, even at high temperatures. The presence of dissolved salts can extend the depth at which water can boil, but only to a few kilometres.

   To illustrate the significance of the relatively small molar volume of water under most crustal conditions, consider the effect of intruding magma into water-bearing sediment either (a) within 1 km of the surface or (b) at moderate depths, say greater than 5 km. In the first instance water will expand rapidly due to boiling (by a factor of 5 or more) very likely resulting in fracturing and bursting apart of the rock. Even if the pores and cracks remain filled with steam, the actual mass of water present will be so small that the sediment has been effectively 'boiled dry'. Such behaviour probably explains why sediments meta-morphosed in sub-volcanic environments are often friable, not hornfelsed. In contrast, at the greater depth heating the water to near-magmatic temperatures will not cause boiling and the increase in volume will be relatively small. A significant amount of water may remain in the pore space of the rock even at very high temperatures and the presence of this water may promote recrystallisation or influence the style of deformation that the rock undergoes. The composition and behaviour of metamorphic fluids are currently the subject of intense research activity.

TERMINOLOGY OF METAMORPHIC ROCKS

All specialist branches of science have their own vocabulary, and metamorphic petrology is no exception. Fortunately the nomenclature is less extensive and capricious than in some other branches of geology.

METAMORPHIC ROCK NAMES

The purpose of a name should be to identify a particular rock and to convey useful information about it. What constitutes useful information depends on your point of view, and so it is quite acceptable to use different names for the same rock, according to the point being made. As an extreme example, geologists working on Archaean rocks may wish at first simply to separate rocks that were once plutonic igneous rocks from those formed at the earth's surface; in this case the whole spectrum of metamorphosed sediments and volcanic rocks can be lumped together as 'supracrustals'.

There are four main criteria for naming metamorphic rocks:

1. The nature of the parent material.
2. The metamorphic mineralogy.
3. The rock's texture.
4. Any appropriate special name.

Names indicating the nature of the parent material

These may be very general, e.g. metasediment, or more specific, e.g. marble. Such names may be used as nouns with or without additional qualification, e.g. diopside marble, or as adjectives qualifying a textural name, e.g. pelitic schist. Some of the common names, and their adjectival forms are as follows:

Original material	Metamorphic rock type (noun/adjective)*
Argillaceous or clay-rich sediment	Pelite/pelitic
Arenaceous or sandy sediment	Psammite/psammitic or quartzofeldspathic (if appropriate)
Clay–sand mixture	Semi-pelite
Quartz sand	Quartzite
Marl	Calc-silicate/calcareous
Limestone	Marble
Basalt	Metabasite/mafic
Ironstone	Meta-ironstone/ferruginous

*In addition, it is acceptable to prefix any igneous or sedimentary rock name by 'meta-' to denote the metamorphic equivalent, as in the last two examples.

Metamorphic Mineralogy

The names of particularly significant metamorphic minerals that may be present are often used as qualifiers in metamorphic rock names, e.g. garnet mica schist, forsterite marble. There are two possible conventions here, the mineral names may be given in order of abundance for the principal metamorphic minerals, to denote the modal mineralogy, e.g. garnet–sillimanite schist, or the names of particularly significant minerals can be given, which indicate specific conditions of metamorphism, irrespective of their abundance, e.g. sillimanite–muscovite schist. The first convention might be more appropriate for a field geologist who wishes to make stratigraphic correlations and can use the modal mineralogy as a rough guide to rock composition. On the other hand a petrologist studying variations in metamorphic grade will specify only those minerals that indicate particular conditions of metamorphism. Some essentially monomineralic rocks are named for their dominant mineral, e.g. quartzite, serpentinite or hornblendite. A number of other names referring to particular mineral associations are described under 'Special names' below.

The rock's texture   Textural terms are very important for naming metamorphic rocks, and indicate whether or not oriented fabric elements are present to dominate the rock's appearance, and the scale on which they are developed. Although mineral **preferred orientations** are best developed in pelites and semi-pelites, they can form in a wide range of rock types if deformation is sufficiently intense. In many regionally metamorphosed rocks, micas develop a preferred orientation aligned perpendicular to the maximum compression direction giving rise to a **planar fabric** or **foliation**. The names used for planar fabrics depend on the grain size and gross appearance of the rock. Deformation and metamorphism of clay-bearing clastic sediments give rise to the following sequence of rocks with characteristic fabrics, in order of increasing grade of metamorphism.

Slate   A strongly cleaved rock in which the cleavage planes are pervasively developed throughout the rock, due to orientation of very fine phyllosilicate grains. The individual aligned grains are too small to be seen with the naked eye, and the rock has a dull appearance on fresh surfaces (Fig. 1.10(*a*)).

Phyllite   Similar to slate but the slightly coarser phyllosilicate grains are sometimes discernible in hand specimen and give a silky appearance to cleaved surfaces. Often, the cleavage surfaces are less perfectly planar than in slates.

Schist   Characterised by parallel alignment of moderately coarse grains, usually clearly visible with the naked eye (Fig. 1.10(*c*)). This type of fabric is known as **schistosity**, and where deformation is fairly intense it may be developed by other minerals, such as hornblende, as well as by phyllosilicates.

Gneiss   Gneisses are coarse, with a grain size of several millimetres, and foliated (i.e. with some sort of planar fabric, such as schistosity or compositional layering). English and north American usage emphasises a tendency for different minerals to segregate into layers parallel to the schistosity, known as gneissic layering; typically quartz- and feldspar-rich layers segregate out from more micaceous or mafic layers. European usage of gneiss is for coarse, mica-poor, high-grade rocks, irrespective of fabric. The term **orthogneiss** is used for gneisses of igneous parentage, **paragneiss** for metasedimentary gneisses (Fig. 1.10(*d*)).

In practice the boundaries between all these types are gradational. It is worth remembering that in cases of doubt the majority of metamorphic rocks can be described as schists!

Sometimes deformation involves extension as well as flattening, and in this case elongate mineral grains (such as amphiboles) may line up to produce a **linear fabric** or **lineation** oriented in the direction of stretching.

Mylonite is a term used for fine grained rocks produced in zones of intense ductile deformation where pre-existing grains have been deformed and recrystallised as finer grains (Fig. 1.10(*e*)). Often the overall appearance of the rock in outcrop is flaggy, even though individual hand specimens show no marked mineral alignment because of the

Fig. 1.10 Examples of metamorphic rock types.
a) Slate. Fractured surface photographed under the scanning electron microscope (SEM), showing the strong alignment of most phyllosilicate grains, and larger interspersed tabular quartz grains. Easdale Slate, Scotland. Photo courtesy G.E. Lloyd. *b*) Hornfels. SEM photomicrograph illustrating randomly oriented micas (some examples are arrowed) interspersed with recrystallised granular quartz grains. Ben Nevis, Scotland, courtesy G.E. Lloyd.

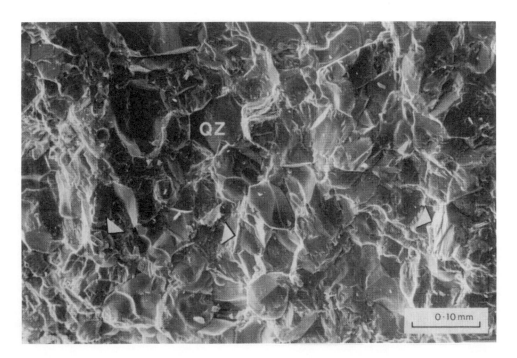

Fig. 1.10   Examples of metamorphic rock types.
c) Garnet mica schist. Outcrop with coarse garnet porphyroblasts in a mica-rich matrix. Note the small veins of segregated quartz. Co. Sligo, Ireland. *d*) Augen–gneiss. This outcrop shows two surfaces in different orientations, each with a very different appearance because of the effects of folding (see inset). The large pale crystals are of K-feldspar, and the rock has been deformed around them to give rise to the distinctive augen structure on the right hand surface. Moldefjord, Norway.

Fig. 1.10 Examples of metamorphic rock types.
e) Mylonite. Intensely deformed rock displaying a characteristic platy appearance. Moine Schist adjacent to the Moine Thrust near Ullapool, Scotland. The pencil lies across a gently-dipping late fault with associated folding of the mylonite fabric. Photo courtesy R.J. Knipe. *f*) Migmatite. Large scale migmatite features typical of high-grade migmatitic basement. Major units of light coloured, injected leucosomes are interbanded with schists that are themselves migmatised on a finer scale. The vertical height of the hillslope shown is about 100 metres. Skagit Gneiss, North Cascades, Washington, USA.

Fig. 1.10 Examples of metamorphic rock types.
g) Marble. Spectacular 'similar-style' folding is typical of medium to high grade regionally metamorphosed marbles. Connemara, Ireland.

thorough recrystallisation. In extreme cases, thin seams of glassy rock, known as **pseudotachylyte** are produced, apparently the result of melting due to frictional heating.
Normally, pseudotachylyte is restricted to dry rocks such as granulites.

Contact metamorphism in the absence of deformation gives rise to a random fabric of
interlocking grains which produces a tough rock known as **hornfels** (Fig. 1.10(*b*)).

Some metamorphic rocks, particularly those relatively poor in sheet silicates, have
textures that are not obviously schistose, even though the rocks are not hornfelses.
Winkler (1976) has proposed the term 'fels' for such rocks, although it has not been
universally adopted. In the older literature the term 'granulite' is used for some such rocks,
particularly psammites with an equigranular texture, but this term is now reserved to
denote particular physical conditions of metamorphism.

Textural names are usually used as nouns, qualified by adjectives indicating the parent
material or the present mineralogy (e.g. garnet schist, pelitic hornfels).

Special Names   Special names are mercifully rare in metamorphic petrology, and most that are used are
also descriptive. However, the mineral associations indicated by the names carry implications for the conditions of metamorphism. Some of the commonest are the following:

Greenschist   Green, foliated metabasite, usually composed predominantly of chlorite,
epidote and actinolite.
Blueschist   Dark, lilac–grey foliated metabasite, owing its colour to the presence of
abundant sodic amphibole, typically glaucophane or crossite: seldom truly 'blue' in hand
specimen.
Amphibolite   An essentially bimineralic dark-green rock made up of hornblende and
plagioclase. A wide range of minerals may occur as accessories. Most amphibolites are

metabasites (**ortho-amphibolite**) but some may be metamorphosed calcareous sediments (**para-amphibolites**).

Serpentinite   Green, black or reddish rock composed predominantly of serpentine. Formed by hydration of igneous or metamorphic peridotites (olivine-rich ultrabasic rocks).

Eclogite   Metabasite composed of garnet and omphacitic clinopyroxene with no plagioclase feldspar. Common accessories include quartz, kyanite, amphiboles, zoisite, rutile or minor sulphides.

Granulite   Rock characterised by both a texture of more or less equidimensional, straight sided (polygonal) grains for all mineral species, and a mineralogy indicative of very high temperature metamorphism, closely akin to the mineralogy of calc-alkaline basic to moderately acid plutonic rocks (feldspar, pyroxene, amphibole). The **charnockite** suite constitutes a distinct variety of K-feldspar and hypersthene bearing granulites.

Migmatite   A 'mixed rock' composed of a schistose or gneissose portion intimately mixed with veins of apparently igneous quartzo-feldspathic material (known as **leucosomes**) (Fig. 1.10(*f*)).

ADDITIONAL TEXTURAL TERMS

Metamorphic textures are described in more detail in Chapter 6, but there are a small number of names in addition to those already defined which are in such common usage as to be essential for the understanding of the intervening chapters.

Porphyroblasts are grains that are significantly larger than those of the matrix (e.g. the staurolite crystals in Fig. 1.1(*a*) or the garnets in Fig. 1.10(*c*)). Many porphyroblasts enclose **inclusions** of matrix minerals; sometimes minerals occur as inclusions even though they are no longer present in the rock matrix, such minerals are termed **armoured relics** or **relic inclusions**. Porphyroblasts with a very high density of inclusions are termed **poikiloblasts**. Where a porphyroblast has well-developed crystal faces it is said to be **idioblastic**, however in general the common term euhedral (for a grain with good crystal form) or anhedral (for an irregularly shaped grain) may be applied to metamorphic minerals.

Sometimes, one or more minerals can replace a pre-existing mineral grain while retaining its shape. Such replacement textures are termed **pseudomorphs**, the original parent mineral can usually only be recognised if traces remain or if it had a particularly distinctive crystal form.

FURTHER READING ON SELECTED TOPICS

Contact Metamorphism

Reverdatto, V.V., 1973. *The Facies of Contact Metamorphism* (translated by D.A. Brown). Australian National University, Canberra.

Regional Metamorphism

See Chapter 7, also:
Ernst, W.G. (ed.), 1977. Metamorphism and Plate Tectonic Regimes. *Benchmark Papers in Geology*, **17**. Dowden, Hutchinson & Ross, Stroudsburg.

Impact Metamorphism

French, B.M. & Short, N.M. (eds) 1968. *Shock Metamorphism of Natural Materials*. Mono Book Co., Baltimore.

Grieve, R.A.F., 1987 Terrestrial impact structures. *Annual Reviews of Earth & Planetary Science*, **15**, 245–70.

Sea-Floor Metamorphism

See Chapter 4, also:

Rona, P.A., Bostrom, K., Laubier, L. & Smith, K.L. (eds) 1983. *Hydrothermal Processes at Seafloor Spreading Centres*. Plenum Press, New York.

Metamorphic Fluids

Fyfe, W.S., Price, N. & Thompson, A.B., 1979. *Fluids in the Earth's Crust*. Elsevier, Amsterdam.

Metamorphic Rocks

Yardley, B.W.D., MacKenzie, W. S. & Guilford, C. 1990. *Atlas of Metamorphic Rocks and their Textures*. Longman, Harlow.

2 CHEMICAL EQUILIBRIUM IN METAMORPHISM

The fundamental principle underlying most studies of metamorphic rocks is that their mineral assemblages reflect the physical conditions, such as pressure and temperature, at the time when the rock recrystallised. This principle is based on numerous field studies that have shown that the mineral assemblages in any particular rock type vary systematically and predictably across an area. Such observations have been taken as the basis for applying the theory of chemical equilibrium to metamorphic rocks. This chapter is concerned with the meaning of chemical equilibrium as applied to rocks, with the information that can be gleaned by treating rocks as equilibrium chemical systems, and with the limitations and pitfalls of so doing. The relevance for geology of the geochemical approach outlined in this chapter is that it allows us in many instances to specify the approximate conditions of temperature, depth of burial, etc., to which a group of rocks have been subjected during metamorphism, on the basis of intelligent comparison with the results of laboratory experiments. In many more cases it becomes possible to deduce relative differences in metamorphic conditions between different parts of a metamorphic belt.

EQUILIBRIUM – AN INTRODUCTION

Suppose we consider the behaviour of the atoms in a lump of rock undergoing metamorphism. The arbitrary and hypothetical lump of rock is a **system** in chemical terms. It is made up of a number of minerals and perhaps also contains an intergranular fluid. Each of these constituents is known as a **phase**: phases are the physically separable constituents of the system and may be solid, liquid or gaseous. Plagioclase and quartz are separate phases in a pelitic schist, but if the plagioclase has a composition intermediate between anorthite and albite, *the anorthite and albite end-members are not phases* because the plagioclase grains could not be further separated into albite particles and anorthite particles.

If we subject our system to some specific conditions of pressure and temperature, and maintain these conditions unchanged for a sufficiently long time, the atoms in the system will group themselves into the most stable configuration possible and the system is then at **equilibrium**. This configuration may involve one or a number of solid, liquid or gaseous phases depending on the atoms available and the conditions to which the system is subjected. Even at equilibrium, atoms are constantly in motion and may exchange between one phase and another. However, there is no overall change in the amount or composition of each phase present over a period of time. In other words, nothing actually happens at equilibrium!

Suppose the temperature (or pressure) of the system is now slowly changed, the coexisting phases will eventually cease to be in equilibrium with one another and some may grow at the expense of others, or new phases may appear. Such changes in rocks are known as **metamorphic reactions** and lead to the production of a new assemblage of phases that is at equilibrium under the new conditions.

All chemical systems with the same chemical composition (i.e. the same types of atoms present in the same proportions) and subjected to the same conditions, will develop the same assemblage of phases if equilibrium is attained. The fact that in regional studies such as those outlined in Chapter 1, all specimens of approximately the same bulk composition collected over an extensive area within a metamorphic zone have the same mineral assemblage, is therefore evidence for the attainment of chemical equilibrium during metamorphism. Equally, variation of mineral assemblage between zones is evidence of different metamorphic conditions in each zone, provided that there is no doubt that the rocks used to define each zone are of essentially the same chemical composition.

THE PHASE RULE

Most metamorphic rocks are chemically quite complex and are made up of a relatively large number of minerals. In order to understand how they formed we first need to know how many minerals can coexist stably at equilibrium in a particular rock, as a first indication of whether the mineral assemblage represents an equilibrium association.

This can be done using the phase rule, which was first applied to rocks by V.M. Goldschmidt. The rule can be demonstrated by imagining a glass of water (at atmospheric pressure) as a simple chemical system.

We can convert the water into ice or steam by changing its temperature only. Ice, liquid water and steam are all physically separable phases as defined above (page 29), however since they have the same chemical formula we can make them all out of one **chemical component**, i.e. H_2O. Components are the chemical constituents needed to make the phases we wish to consider in our system; from the point of view of the phase rule it is the *smallest number* of chemical components needed to define the compositions of all the phases that is significant. For example, a system that contains only andalusite and kyanite has only one component, Al_2SiO_5, whereas a system that contains andalusite, corundum and quartz must have two components, Al_2O_3 and SiO_2, to make all the phases.

Returning to our glass of water, we have a one-component system (H_2O), in which we wish to consider the relationships between three phases at different temperatures and pressures. The question 'what is the temperature of a glass of water?' seems nonsensical because liquid water is stable over a range of temperatures. On the other hand the question 'what is the temperature of water coexisting with steam (i.e. boiling)?' has an obvious answer, 100 °C. To be precise it is of course necessary to know the exact atmospheric pressure in the laboratory where the water is being boiled, but provided this information is forthcoming, the temperature of the boiling water can be obtained from standard tables without recourse to direct measurement.

The conclusions that we can draw from the glass of water, for the behaviour of a one-component system are:

1. One phase can occur alone over a range of temperatures and pressures. This one-phase, one-component system is said to have two **degrees of freedom** because it is possible, within limits, to vary its temperature or its pressure independently without changing the number or nature of the phases present.
2. Two phases can only coexist in equilibrium in the one-component system at a single

temperature at any given pressure. As long as the two phases coexist the system has only one degree of freedom because any change in temperature, for example, leads to a related change in pressure.

These conclusions are summarised for the general case by the phase rule, which can be expressed as:

$$F = C - P + 2$$

where F is the number of independent degrees of freedom for the system; C is the number of chemical components; and P is the number of phases.

This discussion of the behaviour of water clearly does not constitute a proof of the phase rule (see for example Greenwood, 1977), it merely demonstrates that the rule is not entirely unreasonable.

One consequence of the phase rule is that, in very general terms, mineral assemblages with a large number of phases will have only a small number of degrees of freedom. As a result the range of conditions under which the assemblage grew will be relatively precisely constrained, and it will be more easily possible to determine them from the results of experiments.

Compositional variation   Although pressure and temperature are very important variables in metamorphism, the composition of many common metamorphic minerals is also variable due to solid solution effects. While it is often convenient to represent the degrees of freedom of a mineral assemblage in terms of P and T, there is no fundamental reason why this should be so; compositional parameters may be more appropriate. However, no matter how many chemical substitutions are possible within the minerals of a particular rock, the number of degrees of freedom calculated from the phase rule still gives the number of variables that can change *independently*. (or **independent variables**).

As an example, consider an assemblage consisting of the five phases white mica + quartz + kyanite + alkali feldspar + aqueous fluid, in which the mica and feldspar are solid solutions intermediate between the Na and K end-members. This gives us the phases:

Mica	$(Na,K)Al_3Si_3O_{10}(OH)_2$
Quartz	SiO_2
Kyanite	Al_2SiO_5
Feldspar	$(Na,K)AlSi_3O_8$
Fluid	H_2O

which can be made from the five components: Na_2O, K_2O, H_2O, SiO_2, Al_2O_3. From the phase rule, the assemblage has two degrees of freedom, but there are at least four possible variables: P, T, feldspar composition and mica composition. This means that only two variables can be changed *independently*; the two remaining **dependent variables** take on values dictated by those of the independent variables. Hence all mixtures of these five phases that we might choose to make and *separately subject to the same pressure and temperature* (i.e. taking P and T as independent variables) will end up having mica and feldspar of the same composition when equilibrium has been attained.

Suppose we remove kyanite from this system so that there are only four phases, and hence the number of independent degrees of freedom is increased to three. At a specific P and T one degree of freedom remains but both mica and feldspar compositions are variables. This means that the Na/K ratios of mica and feldspar are related, and do not vary independently of one another. For our given value of P and T, only one composition of feldspar can coexist with any specified mica composition, or vice versa.

METAMORPHIC PHASE DIAGRAMS

THE P–T DIAGRAM

One of the commonest ways of representing the stability fields of different metamorphic mineral assemblages is on a plot of pressure against temperature (P–T diagram), because such a diagram allows us easily to visualise the sorts of settings inside the earth where a particular mineral assemblage might form.

It follows from the phase rule that in a rock with C components a mineral assemblage with P phases, where P $\leqslant$ C, will be stable over a distinct region of the P–T diagram because it has two independent degrees of freedom (at least). The assemblage is therefore said to occupy a **divariant field**. Figure 2.1(*a*) shows the divariant fields in which andalusite, sillimanite and kyanite are stable. Any two of these phases can coexist in equilibrium only at a single temperature for each pressure and the system then has one degree of freedom. The conditions for C + 1 phases to coexist stably define **univariant curves** which constitute the boundaries between divariant fields. In Fig. 2.1(*a*) the univariant curves intersect at a single point at which all three phases can occur in equilibrium. From the phase rule, a system with C + 2 phases has no degrees of freedom and can therefore be stable only at a unique pressure and temperature known as an **invariant point**.

Many of the laws that govern metamorphic phase diagrams are purely geometrical, following from the phase rule without further knowledge of geology or chemistry; a fuller account is given in the Appendix.

COMPOSITIONAL PHASE DIAGRAMS

The value of the P–T diagram is that it shows the actual range of physical conditions under which specific assemblages are stable relative to others. However, it is also important to have phase diagrams that show the effects of different rock compositions on the assemblages produced, and hence make it possible to tell if different samples have different mineral assemblages because they were metamorphosed under different conditions or because they differ significantly in their chemical composition.

Compositional phase diagrams have a basis in the phase rule much like P–T diagrams. They are constructed for a specific value of P and T to show the possible mineral compositions and mineral assemblages that may be at equilibrium under the conditions of interest. An example is shown in Fig. 2.1(*b*). Provided the P–T conditions selected do not coincide exactly with a univariant curve, the number of phases present in any rock will equal the number of components in the system, i.e. the conditions lie within a divariant field on the P–T diagram. A major limitation is that while it is possible to plot a three-component chemical system on a triangular diagram, each apex corresponding to a pure component, more complex systems require additional assumptions or simplifications, some examples of which will appear later. Often these are unsatisfactory, sometimes they are untenable.

A wide range of information can however be shown on a compositional phase diagram. Solid solution in minerals is represented by plotting their compositions as lines or areas rather than points. Possible equilibrium mineral assemblages are indicated by lines connecting the phases that coexist; these are known as **tie-lines**. In the case of divariant assemblages on triangular diagrams, the tie-lines define triangles each representing a possible stable assemblage at the P–T conditions for which the diagram is drawn. Each

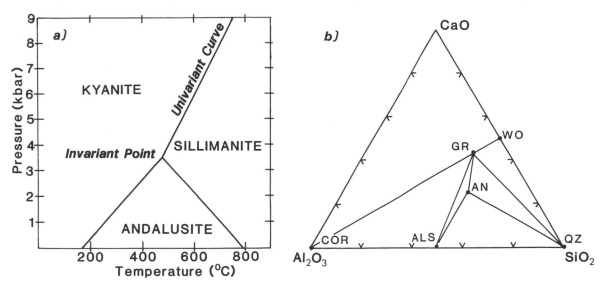

Fig. 2.1   a) Pressure–temperature diagram illustrating the stability fields of andalusite, kyanite and sillimanite, according to Holdaway (1971) (see also Fig. 3.12). *b)* Compositional diagram showing the compositions and stable relationships between corundum (COR), wollastonite (WO), grossular (GR), anorthite (AN), quartz (QZ) and Al-silicate (ALS) in the system $CaO–SiO_2–Al_2O_3$, for some arbitrary pressure and temperature. Only phases directly connected by tie-lines can coexist. Unless lying exactly on a tie-line, any bulk composition in this system will crystallise as a three-phase mixture according to which triangular field it lies within.

of these assemblages forms in a different bulk composition rock, and so they are said to be **compatible assemblages**. Minerals not directly connected by tie-lines cannot stably coexist. Rock compositions can also be plotted on the diagram and this makes it possible to predict what mineral assemblage the rock will develop at the appropriate $P–T$ conditions.

A series of compositional diagrams plotted for successive $P–T$ conditions (from either field or experimental evidence) not only illustrates the changes in mineral assemblage, it also makes it possible to specify what reactions have taken place. This is possible because when two minerals react together the tie-line that previously joined them is removed, while when two minerals grow together for the first time, a new tie-line connecting them appears. This is illustrated for a real example below (page 42).

APPLICATION OF THE PHASE RULE TO NATURAL ROCKS

It is often quite difficult to make the jump from the idealised and simplified systems studied by experimentalists or discussed in textbooks to the natural associations of minerals, often themselves complex solid solutions, found in many of the most common metamorphic rocks.

A chemical analysis of a typical pelitic schist (Table 1.1) contains significant amounts of SiO_2, TiO_2, Al_2O_3, Fe_2O_3, FeO, MgO, MnO, CaO, Na_2O, K_2O and H_2O as well as smaller amounts of P, S, B, F, Sr, Ba, Zr and traces of other elements. Suppose that the phases present are muscovite, biotite, garnet, chlorite, plagioclase, quartz, tourmaline, ilmenite, pyrrhotine, apatite and zircon. This system has at least 16 components but only

11 phases (12 if a pore fluid was present during metamorphism), hence there are at least six degrees of freedom. Does this mean that the assemblage can exist over such a wide range of conditions that it is unlikely to be a useful indicator of conditions of metamorphism, or is it unnecessary to take into account all the minor constituents?

In general, phases such as tourmaline, apatite, zircon and to some extent the sulphides are not useful indicators of metamorphic conditions. They each concentrate a particular minor element, e.g. B in tourmaline, Zr in zircon, that does not readily enter any of the other minerals present. As a result these minerals are unlikely to be involved in metamorphic reactions because there is no other phase for the minor element to enter. To a good approximation they behave as an inert matrix surrounded by other phases that can react together, and it is therefore reasonable to discount both these accessory phases and their distinctive minor element components when applying the phase rule to most metamorphic rocks.

There are other minor element components which do not form their own minerals but occur in small amounts in solid solution in the common silicate minerals. Examples include Sr and Ba in feldspars or Mn in ferromagnesian minerals (though Mn-rich minerals are sometimes found). The concentration of, for example, Sr in feldspar, can vary independently of pressure and temperature depending on how much feldspar and how much Sr are present in the rock, and this variation can be considered as one of the excess independent degrees of freedom. Unless unusually high levels are present, such variation in minor element concentration has little effect on the stability of the major silicate phases and can also be discounted to a first approximation.

As a result, the chemical components which are important for determining whether a mineral assemblage may have formed at equilibrium are those that are *major* constituents of *more than one* of the phases.

METAMORPHIC REACTIONS – SOME FIRST PRINCIPLES

The study of metamorphic reactions and the attainment of equilibrium is governed by the laws of thermodynamics, and a detailed account is beyond the scope of this book. However, some basic principles that are essential to understanding metamorphic rocks are outlined here in a qualitative way, and references for more rigorous study are given at the end of the chapter.

A fundamental concept is that any substance can be assigned a 'free energy' reflecting its energy content. There is more than one way of formulating a free energy, but in geology the **Gibbs free energy** (G) is generally used because the Gibbs free energy of a phase varies with its composition (if it is a solid or liquid solution) and with pressure and temperature, and these are the variables with which geologists are most concerned. G is an **extensive property** – the more material you have the more free energy it contains. In contrast, an **intensive property** such as temperature has the same value irrespective of the size of the system. For extensive properties it is usual to normalise them to 1 mol of the substance (for example the molar volume of quartz is the volume of 1 gram formula weight of quartz).

The molar Gibbs free energy of a discrete phase is a distinct quantity that can, in principle, be measured. However, we often deal with mineral solid solutions where it is the energy of a specific end-member of the solution that is of interest. For example, if the energy of albite in solution in plagioclase is greater than that of albite in solution in an adjacent orthoclase, then we might expect albite to move out of the higher energy environment (the plagioclase crystal) into the lower energy one (orthoclase). The measure used for the Gibbs free energy of end-members in solid solutions (or of components of liquid solutions) is the chemical potential (μ), defined as:

$$\mu_i = \left(\frac{\partial Gp}{\partial n_i}\right)_{T,P,n_j}$$

i.e. for a component i, its chemical potential μ_i is equal to the change in the Gibbs free energy of the phase in which it occurs (G_p) resulting from a change in the number of moles of i (n_i) when temperature, pressure and the number of moles of all other components (n_j) remain constant.

We can define the molar Gibbs free energy of a solution phase containing j components as:

$$Gp = (\sum_{i=1}^{i=j} \mu_i \, n_i)/n_p$$

where n_p is the number of moles of the phase made from $\sum_{i=1}^{i=j} n_i$ moles of the constituent components.

Hence the Gibbs free energy of a system made up of m phases becomes:

$$G_s = \sum_{k=1}^{k=m} G_{p,k}.n_{p,k}$$

where $G_{p,k}$ is the molar Gibbs free energy of the k th phase, of which $n_{p,k}$ moles are present. Clearly, if the same atoms can be combined in a number of different ways involving different assemblages of phases present in different proportions, then the value of G_s will be different in each case.

At equilibrium the Gibbs free energy of a system is at a minimum for the prevailing conditions of pressure and temperature, i.e. it is the combination of atoms that yields the smallest value of G_s that is stable.

The reason why reactions take place is that the value of G for every substance changes with variation in P or T, but for different minerals the magnitude of the change is different. As a result a mineral assemblage at equilibrium under one set of conditions may no longer have the lowest possible free energy after P or T have been changed.

As an example, consider a rock made up solely of andalusite. Under different conditions the same atoms can rearrange themselves into kyanite or sillimanite if these have a lower Gibbs free energy, since the chemical composition of each of these phases is the same, Al_2SiO_5. Andalusite has a lower density ($3.15 \, g/cm^3$) than kyanite (density $3.60 \, g/cm^3$), and therefore high pressure conditions are likely to stabilise kyanite. If temperature is kept constant, the variation in Gibbs free energy with pressure for any phase is given by:

$$\left(\frac{\partial G}{\partial P}\right)_T = V$$

where V is the molar volume of the phase. Because minerals are not very compressible, V is fairly constant over large ranges of pressure, however the notation $V°$ is used for the molar volume measured under standard conditions, usually $25 °C$ and 1 bar. If G is plotted against P for any phase at a particular temperature the result is a line that is nearly straight, with slope $V°$ at 1 bar. This has been done schematically for andalusite ($V° = 51.530 \, cm^3$)

and kyanite ($V^\circ = 44.090\,\mathrm{cm}^3$) in Fig. 2.2($a$). The line representing kyanite has a shallower slope than that for andalusite because the molar volume is less, and hence the lines cross at a point corresponding to $P = P_E$, at which point both phases have the same Gibbs free energy, are therefore equally stable and coexist in equilibrium. At pressures less than P_E, $G_{and} < G_{ky}$ and so any kyanite will tend to react to form andalusite, while at pressures greater than P_E the reverse is true. The actual value of P_E depends on the temperature for which the plot is drawn.

There is a rather similar relationship between G and temperature when P is held constant. In this case it is a property called **entropy** (S), representing the degree of randomness of atomic arrangement in a substance, which determines the way in which G changes with temperature. This is illustrated for andalusite and kyanite in Fig. 2.2(b), and here it is the phase with the larger value of S that becomes stable with increase in T. Again, there is a unique temperature, T_E, at each pressure, at which kyanite and andalusite are equally stable. If kyanite is heated above T_E it will tend to react to andalusite, and vice versa.

The relationships illustrated in Fig. 2.2 are particularly simple because they concern only two phases of the same composition. However, in principle, there is no difference between comparing the Gibbs free energy of two phases of the same composition, and comparing the Gibbs free energy of two mineral assemblages, each of a number of phases, *provided that it is possible to write a balanced chemical reaction between the two assemblages*.

It is possible to apply the phase rule to rocks without any knowledge of the thermodynamic properties of the phases. However, the predictions of the phase rule are consistent with the conclusions drawn from thermodynamic considerations and summarised in Fig. 2.2. Andalusite and kyanite constitute a two-phase, one-component system and therefore if both phases coexist there is only one degree of freedom, as with boiling water. There is only a single T for any given P at which they can coexist stably, and this will be T_E of Fig. 2.2(b).

A final thermodynamic property of fundamental significance for the way in which the earth's crust behaves is **enthalpy** (H). The enthalpy of a phase reflects its heat content and

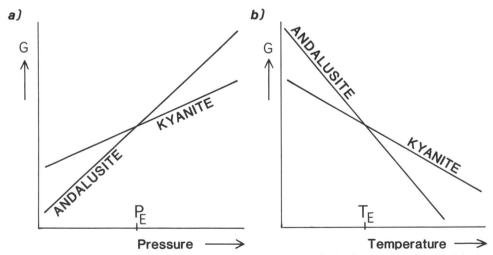

Fig. 2.2   Schematic plots illustrating the variation in the molar Gibbs free energy (G) of andalusite and kyanite, a) with respect to pressure at constant temperature, and b) with respect to temperature at constant pressure. P_E and T_E are the pressure and temperature respectively at which the two phases can coexist in equilibrium.

in the same way as the reactants and products of a reaction may have different volumes, so they may also have different enthalpies. The enthalpy difference between reactants and products is:

$$\Delta H_r = H_{products} - H_{reactants}$$

and ΔH_r is known as the **heat of reaction**. If ΔH_r is positive, additional heat must be put into the system to enable the reaction to proceed even when the equilibrium temperature has been attained. Such reactions are known as **endothermic**, and a good example is the boiling of water. Metamorphic reactions that liberate water from hydrous minerals are similarly endothermic. Other reactions (for example the condensation of steam to liquid water or the conversion of basalt to eclogite) give off heat and are **exothermic**.

THE CONCEPT OF BUFFERING

When the phase rule is applied to a natural mineral assemblage it often transpires that the system has only a small number of independent degrees of freedom, even though there are a relatively large number of possible variables, such as pressure, temperature, fluid pressure and composition of phases. In this case the mineral assemblage effectively fixes the value of some of these variables or **buffers** them at a specific value for a given pressure and temperature. The effect is rather analogous to the way in which the temperature of a beaker of boiling water is always $100\,°C$ at 1 atmosphere pressure, no matter how much it is heated, until one of the phases is completely used up (the beaker boils dry!) when the temperature of the container and the remaining steam inside it will increase rapidly.

A simple but important example of buffering is the way in which an assemblage containing magnetite and hematite can control the concentration of available oxygen in the fluid phase at a constant value, irrespective of the actual amounts of magnetite and hematite present. The relevant equilibrium is:

$$3Fe_2O_3 = 2Fe_3O_4 + \tfrac{1}{2}O_2$$

$$\text{hematite} \quad \text{magnetite} \quad \text{in fluid} \qquad\qquad [2.1]$$

The level of oxygen in the fluid phase is described in terms of its contribution to the total fluid pressure. Since oxygen makes up about 20 per cent of the air, the portion of atmospheric pressure that is due to oxygen molecules is about 0.2 bar; this is known as the **partial pressure** of oxygen in air. For the purposes of chemical equilibrium studies, **fugacity** (f) is used instead of partial presure, being the *thermodynamically effective pressure* of a gas species in the fluid phase; like other measures of pressure, fugacity normally has units of bars or pascals.

Applying the phase rule to the assemblage hematite + magnetite + $H_2O–O_2$ fluid, we have three phases and three components (e.g. $H_2O–FeO–O_2$). There are therefore two degrees of freedom in the system. Hence, although the fugacity of oxygen (f_{O_2}) in the fluid phase is a variable, its value must always be the same for any specific values of P and T to which hematite and magnetite are subjected. In other words if P and T are treated as independent variables, f_{O_2} is a dependent variable with a fixed value at any specified P and T.

Values of f_{O_2} in metamorphism are typically very small (10^{-10} to 10^{-45} bars) since crustal conditions are much more reducing than those at the surface. Myers & Eugster (1983) provide equations for the variation of f_{O_2} with temperature and pressure in equilibrium with a number of important mineral assemblages that serve as oxygen fugacity buffers.

METAMORPHIC REACTIONS – SOME SECOND THOUGHTS

UNIVARIANT CURVES AND CONDITIONS OF REACTION

Univariant curves such as the boundary between the andalusite and kyanite stability fields on Fig. 2.1 are often loosely referred to as representing the conditions of the reaction, e.g.:

$$\text{andalusite} \rightleftharpoons \text{kyanite} \qquad\qquad [2.2]$$

In fact, a moment's reflection (and reference to Fig. 2.2) should suffice to show that the univariant curve represents the only conditions to which a mixture of kyanite and andalusite can be subjected *without* any reaction occurring.

When we boil water the reaction to steam takes place almost exactly at the univariant curve, however most liquids can be cooled below their freezing points without crystallising. It is generally true that reactions that involve the production of new solid phases are unlikely to take place as soon as the conditions of the univariant curve are slightly exceeded, because of the difficulties associated with beginning to grow new solid phases (Ch. 6). If a mineral or mineral assemblage persists beyond the conditions in which it is at equilibrium into the stability field of another assemblage, it is said to be **metastable** and the reaction is said to have been **overstepped**.

It is commonly assumed in studies of metamorphism that most reactions are not overstepped very greatly and take place close to equilibrium, indeed this is a probable consequence of the assumption that metamorphic mineral assemblages have crystallised at equilibrium. Nevertheless, exceptions to this assumption certainly occur, especially with reactions such as the polymorphic transitions between andalusite, kyanite and sillimanite where it is not uncommon to find more than one of these phases close together in the same rock.

METAMORPHIC REACTIONS AND THE PHASE RULE

In the foregoing discussion, reactions have been treated as taking place at a specific temperature and pressure, like the boiling of pure water. The assemblage consisting of both reactants and products has only one degree of freedom, and is in equilibrium only along a univariant curve. Simple reactions of this type are known as **discontinuous** or **univariant reactions**.

Continuous reactions   When the phases involved include solid or liquid solutions, it is possible to have reactions that take place under different conditions according to the composition of the phases. A simple example is the boiling of salt water, which takes place at progressively higher temperatures with increasing salt concentration (Fig. 2.3). Since the steam given off when salt solution boils is essentially pure H_2O, the concentration of salt in the remaining liquid increases progressively along with its temperature as boiling proceeds. Applying the phase rule, the system has two components (NaCl, H_2O) but even when reactants and products coexist there are only two phases (liquid, steam); hence a system with salt solution reacting to steam has two degrees of freedom even while the reaction is taking place. The curve showing the composition of boiling salt solution as a function of temperature in Fig. 2.3 looks like a univariant curve, but this is because the figure has been drawn for a fixed pressure of 1 atmosphere. This means that one of the degrees of freedom has been assigned in specifying pressure, the figure shows that the two remaining

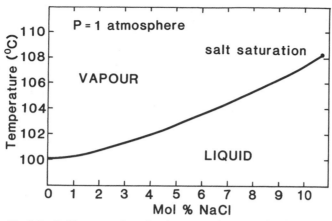

Fig. 2.3 Boiling curve for salt solutions in the system NaCl – H$_2$O, showing the variation in boiling temperature (i.e. liquid–vapour equilibrium temperature) with salt content (expressed as mole per cent NaCl in the liquid phase). The pressure is fixed at 1 atmosphere. This type of plot, showing temperature against a compositional parameter, is known as a *T–X* diagram. Data are from Haas (1976).

variables, temperature and composition of the boiling solution, vary dependently at constant pressure so that by specifying two parameters (e.g. *P* and *T* or *P* and solution composition) the condition of the system is completely specified in accordance with the phase rule.

Reactions of this type that take place progressively so that reactants and products can coexist over an interval of pressure and temperature (divariant field) are known as **continuous reactions**. They invariably involve at least one phase of variable composition and lead to a change in the compositions as well as in the abundances of different phases.

Cation exchange reactions There is a third type of reaction that involves only change in the compositions of phases, with no actual growth or dissolution of any of them; they are known as **cation exchange reactions**. These reactions can take place in assemblages with a large number of degrees of freedom (i.e. few phases but many components); their importance has come to be recognised particularly over the past 20 years when new technology has greatly facilitated the determination of mineral compositions.

Cation exchange reactions occur where two or more minerals in a rock can exhibit the same ionic substitution, for example exchange between $Fe^{2+} \rightleftharpoons Mg^{2+}$ or $K^+ \rightleftharpoons Na^+$. The ferromagnesian minerals in fact provide the most important examples of metamorphic cation exchange reactions. The reason why these reactions occur is that although different minerals may show the same type of ionic substitution, it is normally the case that different lattices show different preferences for the alternative cations (Burns, 1970). Hence, if there are several different ferromagnesian minerals present at equilibrium in a rock, the ratio $Mg^{2+}/(Mg^{2+} + Fe^{2+})$, commonly known as X_{Mg}* or the mole fraction of the Mg end-member, will always be different for each mineral. Garnet shows a strong preference for Fe over Mg, and so has a lower value of X_{Mg} than other, coexisting Fe–Mg silicates. Many studies have produced consistent results about the relative values of X_{Mg} for different coexisting minerals. In the case of pelitic schist minerals it is $X_{Mg}^{gt} < X_{Mg}^{st} < X_{Mg}^{bio} < X_{Mg}^{chl} < X_{Mg}^{cd}$.

* The X_{Mg} value and other compositional ratios are calculated using the **molecular proportions** of the elements (or their oxides) in an analysis, not the weight percentages.

Cation exchange reactions occur between ferromagnesian minerals because the precise degree to which Fe^{2+} is partitioned into a particular mineral relative to Mg^{2+} is dependent on temperature, with smaller differences in X_{Mg} between any two phases at higher temperatures.

In a rock containing only a small number of phascs so that the assemblage has a relatively large number of degrees of freedom (e.g. a schist with garnet, biotite, quartz, plagioclase feldspar and minor accessory minerals), the X_{Mg} values of the ferromagnesian minerals must reflect the proportions of Fe^{2+} and Mg^{2+} in the bulk of the rock as a whole. In Mg-rich garnet–biotite schists, both the garnet and the biotite will have relatively high X_{Mg} values compared with an iron-rich schist in which both minerals will have low values of X_{Mg}. However, the ratio by which Fe and Mg are partitioned between garnet and biotite remains more or less constant, irrespective of the mineral compositions, for all rocks metamorphosed under the same conditions. This ratio is known as the **distribution coefficient** (K_D) and is defined for Fe $\rightleftharpoons$ Mg exchange as:

$$K_D = \frac{\left(Mg/Fe\right)^A}{\left(Mg/Fe\right)^B} = K_D{}^{A-B}_{Fe-Mg}$$

Note that the superscripts are the abbreviated mineral names (see Glossary) used to identify the mineral pair for which the K_D is being calculated, while the subscript indicates the chemical substitution to which the distribution coefficient refers. Similar distribution coefficients can be calculated for any pair of coexisting minerals that have a chemical substitution in common.

The effect of changing the temperature of a garnet–biotite schist is to cause a cation exchange reaction between garnet and biotite, and this changes the value of K_D. The reaction can be written in a schematic fashion as:

$$\text{Fe-garnet} + \text{Mg-biotite} \rightleftharpoons \text{Mg-garnet} + \text{Fe-biotite} \qquad [2.3]$$

although of course there is only a single composition of garnet and biotite present in the rock at any one time. Since the effect of heating is to reduce the tendency of each mineral to distinguish between Fe and Mg, K_D tends towards a value of 1 with increasing temperature.

An important constraint on cation exchange reactions is that they can only occur to any significant extent if the cations are able to diffuse through the mineral lattices to their surfaces. At low grades of metamorphism (certainly below the garnet zone) diffusion rates are probably very slow in most minerals. However, diffusion rates increase exponentially with temperature and under sillimanite zone conditions are probably quite rapid in most minerals (see Ch. 6).

LIMITATIONS TO THE APPLICATION OF CHEMICAL EQUILIBRIUM

Although in theory any rock will react to form an equilibrium assemblage under any conditions, given sufficient time, in practice some reactions are so sluggish that they may not take place to any detectable extent even over geological time. If this were not the case, we would not be able to find rocks at the earth's surface that retain mineral assemblages formed at high temperatures and pressures within the crust. On the other hand many reactions that are important in rocks have been reproduced in the laboratory with run–times of at most a few months, even though times of at least thousands of years are normally available in nature.

Reactions that involve the release of volatiles (e.g. H_2O or CO_2) as a result of heating appear to take place relatively rapidly for the most part, whereas some reactions between solid phases only may be very sluggish, especially if no fluid phase is present in the rock (see Ch. 6, page 185). Much of the fluid liberated by reaction during heating must escape from the rock and is therefore not available to take place in the reverse reaction on cooling. As a result high temperature minerals often remain present as rocks cool, though they may show evidence of having partially reacted during cooling with any small amounts of fluid present.

Because reactions do not invariably go to completion, it is often the case that not all the mineral grains in a rock formed together in equilibrium. However, reactions during heating appear to be rapid, and remnants of the reactants are unlikely to remain unless they are protected by growth of a product mineral grain around them. In contrast, reactions during cooling are often incomplete unless fluid was reintroduced to the rock at the same time. It is often readily apparent in thin section that not all the minerals in a rock grew together at equilibrium, for example one mineral may occur only as a replacement for a specific precursor, as in the case of chlorite replacing biotite in retrogressed schists.

THE INFLUENCE OF FLUIDS ON METAMORPHIC PHASE EQUILIBRIA

Many metamorphic reactions consume or release fluid, most commonly H_2O but also frequently CO_2 and at low grades CH_4. More rarely reactions may release N_2 or H_2. These reactions are known as **devolatilisation reactions**, and the conditions under which they occur depend not only on pressure and temperature, but also on pore fluid pressure. At any given lithostatic pressure, the temperature needed for a dehydration reaction to occur will be greater the greater the partial pressure of H_2O (P_{H_2O}) in the pore space. Similarly a high value of P_{CO_2} inhibits decarbonation reactions on heating (Fig. 5.2).

Although there is still considerable controversy about the composition, quantity and behaviour of metamorphic fluids, there are at least two important natural constraints on values of P_{H_2O} or P_{CO_2} during metamorphism. Firstly, the effect of devolatilisation reactions must be to increase pore fluid pressure by generating more fluid. The upper limit to which the fluid pressure can rise is controlled by the overall lithostatic pressure, P_l and the tensile strength of the rock, beyond which the rock will be cracked by hydraulic fracturing, preventing any further rise in fluid pressure (Ch. 1, page 18). It seems likely that this extreme situation is not always attained, some rocks becoming sufficiently permeable for the newly generated fluid to escape without fracturing as P_{fluid} approaches P_l. Because of this mechanical limit to fluid pressure, it is commonly assumed that metamorphism takes place with $P_{fluid} = P_l$ and this is also the condition used in many experimental studies.

The second important constraint concerns the composition of the fluid phase. Dehydration reactions take place at lower temperatures if the pressure of H_2O in the pores is low. However, they also release H_2O which will cause P_{H_2O} to rise. It is generally agreed that the porosity of rocks undergoing metamorphism must be small, probably less than 1 per cent under many conditions, so that only a small proportion of the hydrous phases in the rock need break down before the pores are filled with water at high pressure. This means that it will only rarely be possible for dehydration or decarbonation reactions to proceed to any great extent with P_{H_2O} or P_{CO_2} remaining significantly less than P_l. This subject is returned to in Chapter 5.

REDOX REACTIONS

The oxygen fugacity in metamorphic rocks (above, page 37) is so low that the amounts of oxygen available to participate in reactions or diffuse through rocks are negligible. However, where oxygen- or hydrogen-bearing fluids are being generated by reactions within a rock and escaping from it, or are infiltrating the rocks from elsewhere, it does become possible to change the oxygen content by redox reactions.

For example, Shaw (1956) found that high grade metapelites have comparable Fe-contents to similar unmetamorphosed sediments, but usually a higher proportion of it occurs as ferrous, rather than ferric, iron (Table 1.1). At the same time the amount of carbon present as graphite in the high grade rocks is often much less than that present in decayed organic matter in the original sediment. This can be explained if, during metamorphism, carbon combines with ferric iron minerals to produce CO_2, which then leaves the rock with other pore fluids, leaving behind ferrous iron minerals, i.e.:

$$C + 2'Fe_2O_3' = CO_2 + 4'FeO' \qquad [2.4]$$

where $'Fe_2O_3'$ and $'FeO'$ denote ferric and ferrous iron in unspecified minerals, rather than occurring as free oxides (Miyashiro, 1964).

The most extreme example of reduction during metamorphism is the serpentinisation of olivine-rich periodites. Here, magnesium and silicon in the original olivine react with infiltrating water to produce serpentine, but the ferrous iron from the olivine combines with oxygen from the water to produce magnetite. This liberates hydrogen, and can result in waters with an appreciable hydrogen content and conditions so reducing that Fe–Ni metal alloys are able to form. Frost (1985) gives an account of this reaction which is a comprehensive explanation of redox processes in metamorphism.

APPLICATION OF CHEMICAL EQUILIBRIUM TO NATURAL ROCKS : AN EXAMPLE

Before attempting to use the principles of chemical equilibrium to interpret metamorphic rocks in general, it is worth demonstrating that the approach is valid, using an example of metamorphism in rocks whose chemistry is sufficiently simple that they can be very closely approximated by a three-component chemical system, enabling the use of simple compositional phase diagrams.

The example chosen is of metamorphosed ultrabasic rocks, since these are composed predominantly of MgO, SiO_2 and H_2O, although CaO, Al_2O_3 and some FeO are also usually present in minor amounts. Bodies of dunite with wehrlite, lherzolite and related types are common in orogenic belts and have often been extensively hydrated. Subsequent regional or contact metamorphism may cause breakdown of some of the hydrous phases and even lead to regeneration of the original igneous minerals growing as new metamorphic grains with distinctive texture.

Trommsdorf & Evans (1972) have described the effects of contact metamorphism in the aureole of the Bergell tonalite of the Italian Alps (whose location is shown on Fig. 7.10) on the serpentinite body of Val Malenco. The regional assemblage in the serpentinite is antigorite + forsterite + diopside + minor chlorite and magnetite. Trommsdorf and Evans found a sequence of contact metamorphic zones on approaching the intrusion, each characterised by a particular mineral association:

Zone A: antigorite+olivine+diopside (regional assemblage).
Zone B: antigorite+olivine+tremolite.
Zone C: talc+olivine+tremolite.
Zone D: anthophyllite+olivine+tremolite (enstatite is present at one locality).

The distribution of the zones is shown in Fig. 2.4. The metamorphic reactions that take place at the isograd between Zones B and C and Zones C and D involve only phases that can be made from the components MgO, SiO_2 and H_2O. Other minerals present do not participate in the reactions and are effectively inert; they are analogous to the vessel in which a chemist performs an experiment. The compositions of the reacting phases, antigorite, olivine, talc, anthophyllite, enstatite and the H_2O fluid assumed to have been given off by the dehydration reactions, are plotted on a triangular diagram with MgO, SiO_2 and H_2O as the corner compositions in Fig. 2.5(*a*). Note that this diagram has been constructed using the **molecular proportions** of the components so that enstatite ($MgSiO_3$) plots at the midpoint of the MgO–SiO_2 side of the triangle. The mineral assemblage of Zone B (or at least that part of it which is involved in the reactions) is shown in Fig. 2.5(*b*) by the triangle of tie-lines connecting the phases present. Although only

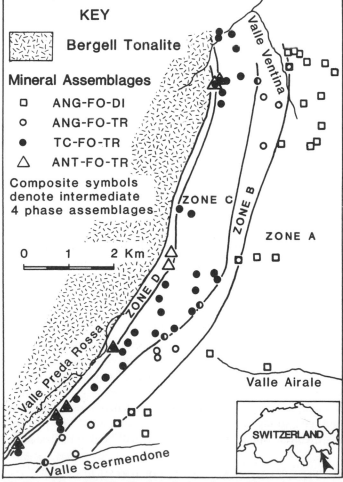

Fig. 2.4 Map of the distribution of mineral assemblages in metamorphosed ultrabasic rocks in the aureole of the Bergell tonalite, Italy (see also Fig. 7.10). After Trommsdorff & Evans (1972). Mineral abbreviations are given in the glossary.

antigorite and olivine are present today in rocks from Zone B, it is assumed that an aqueous fluid phase was also present during metamorphism. Since this assemblage has three components and three phases it follows from the phase rule that it has two degrees of freedom, and can thus exist over a range of temperature and pressure.

The exercise is repeated for the Zone C assemblage in Fig. 2.5(*c*). The difference from Zone B is that antigorite is no longer present and its composition lies within the talc–olivine–H_2O triangle. In other words antigorite cannot be stable under the conditions of Zone C and a rock with the *chemical composition* of pure antigorite would recrystallise to talc + olivine + H_2O. This tells us that the reaction that takes place in passing to Zone C is:

$$\text{antigorite} \rightarrow \text{olivine} + \text{talc} + H_2O \qquad [2.5]$$

which can be balanced as:

$$5Mg_3Si_2O_5(OH)_4 \rightarrow 6Mg_2SiO_4 + Mg_3Si_4O_{10}(OH)_2 + 9H_2O$$

and represents the ultimate high temperature stability limit of serpentine. Comparing the assemblage of Zone D (Fig. 2.5(*d*)) with that of Zone C, a rather different picture is present. The composition of talc does not lie within the anthophyllite–olivine – H_2O triangle, and so it is possible that talc could survive under Zone D conditions in rocks that were more siliceous than the Val Malenco body. What we can be sure of is that talc can no longer coexist with olivine in Zone D, because talc and olivine lie on opposite sides of the new anthophyllite – water tie-line which characterises Zone D. Phases can only coexist over a range of P and T (divariant assemblages) if they are connected by a tie-line on a diagram of this type. The reaction that takes place between Zones C and D results in the replacement of the association talc + olivine, stable in Zone C, with the association anthophyllite + H_2O (found in Zone D and by implication not possible in Zone C because they lie on opposite sides of the olivine–talc tie-line). Hence the reaction must be:

$$\text{olivine} + \text{talc} \rightarrow \text{anthophyllite} + H_2O \qquad [2.6]$$

or: $4Mg_2SiO_4 + 9Mg_3Si_4O_{10}(OH)_2 \rightarrow 5Mg_7Si_8O_{22}(OH)_2 + 4H_2O$

A reaction of this type is said to involve a '**tie-line flip**' on the phase diagram; it does not represent the ultimate limit of stability of any of the phases. The reason why talc is not found in Zone D is that there is always more olivine available than is required to react with the amount of talc present initially, and so reaction [2.6] continues until all the talc has been used up.

The rocks of the Bergell aureole provide a clear demonstration of a progressive sequence of contact metamorphic zones, each characterised by a divariant assemblage, with univariant assemblages occurring near the zone boundaries. The principle of interpreting metamorphic rocks in terms of chemical equilibrium is therefore well founded in this case.

This example is particularly straightforward because solid solution effects are not important, and so the reactions are all discontinuous. The plotting of assemblages and reactions involving solid solution minerals will be demonstrated in Chapter 3.

A FURTHER REFINEMENT: THE PROJECTION

Representing the reactions that take place at Val Malenco by the system $MgO–SiO_2$–H_2O is a simplification because other components and phases are also present and one

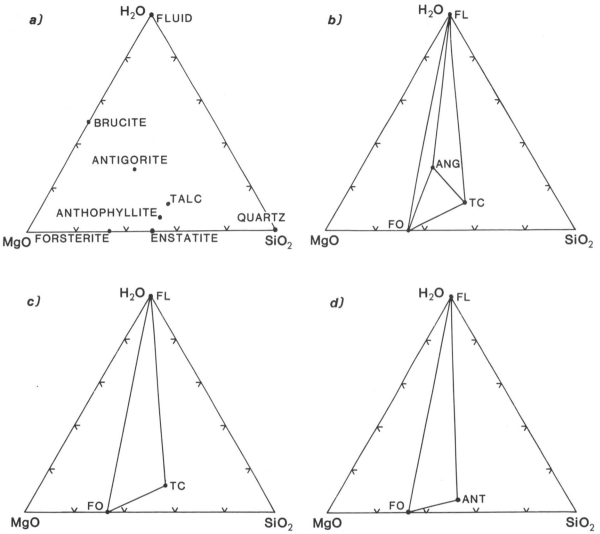

Fig. 2.5 MgO–H$_2$O–SiO$_2$ triangular diagrams showing: *a*) the compositions of the major phases of metamorphosed ultrabasic rocks; *b*)–*d*) phase compatibilities in successive zones of the Bergell aureole (Fig. 2.4).

isograd involves the calcic phases diopside and tremolite. This introduces the chief problem with metamorphic phase diagrams: how to represent systems of more than three components on a flat piece of paper. The problem can often be overcome by using a **projection** which shows only a particular subset of the total mineral assemblage. For example, all the assemblages shown in Fig. 2.5(*b*)–(*d*) contain H$_2$O as a phase. We could therefore represent them all by just showing the pair of minerals that coexist with H$_2$O in each zone, plotting their compositions in terms of their proportions of MgO to SiO$_2$. This is done by drawing a line from the H$_2$O apex of the H$_2$O–MgO–SiO$_2$ triangle through the composition of the mineral until it intersects the SiO$_2$–MgO side at a point corresponding to the **projected composition** of that mineral. The procedure is shown for talc in Fig. 2.6(*a*). Using this technique we can represent all mineral assemblages *that include H$_2$O* on a one-dimensional MgO–SiO$_2$ diagram, and this can be used to deduce the reactions in exactly the same way as the triangular diagrams of Fig. 2.5, by assuming that H$_2$O is always

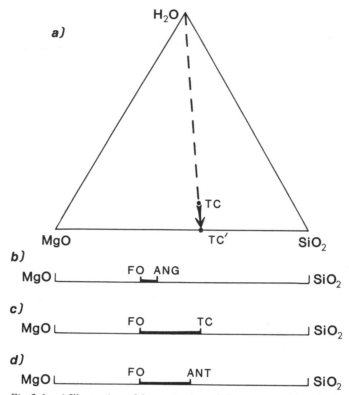

Fig. 2.6   *a*) Illustration of the projection of phase compositions in the system $MgO-H_2O-SiO_2$ on to the $MgO-SiO_2$ join. *b*)–*d*) Mineral assemblages (including an aqueous fluid phase) from the Bergell aureole (Fig. 2.5 *b*)–*d*)) shown projected on to an $MgO-SiO_2$ binary diagram from H_2O.

available to balance reactions as necessary. The $MgO-SiO_2$ projections for Zones B, C and D in the Bergell aureole are shown in Fig. 2.6(*b*)–(*d*).

Having reduced the three-component $MgO-SiO_2-H_2O$ system to one line introduces the possibility of representing a fourth component with a triangular diagram. To illustrate the reaction from Zone A to Zone B, CaO is an essential component, and in Fig. 2.7(*a*) all the participating phases are shown on a $CaO-MgO-SiO_2$ projection, projected from H_2O. Again, the projection means that only assemblages containing H_2O as a fourth phase can be represented, and the phases are plotted according to the relative proportions of CaO, MgO and SiO_2 in their formulae, neglecting any H_2O. Some projections used in subsequent chapters are more difficult to plot than this, although the underlying principle is the same. As an exercise, deduce the reaction between Zone A and Zone B using the $CaO-MgO-SiO_2$ diagrams of Fig. 2.7(*b*), (*c*).

EVIDENCE FOR EQUILIBRIUM IN METAMORPHISM: SUMMARY AND CRITIQUE

The best evidence that mineral assemblages once existed in thermodynamic equilibrium comes from regional studies that demonstrate: (a) uniformity of mineral assemblage for all rocks of specific chemical composition within well-defined areas; and (b) systematic

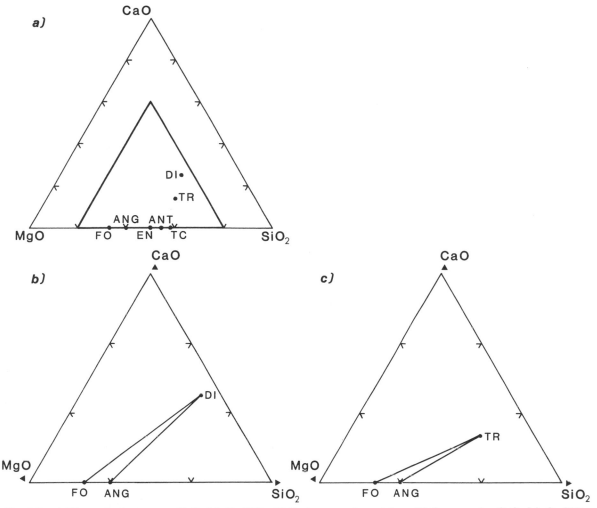

Fig. 2.7   *a*) Phases in the system $CaO-MgO-SiO_2-H_2O$ shown projected from H_2O on to the $CaO-MgO-SiO_2$ triangle. *b*), *c*) Phase compatibilities in Zones A and B of the Bergell aureole respectively, shown within the portion of the $CaO-MgO-SiO_2$ triangle outlined in *a*).

variations in mineral assemblage for rocks of a particular chemical composition over a larger distance.

A somewhat different problem is presented by an individual rock or thin section. Almost every metamorphic rock that one studies will have had a more or less prolonged history of recrystallisation, with probably one episode dominant. Usually, but by no means invariably, this will be the metamorphic episode that took place at the highest temperature attained by the rock. However, the rock may retain minerals that were present at an earlier stage in the metamorphism and have not broken down completely (relics) or grains that grew later than the main episode during partial recrystallisation, e.g. retrograde metamorphism. How can these be distinguished?

In an ideal rock each mineral will be found to occur in contact with every other mineral within a single thin section, and the grains will show no evidence of breakdown. This ideal textural arrangement is approached in some carbonate-rich rocks, or rocks metamorphosed at very high temperatures, but usually textures are more equivocal. The following criteria can be useful pointers, but should not be followed slavishly.

1. *Is the rock obviously layered?* Compositional variation between layers may be observed even on the scale of a single thin section, and it would clearly be inappropriate to assume that minerals from different layers were necessarily in equilibrium with one another.

2. *Does a particular mineral occur in contact with the grain boundary network of the rock?* Most movement of material between reacting grains in a rock takes place by diffusion along the grain boundaries rather than through mineral lattices, except at the very highest temperatures. As a result grains that become isolated from the continuous network of grain boundaries in the rock may be preserved as relics because other components are unable to react with them. There are two modes of occurrence of such relics: (a) as cores to zoned grains of a single mineral species, such as sodic cores to calcic amphiboles or the conspicuous coloured zones seen in tourmaline in many thin sections of schist; or (b) as inclusions within a different mineral species, for example chloritoid or staurolite may occur as relic inclusions in garnet or plagioclase in pelitic schists from higher grade zones than those where they are normally stable (Ch. 6, page 163). A cautionary note however: a grain that appears in thin section to be an inclusion may sometimes have been in connection with the rock matrix in the third dimension.

3. *Is there a strong association between a particular pair of minerals?* This is a difficult criterion to apply but many retrograde minerals tend to replace a specific higher temperature mineral, which helps distinguish them from the peak-metamorphic assemblage. For example chlorite often fringes and replaces garnet or biotite, and sericitic white micas often replace Al_2SiO_5 minerals, staurolite or feldspar. Fortunately, pseudomorphs of fine grained retrograde alteration products such as sericite are usually quite conspicuous. On the other hand, if the metamorphic episode that gave rise to the most abundant minerals in a rock was itself retrograde, this process of selective replacement will give rise to a non-random distribution of different retrograde metamorphic minerals even if they grew in equilibrium! The problem is particularly acute for low grade metamorphism of igneous rocks, where each different igneous mineral may be replaced by a different metamorphic mineral. There are some cases where even at high temperature, specific minerals tend to be associated with one another. In particular, sillimanite often grows on biotite, even though it may be in equilibrium with other minerals in the rock. Yardley (1977b) described staurolite–sillimanite schists in which staurolite and sillimanite never occurred in contact with one another although there was every reason to suppose that they were once in equilibrium.

4. *Are there theoretical reasons for supposing that all minerals were not in equilibrium with one another?* This is a line of argument that must be applied only with great caution and requires a good understanding of the ranges of conditions over which particular mineral assemblages may be stable. However, it is sometimes apparent that, from phase rule considerations, a rock contains too many phases. A rock in which andalusite, kyanite and sillimanite are all present might just conceivably have been metamorphosed at the unique value of P and T at which all these three can coexist, but it is much more probable that conditions of metamorphism changed as it recrystallised so that at different times each polymorph formed in its own stability field, but failed to break down completely as it became unstable.

Even where an assemblage does not overtly contravene phase rule criteria for equilibrium, it can sometimes be argued that one mineral in a rock would not have been stable at the same conditions as the others, although it is an argument that is easily abused. For example chlorite can be stable to very high temperatures (sillimanite zone) in rocks of suitable composition, and is not necessarily a retrograde phase in high grade rocks. On the other hand the association of chlorite + K-feldspar is stable only under very low grade (chlorite zone) conditions and chlorite in, say, a garnet-bearing schist with K-feldspar is almost certain to be of retrograde origin.

Where chemical analyses of minerals are available it is possible to use the partitioning of elements between mineral solid solutions as a criterion for equilibrium. Distribution coefficients (K_D) for partitioning of Fe and Mg between pairs of ferromagnesian minerals are quite well known for most metamorphic conditions. When very different results are obtained for a particular mineral pair in an individual sample it is likely to mean that those two minerals did not in fact coexist in equilibrium.

5. *Textures indicative of reaction*   It has been pointed out earlier in this chapter that many metamorphic reactions take place over a range of conditions, and within that range products and reactants can coexist in equilibrium even though one is tending to grow at the expense of the other. For this reason *the fact that a particular mineral appears to be corroded and breaking down does not necessarily indicate that it was out of equilibrium with the other phases in the rock*.

METAMORPHIC FACIES

Once the case is made that some metamorphic mineral assemblages represent associations of phases that were once in equilibrium, these assemblages can be used to draw conclusions about the sorts of pressures and temperatures that prevailed during metamorphism. Until fairly recently it was only possible to compare the relative temperatures and pressures of formation of different metamorphic rocks, but now absolute determinations can be made in favourable circumstances.

The principal problem that arises when comparing metamorphic conditions between different areas is that of relating the grades of metamorphism of different types of rocks. For example, although it is possible to recognise separate staurolite, kyanite and sillimanite zones in the Dalradian rocks of Scotland (Ch. 1), the associated metabasites have the same assemblage (dominated by hornblende and plagioclase) in all three zones. In regions where pelitic schists are absent but metabasites occur, it would not be possible to recognise whether all these zones, or only one of them, were present.

Zonal schemes have been devised for different regions based on a range of rock types including pelitic schists, metabasites, calc-silicate rocks, marbles and even cherts. The precise correlation of $P–T$ conditions between zones based on different rock types is, however, difficult. To overcome this problem Eskola (1915 and subsequent publications), devised a scheme of **metamorphic facies** which represents broader $P–T$ subdivisions than most specific zones recognised in individual rock types.

A metamorphic facies can be defined as a range of pressure–temperature conditions over which a particular common mineral assemblage or range of mineral assemblages are stable. The individual facies are also defined so that rocks of a wide range of compositions can be assigned to a facies, whereas they could not necessarily be correlated with a specific zone identified in a different rock type.

Eskola recognised facies principally on the characteristic mineral assemblages of metabasic rocks. He also correctly deduced the relative pressures and temperatures represented by the different facies, basing his deductions on the facts that: (a) rocks formed at higher pressures will tend to contain denser minerals than rocks of the same composition metamorphosed at lower pressures; and (b) devolatilisation reactions almost invariably take place with increased temperature, so that progressively higher temperature rocks will have progressively lower volatile (H_2O and CO_2) contents.

A scheme of metamorphic facies, based on that of Eskola and approximately calibrated in the light of more recent experimental work, is illustrated in Fig. 2.8; details of the assemblages typical of each facies are discussed in Chapter 4 and are summarised in Table 4.1. It is convenient to consider the facies in four groups:

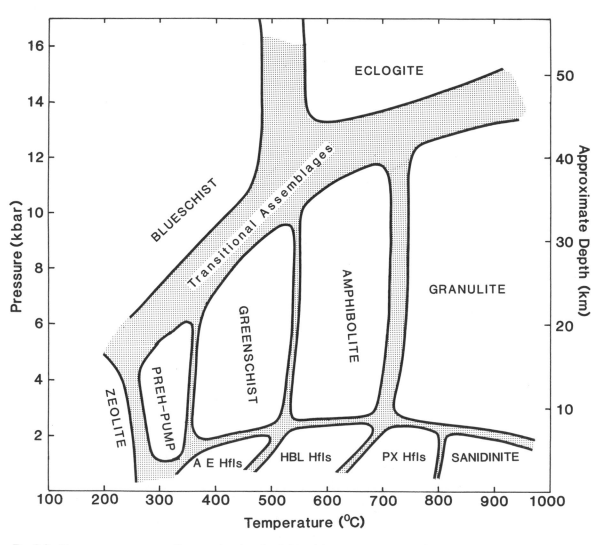

Fig. 2.8 Pressure–temperature diagram showing the fields of the various metamorphic facies. Details of the characteristic assemblages are given in Chapter 4 (Table 4.1). Abbreviations used are: Hfls = hornfels, AE = albite–epidote, HBL = hornblende, PX = pyroxene, PREH–PUMP = prehnite–pumpellyite.

1. *Facies of moderate pressure and moderate to high temperature* The greenschist, amphibolite and granulite facies probably account for the bulk of metamorphic rocks. The pelite zones of the Scottish Highlands, described in Chapter 1, belong to the greenschist and amphibolite facies; granulite facies rocks form at still higher grades.
2. *Facies of low grades* These facies were recognised later than the others, principally through the work of Coombs and colleagues in New Zealand. In this book, low grade rocks are divided into the zeolite facies and the prehnite–pumpellyite facies, but more complex subdivisions have been proposed by most recent workers (Turner, 1981; Liou, Maruyama & Cho (1987). In the present writer's opinion these further subdivisions only have the status of zones since they can only be distinguished in a small range of rock types.
3. *Facies of contact metamorphism* At the low pressures and high temperatures of metamorphic aureoles rather distinctive assemblages are often produced, although there is considerable overlap and gradation between assemblages of the albite–epidote hornfels facies, hornblende hornfels facies and pyroxene hornfels facies, and those of the greenschist, amphibolite and granulite facies respectively. For example, in pelites at these low pressures, almandine garnet is absent, and the Al_2SiO_5 phase is normally andalusite.
4. *Facies of high pressures* The blueschist and eclogite facies represent distinctive conditions characterised by the production of relatively dense phases under conditions of unusually high pressure. Blueschist facies metabasites obtain their characteristic 'blue' colour (most commonly a lilac–grey in hand specimen) from the presence of the sodic amphibole glaucophane rather than the more common green calcic amphiboles. Similarly the eclogite facies is characterised by the occurrence of Na-rich clinopyroxene (omphacite) instead of calcic clinopyroxene in metabasites.

DETERMINATION OF PRESSURE – TEMPERATURE CONDITIONS OF METAMORPHISM

When the scheme of metamorphic facies was first developed, geologists had little idea of the absolute values of pressure and temperature to which rocks had been subjected. With the development of experimental techniques for the study of mineralogical phase equilibria at high pressures and temperatures in the 1950s and 1960s, the conditions of formation of many naturally occurring mineral assemblages have been defined, and we now have a reasonably good idea of the sorts of pressures and temperatures required for the formation of many common rock types. Mineral indicators of metamorphic P or T are known as **geobarometers** and **geothermometers**.

GEOTHERMOMETRY AND GEOBAROMETRY

There are several different approaches to determining the $P–T$ conditions of formation of a mineral assemblage *once it has been established that the assemblage formed at equilibrium*. All the most widely used ones are based on mineralogical reactions, which may be univariant, continuous or cation exchange reactions. Obviously the most useful reactions for determining pressure of metamorphism will be those which take place at nearly the same pressure over a wide range of temperature, while a reaction with the opposite characteristics will make a good temperature indicator. It is possible to predict reasonably well which reactions will best fit these requirements.

At the beginning of this chapter it was noted that the variation in the Gibbs free energy of a mineral (or assemblage of minerals) with changing pressure depends on the molar volume of the mineral, whereas variation in Gibbs free energy with temperature depends on entropy. We can define the volume change of a reaction:

$$\Delta V_r = V_{products} - V_{reactants}$$

Similarly the entropy change is:

$$\Delta S_r = S_{products} - S_{reactants}$$

A reaction that makes a good pressure indicator will have a large ΔV_r but small ΔS_r. This means that variations in pressure will lead to large changes in the relative free energies of reactants and products, making reaction more likely (Fig. 2.2(*a*)). Fluctuations in temperature will have relatively little effect. In contrast, reactions with a large ΔS_r but small ΔV_r are predominantly temperature sensitive. Some reactions are almost exclusively sensitive to temperature, especially cation exchange reactions, because the volume change accompanying such reactions is very small. Devolatilisation reactions are also predominantly temperature sensitive under most conditions of pressure and temperature because of the large increase in entropy that accompanies release of volatiles from ordered mineral structures. Most pressure-sensitive reactions are, however, also somewhat sensitive to temperature and so good geobarometers are much rarer than good geothermometers.

The following section is intended only to provide an outline of the types of geothermometers and geobarometers that have been used in recent years. Further details are given in subsequent chapters, while the article by Essene (1982) is an extremely readable and comprehensive introduction to most of the geothermometers and geobarometers in current use, despite being already outdated in its details.

A petrogenetic grid — The simplest, and often the most reliable, way of determining the *P–T* conditions of formation of a particular mineral assemblage is to compare it with a *P–T* diagram showing the univariant curves for the reactions that control the appearance and breakdown of the phases present. This sort of diagram was conceived by N.L. Bowen, who termed it a **petrogenetic grid**. Today, experiments have been carried out to determine the location of many univariant curves on the petrogenetic grid, providing limits to most of the common assemblages of metamorphism, and there will be examples in the following chapters. The *P–T* conditions of other univariant curves that have not been studied experimentally, can instead be calculated from thermodynamic data in many instances.

The majority of metamorphic minerals form solid solutions of variable composition, and the different end-members may have different stability limits. This means that experiments performed on a specific mineral composition may not accurately define the conditions of formation of the same assemblage in a rock in which the solid solution mineral has a different composition. For example, the white micas found in pelitic schists metamorphosed under moderately high grade conditions, such as the staurolite zone, are solid solutions between paragonite and muscovite. At high grades white mica breaks down to produce alkali feldspar which is also a solid solution between Na and K end-members. The breakdown reactions for the two white mica end-members are:

$$NaAl_3Si_3O_{11}(OH)_2 + SiO_2 \rightarrow NaAlSi_3O_8 + Al_2SiO_5 + H_2O \qquad [2.7]$$

$$KAl_3Si_3O_{11}(OH)_2 + SiO_2 \rightarrow KAlSi_3O_8 + Al_2SiO_5 + H_2O \qquad [2.8]$$

The univariant curves for these reactions (with $P_{H_2O} = P$) have been determined experimentally by Chatterjee (1972) and Chatterjee & Johannes (1974) and are shown in

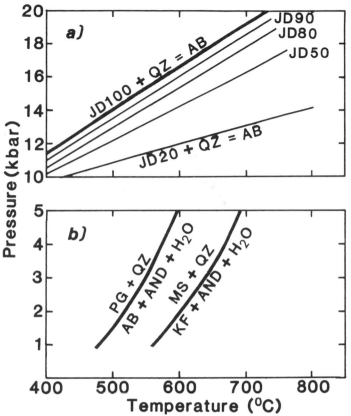

Fig. 2.9 Examples of the influence of solid solution on mineral stabilities. *a*) Effects of solid solution in jadeite on the equilibrium between jadeite-rich pyroxene, albite and quartz. Equilibrium curves are shown for various pyroxene compositions intermediate between jadeite and diopside, labelled according to mole per cent jadeite in the pyroxene (after Holland, 1983). *b*) Univariant curves for the high temperature breakdown of the end-member white micas, paragonite and muscovite, in the presence of quartz. Data from Chatterjee (1972) and Chatterjee & Johannes (1974).

Fig. 2.9(*b*). Natural white micas will break down at intermediate conditions according to composition. The difference between the *P–T* conditions of the two curves is not particularly great in this case, and both the product and the reactant assemblages involve Na–K solid solution minerals. As a result, it is reasonable to use the experimentally-determined muscovite breakdown reaction on a petrogenetic grid to indicate the approximate conditions under which muscovite–quartz assemblages react to K-feldspar–Al-silicate assemblages, irrespective of the presence of Na in the natural minerals, because the Na-content of the muscovites in appropriate rocks is usually small.

In their study of the Bergell aureole outlined earlier in this chapter, Trommsdorf and Evans (1972) were able to show quite rigorously that the *P–T* conditions of the reactions that took place among $CaO–MgO–SiO_2–H_2O$ phases in the aureole were not significantly affected by the small amount of Fe substituting for Mg in the natural minerals. A petrogenetic grid, indicating the approximate temperatures attained at the different isograds in the Bergell aureole, is shown in Fig. 2.10.

There has been a tendency to underestimate the value of the petrogenetic grid in some recent studies. In fact, a number of important metamorphic mineral assemblages are

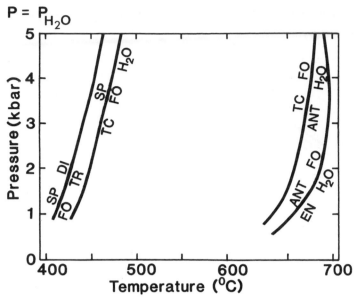

Fig. 2.10 Petrogenetic grid showing equilibrium curves for the reactions encountered in the Bergell aureole. The curves have been calculated from the data set of Holland & Powell (1985). Note that the serpentine phase (SP) for which data are available is chrysotile, whereas it is antigorite that predominates in the Bergell rocks.

stable over such a restricted range of P–T conditions that their presence defines metamorphic conditions just as well as superficially more sophisticated calculation procedures.

Markedly continuous reactions Many important natural metamorphic reactions can be approximated by univariant curves on a petrogenetic grid, even where there is some solid solution in both products and reactants.

However, there are other common reactions in which product and reactant minerals may display different types of solid solution. For example the reaction:

$$NaAlSi_3O_8 \rightarrow NaAlSi_2O_6 + SiO_2 \qquad [2.9]$$
$$\text{albite} \qquad\quad \text{jadeite} \qquad \text{quartz}$$

found in blueschist facies rocks, is univariant in the pure end-member system. However, because jadeite is a pyroxene it can undergo markedly different atomic substitutions from albite, e.g. towards diopside, $CaMgSi_2O_6$ (i.e. $(Ca, Mg) \rightleftharpoons (Na, Al)$) or aegirine, $Na Fe^{3+}Si_2O_6$ ($Fe^{3+} \rightleftharpoons Al$). Neither of these types of substitution can occur in albite, and so for example Fe^{3+} and Al^{3+} cannot be considered to behave as a single component because albite effectively discriminates between them. In natural rocks containing Ca, Mg, Fe^{2+} and Fe^{3+}, the assemblage albite (pure) + quartz + Na-rich pyroxene solid solution is not uncommon because the reaction is continuous and reactants and products can coexist with several degrees of freedom over a large P–T interval.

The breakdown of albite by reaction [2.9] has been studied experimentally by several workers (most recently, Holland, 1980) and a univariant curve for the end-member reaction is shown in Fig. 2.9(*a*). The assemblage jadeite + quartz is restricted to unusually high pressures in the pure system. However, natural jadeite-rich pyroxenes can occur coexisting with albite and quartz in rocks formed at much lower pressures. Curves for the coexistence of a range of pyroxene compositions with albite and quartz are also shown in Fig. 2.9(*a*). It is clear that such pyroxenes are stable over a much wider range of pressures than pure jadeite. A simple rule of thumb is that, when comparing a natural assemblage with experiments in a pure end-member system, *the position of the equilibrium curve for coexistence of products and reactants will be shifted in such a way as to extend the stability field of the phases which are most impure.*

A comparable equilibrium, also widely used as an indicator of metamorphic pressure, is based on the breakdown of anorthite with increasing pressure:

$$3CaAl_2Si_2O_8 = Ca_3Al_2Si_3O_{12} + 2Al_2SiO_5 + SiO_2 \qquad [2.10]$$
$$\text{anorthite} \qquad\quad \text{grossular} \qquad\quad \text{kyanite} \qquad \text{quartz}$$

Here, both the anorthite and grossular normally occur as dilute components of solid solutions (Newton & Hasleton, 1981).

Markedly continuous reactions of this sort are difficult to interpret in terms of $P-T$ conditions of metamorphism from the petrogenetic grid. However, if the compositions of the natural phases are known it is often possible to use thermodynamic calculations to estimate the $P-T$ conditions of formation of a natural assemblage (see the adjacent box for an outline of the principles). The details of such calculations are beyond the scope of this book, but are given in textbooks of thermodynamics for petrology, listed at the end of the chapter. The most important use of continuous reactions in this way is as geobarometers, rather than geothermometers, and Newton (1983) has provided a review of the most important ones.

HOW TO MAKE THE JUMP BETWEEN EXPERIMENTS ON PURE SYSTEMS AND EQUILIBRIA BETWEEN COMPLEX NATURAL MINERALS: THE CONCEPT OF ACTIVITY

More often than not, natural solid solution minerals have compositions that differ from those of the pure end-members used in experiments. To calculate $P-T$ conditions for rocks from continuous equilibria involving solutions it is therefore necessary to correct for these compositional differences.

The way in which this is done is to determine the **activities** of the end-member substances as they occur in solution in the natural minerals. Activity, as normally used in geology, amounts to the *thermodynamically effective concentration* of the component in the solution. For example, in a pure H_2O fluid, the activity of water (a_{H_2O}) is 1, whereas if the fluid is a mixture of H_2O and CO_2, then $a_{H_2O} < 1$. Similarly, in a pure jadeite pyroxene, $a_{jd} = 1$, whereas in an intermediate clinopyroxene the activity will be less, but will depend on the precise composition of the mineral. Knowing the activities of end-members occurring as components in natural minerals, it becomes possible to calculate precisely how much the equilibrium curve will be displaced from its position in the pure system. The catch is that the relationships between activity and composition are often complex, and are only poorly understood for many of the most important mineral solid solutions.

Strictly speaking, activity is a dimensionless ratio whose value varies according to the way in which the ratio is defined. This is because activity is the ratio between the fugacity of the component in the natural solution and its fugacity in a reference condition known as a **standard state**. The fugacity of the component can be thought of as its vapour pressure in a gas phase coexisting with the mineral in its natural or standard state, and the standard state is usually defined as the component occurring as a pure phase at the metamorphic P and T, so that the pure phase always gives an activity of 1.

Cation exchange reactions

Because these reactions do not involve growth or breakdown of mineral lattices, merely exchange of cations between phases, the volume change due to the ion exchange in one phase is usually almost exactly compensated by the volume change accompanying the reverse exchange in the other phase. As a result $\Delta V_r \simeq 0$ and cation exchange reactions are almost solely temperature dependent. The study of cation exchange reactions was pioneered by Kretz (1959) and Albee (1965) in the U.S., and by Perchuk in the U.S.S.R., but only became routine when the development of the electron probe microanalyser made it possible to obtain rapid and accurate analyses of coexisting solid solution minerals in a rock. As cation exchange proceeds, the parameter K_D (defined above, page 40) changes. There are thermodynamic reasons for predicting that a plot of $\ln K_D$ versus $1/T$ will yield a straight line; and hence variation in K_D, if large enough, can make a useful geothermometer.

One of the best known cation exchange reaction geothermometers is the exchange of Fe and Mg between garnet and biotite (reaction [2.3] page 40). This has been studied experimentally by Ferry & Spear (1978) who reacted biotite and garnet crystals of different compositions at a variety of temperatures and measured the final mineral compositions at the end of the run. Their results are plotted in the form of $\ln K_D$ against $1/T$ in Fig. 2.11(*a*). Such a diagram provides a quantitive calibration of the temperature sensitivity of the reaction. If a K_D value is calculated from mineral analyses in a particular rock, Fig. 2.11(*a*) can be used to estimate the temperature at which the mineral assemblage equilibrated.

In practice, the value of K_D determined for natural rocks is influenced by the presence of additional components that were absent from Ferry & Spear's experiments; for example Ca and Mn in garnet, Ti in biotite. In recent years a host of refined calibrations of this thermometer have been proposed which purport to overcome this problem, but none is of undoubted universal applicability.

Cation diffusion (described in more detail on page 150) takes place quite readily in many minerals at the temperatures of high grade metamorphism but becomes rapidly less effective as temperature drops. As a result minerals may change their composition as the rock cools, so that the temperature indicated by cation exchange equilibrium may be fact be a 'blocking temperature', merely representing the temperature at which cation diffusion becomes ineffectual at geological cooling rates.

Oxygen isotope thermometry   In much the same way that solid solution minerals can exchange cations, so can many rock-forming minerals and fluids exchange oxygen atoms of different masses. There are three isotopes of oxygen found in nature: ^{16}O, ^{17}O and ^{18}O, and all are stable, i.e. they do not undergo radioactive decay. Different minerals can show different preferences for oxygen atoms of different masses in their structure in much the same way that ferro-magnesian minerals can show different preferences for Fe relative to Mg. A fractionation factor, α, can be defined which is analogous to the K_D parameter used to monitor the progress of cation exchange reactions.

$$\alpha_{A-B} = \frac{(^{18}O/^{16}O)_A}{(^{18}O/^{16}O)_B}$$

where the A and B suffixes denote different minerals.

Fractionation factors usually vary with temperature in a comparable way to K_D.

The application of oxygen isotopic exchange to determining metamorphic temperatures has been reviewed by Clayton (1981). It appears to give excellent results at low temperatures (where fractionation is also greatest), but oxygen diffusion in most mineral lattices becomes significant at higher grades, so that the temperatures obtained from amphibolite facies or higher grade rocks 'blocking temperatures', merely recording the temperature at which oxygen diffusion became ineffectual. Nevertheless, various studies

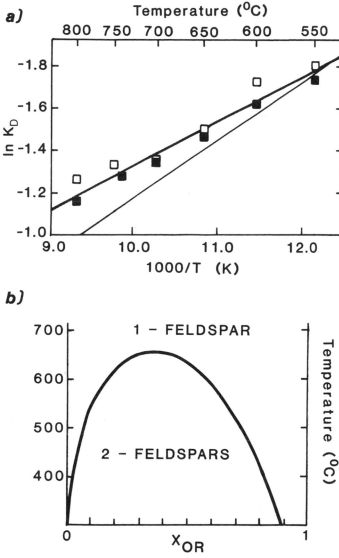

Fig. 2.11   Examples of metamorphic geothermometers. *a*) Plot of ln K_D versus $1/T$ for the Fe-Mg cation exchange reaction between biotite and garnet. Square symbols are experimental points of Ferry & Spear (1978), at 2.07 kbar. Filled squares are experiments using Fe-rich biotite; open squares are experiments with Mg–Fe biotite. The heavy line is a fit to the experimental points, whereas the light line is the empirical calibration of Thompson (1976). *b*) Alkali feldspar solvus, $\times_{OR}$ denotes the mol fraction of orthoclase in alkali feldspar. After Smith & Parsons (1974).

report systematic variation in α with grade from amphibolite facies rocks (e.g. O'Neill & Ghent, 1975).

Extent of mutual solid solution between mineral pairs   Some mineral pairs exhibit limited mutual solid solution at low temperatures, but are completely miscible at high temperature. The phase diagram for this sort of system exhibits a **solvus curve** above which only one phase of variable composition is found, but below which two coexisting immiscible phases occur (Fig. 2.11(*b*)). The solvus curve indicates the amount of mutual solid solution that can occur at any particular temperature, and this will be at a maximum when both phases coexist. Thus measurement of the composition of the coexisting minerals should define the temperature at which they crystallised together. Solvus geothermometers become more sensitive temperature indicators as temperature increases. This type of geological thermometer was first proposed by T.F.W. Barth for coexisting K-feldspar and albite (Fig. 2.11(*b*)), although it has seldom been successfully applied to this common assemblage in metamorphic rocks because of the complications introduced by additional solid solution between anorthite and albite, and by the ease with which cations diffuse in alkali feldspars down to quite low temperatures, changing the mineral's composition as it cools.

The most widely used thermometer of this type is based on mutual solution between Ca-rich clinopyroxenes or pigeonite and orthopyroxene. This was first developed by Davis and Boyd (1966) and is applicable principally to very high grade metamorphic rocks. It has been further refined by Wood and Banno (1973), who presented a thermodynamic treatment for natural pyroxene solid solutions, and by Wells (1977). Recent developments in the use of this thermometer are reviewed by Lindsley (1983).

The extent of solid solution between calcite and dolomite also provides a useful geothermometer, and is described further in Chapter 5.

Some mineral pairs with unlike structures exhibit incomplete mutual solid solution and are also of use as indicators of metamorphic conditions, e.g. solution of garnet in orthopyroxene results in significant Al-contents for some natural orthopyroxenes. This solid solution is favoured by high pressures and the Al-content of orthopyroxene coexisting with garnet has been studied experimentally for use as a geobarometer by Harley (1984a, b).

SUMMARY

This chapter has introduced the basic rationale behind many of our interpretations of metamorphic rocks, which are frequently based on the assumption that all or part of the mineral assemblage originally grew at equilibrium under some specific physical conditions to which the rock was subjected. Changing conditions of pressure and temperature may lead to reaction and the production of new assemblages, but while some reactions take place abruptly at a particular temperature and result in the disappearance of certain minerals and the growth of new ones, other reactions may take place over a range of conditions within which reactants and products coexist but change progressively in composition and abundance.

If assemblages of minerals that coexisted in equilibrium during metamorphism can be recognised, they can be used to estimate the depths and temperatures at which metamorphism occurred. Relative pressures and temperatures can usually be readily determined by assigning rocks to a metamorphic facies, and a petrogenetic grid can often be used to assign approximate numerical values of pressure and temperature. More precise estimation of conditions requires chemical or isotopic analyses of the minerals present.

FURTHER READING

Ernst, W.G., 1976. *Petrologic Phase Equilibria*. W.H. Freeman & Co., San Francisco.
Ferry, J.M. (ed) 1982. Characterization of metamorphism through mineral equilibria. *Reviews in Mineralogy*, **10**.

Fyfe, W.S., Turner, F.J. & Verhoogen, J., 1958. *Metamorphic Reactions and Metamorphic Facies*. Geological Society of America Memoir 73.

Greenwood, H.J. (ed) 1977. *Applications of Thermodynamics to Petrology and Ore Deposits*. Mineralogical Association of Canada Short Course Handbook 2.

Powell, R., 1978. *Equilibrium Thermodynamics in Petrology*. Harper & Row, London.

Wood, B.J. & Fraser, D.G., 1976. *Elementary Thermodynamics for Geologists*. Oxford University Press, Oxford.

3 METAMORPHISM OF PELITIC ROCKS

Pelitic rocks are derived from clay-rich sediments and are of particular importance in studies of metamorphism because they develop a wide range of distinctive minerals.

The term 'pelitic rocks' is sometimes used loosely as a field term to signify all slaty or schistose rocks with a high proportion of micas or other phyllosilicates. However, the distinctive minerals that characterise the metamorphic zones in pelites at high grades can only develop in a much more restricted range of compositions, rich in A1 and poor in Ca (Table 1.1), and it is with these 'true pelites' that this chapter is principally concerned.

REPRESENTATION OF PELITE ASSEMBLAGES ON PHASE DIAGRAMS

In chemical terms, reactions in pelitic rocks principally involve the components SiO_2, Al_2O_3, FeO, MgO, K_2O and H_2O, and most theoretical and experimental studies attempt to model natural rocks using this simplified system, which has become known as the KMFASH system from the initial letters. Other components, especially Fe_2O_3, TiO_2, MnO, CaO, Na_2O and C may also be present to a significant extent, but with some important exceptions they do not usually play a major role in the reactions that produce the key metamorphic index minerals.

In order to represent even the idealised six-component pelite system on a compositional phase diagram requires considerable simplification, and this can be done by means of projection, in much the same way as was done to represent phases in the CaO–MgO–SiO_2–H_2O system on a triangular phase diagram in Chapter 2. Almost all metapelites contain quartz, and we can also make the assumption that during progressive heating an aqueous fluid phase is present, since most of the reactions release H_2O. It is therefore reasonable to restrict our study to such rocks, and show pelite assemblages projected from quartz and H_2O. This means that the remaining minerals of the assemblage are plotted in a three-dimensional tetrahedron whose corners correspond to the remaining components: Al_2O_3, K_2O, FeO and MgO. It should be apparent from Fig. 3.1(*a*) that this is not an easy diagram to work with, since specific compositions cannot be plotted on it uniquely,

Fig. 3.1 opposite The Thompson AFM projection for representing mineral assemblages of pelites. *a*) Graphical representation of the projection of the composition of a biotite from muscovite on to the AFM plane. *b*) AFM diagram showing the major compositional variation of the most common pelite minerals (based on Winkler, 1976). Numbers along the left side are values of the A co-ordinate, those along the base are the M co-ordinate (equivalent to X_{Mg}).

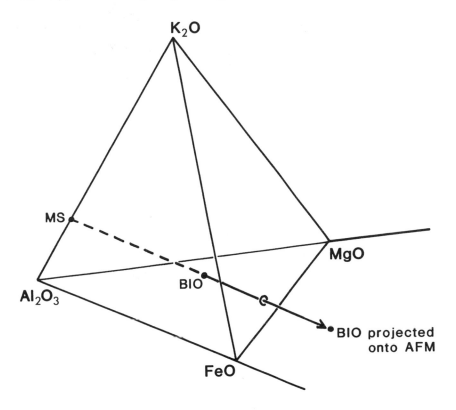

a)

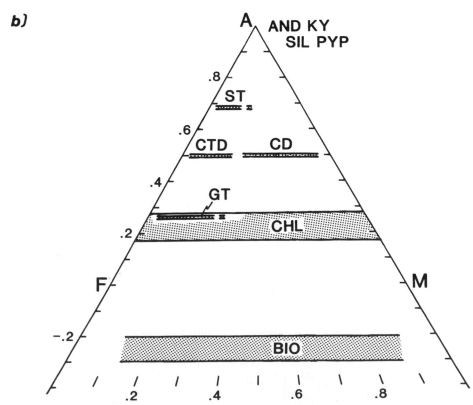

b)

although at least the problem has been reduced to one that can be represented on a sheet of paper!

Eskola, who pioneered the representation of metamorphic mineral assemblages on phase diagrams, attempted to solve the problem of producing a manageable triangular diagram for pelites by combining FeO and MgO as a single component to produce a triangular **AKF diagram** with Al_2O_3, K_2O and FeO + MgO as the corner components. Unfortunately, as we have already seen, when several ferromagnesian minerals coexist in a rock they commonly have different X_{Mg} values and are therefore effectively able to distinguish FeO and MgO as separate components. As a result many metapelite assemblages have four phases that can be plotted on the AKF diagram and this precludes the possibility of using the diagram in a rigorous way to decide whether different assemblages result from different metamorphic conditions or from different rock compositions, and to deduce the reactions that have taken place between zones.

A much more elegant solution to the problem is provided by the **AFM projection** of J.B. Thompson Jr. (1957). This projection is based on the fact that most metapelites contain muscovite, and involves projecting from the composition of muscovite on to the Al_2O_3–FeO–MgO (AFM) face of the Al_2O_3–K_2O–FeO–MgO (AKFM) tetrahedron, in addition to projecting from quartz and water. This procedure is illustrated in Fig. 3.1(*a*), which shows a biotite composition projected on to the AFM face by drawing a line from muscovite through biotite until it intersects the AFM plane. Note that for biotite, the projected composition on the Thompson AFM diagram lies outside the original AKFM tetrahedron.

The numerical procedure for calculating where a rock or mineral composition lies has to take into account the fact that muscovite contains Al, and therefore the AFM co-ordinates cannot be given simply by the relative (molecular) properties of Al_2O_3, FeO and MgO. Instead, for analyses containing K, the analysis must be recast to give its composition in terms of the components $K_2Al_6Si_6O_{20}(OH)_4$ (muscovite), Al_2O_3, FeO, MgO. The amount of muscovite component is dictated by the amount of K in the analysis, but for every mole of K_2O assigned to muscovite, 3 moles Al_2O_3 must also be assigned, reducing the amount of the Al_2O_3 component itself. Hence the AFM co-ordinates of any analysis of an AKFM mineral for plotting on the AFM projection are given by:

$$A = \frac{[Al_2O_3] - 3\,[K_2O]}{[Al_2O_3] - 3[K_2O] + [MgO] + [FeO]}$$

$$M = \frac{[MgO]}{[MgO] + [FeO]}$$

where the square brackets denote the number of moles of the oxide concerned in the analysis. To plot a rock analysis, the A value must be further reduced to allow for Al combined with Na and Ca in paragonite, plagioclase or epidote.

The A values obtained define the position of a horizontal line in terms of its distance from the F–M join towards A, while the M values define a line radiating from A, lying between the A–F join (M = 0) and the A–M join (M = 1). The analysis plots at their intersection. The compositions of the common pelite minerals are shown plotted on an AFM projection in Fig. 3.1(*b*).

Despite the greater complexity of the AFM projection, the principle is exactly the same as was demonstrated for projection of the MgO–SiO_2–H_2O system in Fig. 2.6 and AFM diagrams can be used to deduce reactions between zones. Only the assemblages of rocks that also contain muscovite and quartz, and can be assumed to have had an aqueous fluid phase during metamorphism, can be plotted on the AFM projection, and in attempting to balance reactions deduced from AFM projections it should be remembered that these phases may also have been involved.

In this chapter, AFM diagrams are used to illustrate some of the major changes that take place during metamorphism of pelitic rocks. A very thorough analysis of continuous and discontinuous reactions in pelites using this and other projection schemes is given by A.B. Thompson (1976).

PELITIC ROCKS AT LOW GRADES

Clay-rich sediments may undergo extensive changes during diagenesis, and there is no sharp distinction between diagenetic and metamorphic processes. Low temperature recrystallisation has been the subject of considerable research effort in recent years because of its significance for the development of hydrocarbon reservoirs, and for some types of geothermal field, but it requires rather specialist techniques because the very fine grain size of the materials precludes easy identification of the phases. The transition from diagenesis to low grade metamorphism in clastic sediments has been reviewed by Frey (1987) and Dunoyer de Segonzac (1970), while Curtis (1985) specifically reviews diagenetic to early metamorphic changes in clays.

During the advanced stages of diagenesis many clays become unstable and pelitic sediments are converted to mixtures of chlorite and illite, with some of the kaolinite group minerals possibly also present. These changes are probably not isochemical; there may be considerable exchange of ions with the pore fluid which is driven out as the clays compact. Some authors use the term **anchizone** for the zone of chlorite–illite rocks transitional into metamorphism, and the term **epizone** for the succeeding lowest grades of metamorphism in which illite is replaced by white mica. The very fine white micas of epizonal slaty rocks are known as **sericite** and the dominant constituent is phengite, a variety of muscovite in which there has been a coupled substitution of Si^{4+} and Fe^{2+} (or Mg^{2+}) for $2Al^{3+}$. In muscovite Al occurs in both octahedral co-ordination (Al^{VI}) and tetrahedral co-ordination (Al^{IV}) whereas Si occurs in tetrahedral sites, Fe and Mg in octahedral sites. Hence the phengite substitution can be written:

$$(Si^{IV}, Fe^{VI}) \rightleftharpoons (Al^{IV}, Al^{VI})$$

In some areas where detailed studies have been carried out, for example the work of Frey and co-workers in the central European Alps, it has been found that as well as phengite the sericite may also contain the Na-mica paragonite, and pyrophyllite, a hydrous aluminium silicate.

It is not feasible to map conventional isograds in very low grade pelites, but two techniques have been developed to study the variations in grade. Both depend on the progressive recrystallisation that occurs during heating. One method is to use X-ray diffraction to measure the crystallinity of illite, which increases as it recrystallises, coarsens and passes into phengite, resulting in sharper peaks on the diffraction trace at higher grades. The other method is to measure the reflectance of original organic matter on polished surfaces, since this also increases with recrystallisation to graphite. Examples of regional studies utilising these techniques are the work of Frey *et al.* (1980) in the central Alps, of Kisch (1980) in western Sweden, and of Weaver (1984) in the southern Appalachians, USA.

METAMORPHISM OF PELITE IN THE BARROVIAN ZONAL SCHEME

The classic zonal scheme found in the Scottish Highlands and many other parts of the world and outlined in Chapter 1 provides an excellent example of the way in which the mineralogy of pelites can vary with increasing temperature and pressure, and is therefore described here in more detail. The assemblages found are also summarised in Table 3.1. Different assemblages can be produced if heating took place at higher or lower pressures than in Barrow's area, or attained higher temperatures, and these variations are described subsequently.

Chlorite zone:

In the chlorite zone of Scotland, pelitic rocks are fine grained slates, often with graphite. The fine grain size makes it difficult to study these rocks under the microscope, but they typically contain chlorite and phengitic muscovite with variable amounts of quartz, albite and accessories such as pyrite. Some pelites or associated semi-pelites may contain K-feldspar, stilpnomelane or minor calcite. In some other regions, chlorite zone rocks are coarser grained schists and the minerals are more readily identified.

Biotite zone:

The definition of the biotite zone is not as straightforward as was originally supposed because it has been pointed out by Mather (1970) that the rocks in which biotite develops first are not strictly pelites but are greywackes with detrital K-feldspar, in which biotite can form by the reaction:

$$\text{K-feldspar} + \text{chlorite} \rightarrow \text{biotite} + \text{muscovite} + \text{quartz} + H_2O \qquad [3.1]$$

Although this is a continuous reaction, involving six phases and the six KMFASH components, it appears to go to completion over a fairly narrow temperature range since the assemblage K-feldspar + chlorite + biotite is a rare one except in partially retrogres-

Table 3.1   Characteristic assemblages of pelitic rocks from the Barrovian zones of the Scottish Highlands. (Compiled principally from Atherton, 1977; Chinner, 1965, 1967; and Harte & Hudson, 1979)

Zone	Typical assemblage
Chlorite	Chlorite + phengitic muscovite + quartz + albite ± calcite ± stilp-nomelane ± paragonite
Biotite	Biotite + chlorite + phengitic muscovite + quartz + albite ± calcite
Garnet	Garnet + biotite + chlorite + muscovite + quartz + albite + epidote (chloritoid is very rarely present in this region, but is common at comparable grades elsewhere)
Staurolite	Staurolite + garnet + biotite + muscovite + quartz + plagioclase (possibly chlorite at low grades)
Kyanite	Kyanite ± staurolite + garnet + biotite + muscovite + quartz + plagioclase
Sillimanite	Sillimanite ± staurolite + garnet + biotite + muscovite + quartz + plagioclase ± kyanite relics
Common accessories:	Ilmenite, magnetite, hematite, rutile (principally in the kyanite zone), pyrite, tourmaline, apatite, zircon, graphite

sed rocks. True pelites lack K-feldspar, and in them biotite is formed at somewhat higher temperatures through another continuous reaction:

$$\text{phengite} + \text{chlorite} \rightarrow \text{biotite} + \text{phengite-poor muscovite} + \text{quartz} + H_2O \quad [3.2]$$

Note that phengite and muscovite appear on opposite sides of the reaction but this is actually just for convenience in writing the equation. In reality only one potassic white mica phase is present and this changes in composition from phengite towards pure muscovite as the reaction proceeds. As a result, the reacting assemblage still has three degrees of freedom. In other words different chlorite–biotite–muscovite schists equilibrated at the same P and T (accounting for two degrees of freedom) may still display different mineral compositions, but the compositions of each individual mineral cannot vary independently. The temperature at which reaction [3.2] commences depends on the extent of phengite substitution in the initial mica, and the X_{Mg} values of the chlorite and mica. High phengite content and low X_{Mg} favour biotite growth at lower temperature. The association chlorite + muscovite + biotite is stable over a wide temperature interval, emphasising the markedly continuous nature of this reaction.

Garnet zone:

The next mineral to appear in the Barrovian sequence is garnet, and at this grade pelitic rocks are typically schists whose minerals are coarse enough to be readily identified in thin section, even though the actual grain size and textures may vary between different areas. The garnet isograd is usually easy to trace in the field because garnet appears in a wide range of rock composition as well as in pelites at about the same grade. Frequently it forms conspicuous porphyroblasts and this helps to distinguish almandine-rich garnets typical of the garnet zone from spessartine garnets which may develop at lower grades in Mn-rich sediments, but are usually very fine grained.

A typical garnet zone pelitic rock has the assemblage: garnet + biotite + chlorite + muscovite + quartz + plagioclase (albite). Ilmenite or magnetite may be present and sometimes also epidote; possible accessory constituents, usually sufficiently coarse grained to be identified in thin section, may include apatite, tourmaline and zircon. Minor sulphides are often present and may be pyrrhotine or pyrite, sometimes with traces of chalcopyrite; graphite may also occur.

The characteristic garnet of the garnet zone is rich in almandine and probably grows by a continuous reaction such as:

$$\text{chlorite} + \text{muscovite} \rightarrow \text{garnet} + \text{biotite} + \text{quartz} + H_2O \quad [3.3]$$

It was noted in Chapter 2 (page 39) that where chlorite, garnet and biotite coexist, Fe and Mg are distributed between them so that $X_{Mg}^{chl} > X_{Mg}^{bio} > X_{Mg}^{gt}$. Since chlorite has a stronger preference for Mg than the phases on the right-hand side of the reaction, the effect of this continuous reaction will be to deplete the remaining chlorite in Fe as the reaction proceeds.

These continuous changes can be demonstrated conveniently by means of a series of AFM diagrams, shown in Fig. 3.2. The biotite zone assemblage muscovite + quartz + biotite + chlorite with its three degrees of freedom is represented in Fig. 3.2(*a*) by lines representing a range of chlorite compositions that can coexist with biotite, and a range of biotite compositions that coexist with chlorite. The diagram is constructed for a constant value of P and T, and although a wide range of chlorite compositions is possible (in terms of X_{Mg}), only one specific composition of biotite can then coexist with any chosen chlorite at the particular P–T conditions. This is a consequence of the assemblage having three degrees of freedom and is represented in Fig. 3.2(*a*) by tie-lines connecting specific

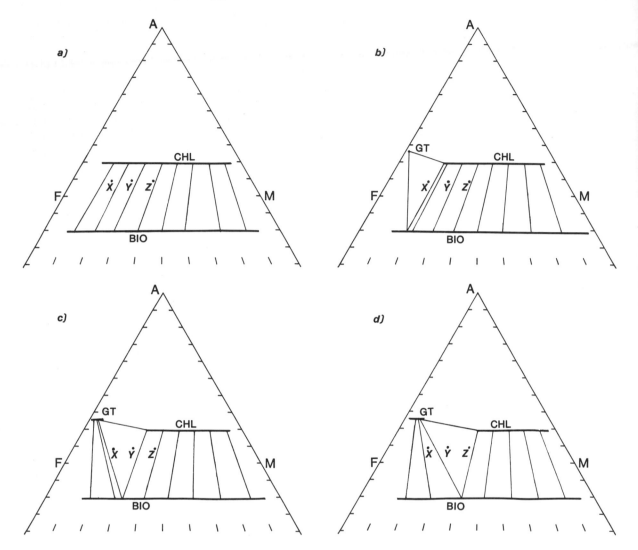

Fig. 3.2   AFM projections for the biotite and garnet zones of the Scottish Highlands. *X*, *Y* and *Z* are possible rock compositions. All assemblages also include muscovite, quartz and H_2O. *a*) Biotite zone, showing the trivariant chlorite-biotite field with representative tie-line connecting coexisting compositions. *b*) First appearance of garnet, restricted to Fe-rich compositions only (rock *X*). *c*) With further heating, garnet appears in a wider range of rocks as reaction [3.3] progresses, but is still absent from rock *Z*. *d*) Garnet is now present in most normal pelite compositions, and chlorite is absent from the more Fe-rich rock, *X*.

chlorite and biotite compositions. Although an infinite number of tie-lines are possible, they must all slope in such a way that:

$$K_{D_{Fe-Mg}}^{bio-chl} = \left(^{Mg}/_{Fe}\right)_{bio} / \left(^{Mg}/_{Fe}\right)_{chl} = \text{constant}$$

This is because the distribution coefficient K_D varies with temperature but ideally is independent of mineral composition. A consequence of this constraint is that no two tie-lines, however closely drawn, may actually cross over one another if they represent equilibrium at the same pressure and temperature.

   The typical garnet zone assemblage, muscovite + quartz + chlorite + biotite + garnet, is shown in Fig. 3.2 (*b*)–(*d*) as a triangle whose corners represent the unique compositions of the phases that coexist at the *P–T* conditions for which the diagram is drawn. This is a consequence of the fact that the assemblage has only two degrees of freedom and so if *P* and *T* are taken as the independent variables, the mineral compositions cannot vary at any specified *P–T* condition. Because garnet has a marked preference for Fe, it develops first in the most Fe-rich rocks (Fig. 3.2(*b*)). With heating, the continuous reaction (reaction [3.3]) progresses, further restricting the range of possible chlorite compositions (compare Figs 3.2(*b*) and (*c*)). In addition, garnet develops in progressively more magnesian rock types. Points *X*, *Y* and *Z* in Fig. 3.2 represent the compositions of different pelitic rocks. At the *P–T* conditions of Fig. 3.2(*b*), only rock *X* will develop garnet, whereas by the higher temperature conditions represented by Fig. 3.2(*d*), all three rocks contain garnet and in rock *X* all chlorite has been consumed and the rock has the assemblage muscovite + quartz + garnet + biotite. Figure 3.2(*c*) represents an intermediate temperature. The significance of Fig. 3.2 is two-fold; it shows schematically how the compositions of chlorite, biotite and garnet evolve with temperature in the garnet zone, with Fe-rich chlorites becoming unstable, and secondly it demonstrates the influence of bulk rock composition on the precise temperature at which index minerals appear.

In some true pelites chloritoid occurs in the garnet zone, and may indeed appear at lower grades in favourable circumstances. Chloritoid is however very restricted in the range of rock compositions in which it can develop during Barrovian metamorphism. It very rarely occurs with biotite, (an association which does, however, become possible at lower pressures), but is more commonly present in the assemblages garnet + chloritoid + chlorite + muscovite + plagioclase + quartz or chloritoid + chlorite + muscovite + paragonite + plagioclase + quartz.

Figure 3.3(*a*) is an AFM diagram showing a full range of garnet zone assemblages including those with chloritoid. Note that the fact that garnet and chlorite can coexist means that biotite does not develop in the most aluminous rocks and cannot occur with chloritoid. This is in accord with what is found in nature. Chloritoid is restricted to very aluminous rocks whose compositions plot above the garnet–chlorite tie-line, and it is perhaps unfortunate that such rocks are almost entirely lacking in Barrow's type area.

The garnet zone is also often marked by a change in plagioclase compositions, seen not only in pelites but also in other metasediments and in metamorphosed igneous rocks. At lower grades the only plagioclase found is albite, and Ca may be present in the rock in accessory epidote or other phases. Within the garnet zone oligoclase or andesine appear, and rocks have been described from some parts of the world that have rather pure albite coexisting with oligoclase (An20 to An25) as separate phases (e.g. Crawford, 1966). This is taken to mean that the plagioclase solid solution series is not continuous at these temperatures, and there is a break in the compositions possible known as the **peristerite gap**. By the upper part of the garnet zone a complete range of plagioclase compositions between albite and oligoclase is found, and this change has been mapped as an isograd in some studies.

Staurolite zone:

It is only in this zone that the distinction between 'true pelites' and other mica schists becomes apparent because staurolite grows in a range of Al-rich, Ca-poor pelitic rocks but only rarely in other lithologies. Many garnet–mica schists prove to be of unsuitable composition for staurolite to form, and should mostly be referred to as semi-pelites. A critical factor in this is the stabilisation of plagioclase at this grade, which allows available Ca (e.g. in epidote) to combine with additional Al to produce anorthite component, effectively reducing the amount of Al available to form other aluminosilicate minerals.

The effect is analogous to moving the rock analysis downwards on the AFM diagram at constant '*M*' value, to lie in the garnet–biotite field (Fig. 3.3(*b*)). In addition, the range of X_{Mg} values over which staurolite can form is also quite restricted, but Mg-rich 'true pelites' develop other distinctive Al-rich minerals instead at this grade (e.g. kyanite).

Typically, staurolite zone pelites contain the assemblage staurolite + garnet + biotite + muscovite + quartz + plagioclase. The minor and opaque phases found are similar to those present in the garnet zone. Chlorite is also sometimes present.

In areas where suitably aluminous pelites occur, staurolite is produced from breakdown of chloritoid by a reaction such as:

$$\text{chloritoid} + \text{quartz} \rightarrow \text{staurolite} + \text{garnet} + H_2O \qquad [3.4]$$

Textural evidence for a reaction of this type is found in some staurolite schists that contain relics of chloritoid preserved as inclusions in garnet. This is a texture that may be produced if the new-formed garnet grows around a chloritoid grain as it breaks down, eventually armouring it so that it is no longer able to react with quartz in the rock matrix, and therefore persists stably to higher temperatures.

Staurolite is also found in pelitic rocks that are of unsuitable composition to have ever developed chloritoid (i.e. they plot below the garnet–chlorite join). Here, staurolite may have been produced by the discontinuous reaction:

$$\text{garnet} + \text{muscovite} + \text{chlorite} \rightarrow \text{staurolite} + \text{biotite} + \text{quartz} + H_2O \quad [3.5]$$

Chlorite is relatively rare in staurolite zone rocks, except as a product of retrograde alteration. However, a number of studies have reported primary chlorite coexisting with staurolite, biotite, muscovite and quartz. This assemblage is shown on the AFM projection in Fig. 3.3(*b*), together with the assemblage garnet + biotite + staurolite + muscovite + quartz. The difference between Figs 3.3(*a*) and 3.3(*b*) is the replacement of the garnet–chlorite tie-line with a staurolite–biotite tie-line. As a result, staurolite is found in a wider range of rocks than chloritoid because it can be present in less aluminous layers whose composition projects below the garnet–chlorite tie-line, e.g. compositions such as *X*, *Y* and *Z* in Fig. 3.2.

Reaction [3.5] takes place at a fixed temperature for any given pressure since it is a discontinuous reaction, and proceeds until one of the three reactants has been consumed. When this reaction has ceased some further staurolite may be produced by continuous reaction involving the remaining phases, for example:

$$\text{chlorite} + \text{muscovite} \rightarrow \text{staurolite} + \text{biotite} + \text{quartz} + H_2O \qquad [3.6]$$

Kyanite zone:

The Barrovian kyanite zone is typified by a range of assemblages including the staurolite zone assemblage of garnet + staurolite + biotite (+ muscovite + quartz) as well as those with kyanite: kyanite + staurolite + biotite or kyanite + biotite (+ muscovite + quartz). These assemblages are illustrated in Fig. 3.3(*c*), which differs from Fig. 3.3(*b*) in that the staurolite–chlorite tie-line is replaced by a kyanite–biotite tie-line. This change corresponds to a discontinuous reaction:

$$\text{muscovite} + \text{staurolite} + \text{chlorite} \rightarrow \text{biotite} + \text{kyanite} + \text{quartz} + H_2O \quad [3.7]$$

Reaction [3.7] can only take place in pelites with relatively Mg-rich minerals since it does not affect rocks with the assemblage garnet + staurolite + biotite + muscovite + quartz (compare Figs. 3.3(*b*) and (*c*)). Once the kyanite–biotite association has been stabilised by reaction [3.7], further growth of kyanite can occur by the continuous reaction:

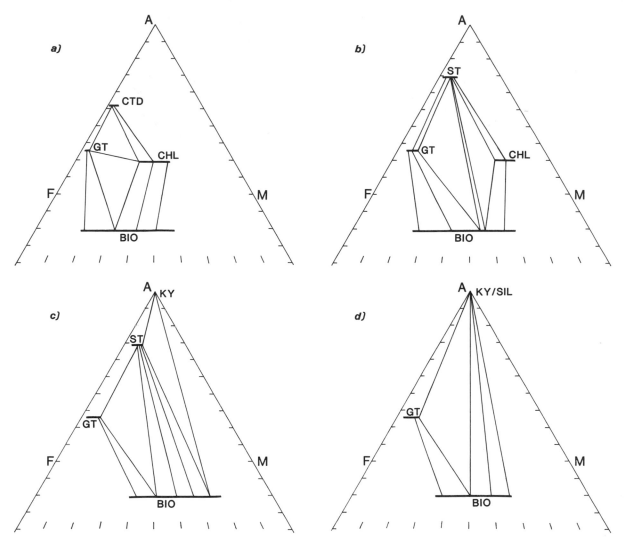

Fig. 3.3 AFM projections showing assemblages coexisting with muscovite, quartz and H_2O in Barrovian-type metamorphism: *a*) garnet zone showing divariant (two-phase) and trivariant (three-phase) fields, including chloritoid-bearing assemblages; *b*) staurolite zone; *c*) kyanite zone; *d*) sillimanite zone.

$$\text{staurolite} + \text{muscovite} + \text{quartz} \rightarrow Al_2SiO_5 + \text{biotite} + H_2O \qquad [3.8]$$

This has the effect on the AFM diagram of enlarging the kyanite–biotite field and reducing the range of rock compositions that retain staurolite (Fig. 3.3(*d*)). It therefore causes the gradual development of kyanite in rocks that did not contain chlorite under staurolite zone conditions.

Sillimanite zone:
This zone differs from the kyanite zone only in the presence of sillimanite, and kyanite may also still be present. The sillimanite normally occurs in the form of very fine needles which may be matted together or penetrate grains of biotite or quartz and is known as **fibrolite**. Coarse prismatic sillimanite is largely restricted to the granulite facies, except where it occurs pseudomorphing andalusite. The reaction that represents the transition from the kyanite zone is the polymorphic transition:

$$\text{kyanite} \rightarrow \text{sillimanite} \qquad\qquad [3.9]$$

However, the fact that some kyanite commonly remains, suggests that this reaction is very sluggish. Instead it is probable that much of the sillimanite is produced from breakdown of other minerals. For example, reaction [3.8] appears to take place over a range of temperatures spanning the boundary between the stability fields of kyanite and sillimanite; at higher temperatures sillimanite replaces kyanite as a product and is generated directly. Within the sillimanite zone staurolite disappears from muscovite–quartz pelites as a result of the discontinuous reaction:

$$\text{staurolite} + \text{muscovite} + \text{quartz} \rightarrow \text{garnet} + \text{biotite} + \text{sillimanite} + H_2O \,[3.10]$$

Although this causes staurolite to disappear from the AFM diagram (Fig. 3.3(d)), it is not uncommon to encounter pelites that lack muscovite at this grade, because muscovite has appeared as a reactant in most reactions from the biotite isograd and may have been

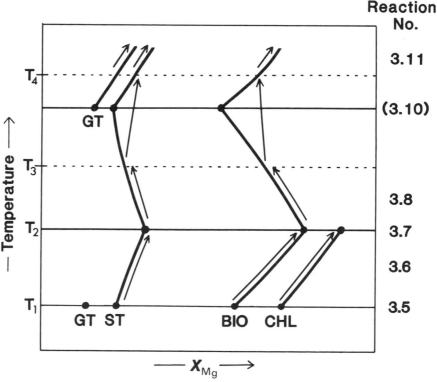

Fig. 3.4 T–X_{Mg} diagram illustrating the sequence of continuous and discontinuous reactions as a pelitic schist is heated above the staurolite isograd. Bold lines represent mineral compositions changing as a result of continuous reactions (labelled on the right). Horizontal solid lines denote the temperature of discontinuous reactions in the series. The arrows monitor the changes in mineral composition for a specific pelite. The initial assemblage is staurolite + biotite + chlorite + muscovite + quartz + H_2O. Continuous breakdown of chlorite makes all Fe–Mg minerals more magnesian until chlorite disappears at T_2 due to discontinuous reaction [3.7]. Further continuous breakdown of staurolite drives mineral compositions to lower X_{Mg} values until the reaction ceases at T_3 because all muscovite in the rock has been consumed. With further heating only minor adjustments take place due to cation exchange reaction, and reaction [3.10] cannot occur (no muscovite). At T_4, garnet appears due to staurolite breakdown by reaction [3.11], and since garnet is more Fe-rich than staurolite, staurolite breakdown now drives mineral compositions to higher X_{Mg} values.

entirely consumed at lower grades. In these muscovite-free pelites, staurolite may still persist although the assemblages cannot be shown on the AFM projection. Yardley, Leake & Farrow (1980) have described the assemblages and reactions found in a sequence of pelites of this type; this is summarised below, and in Fig. 3.4.

SOME LIMITATIONS TO THE APPLICABILITY OF THE AFM DIAGRAM

It ought to be an easy practical exercise to identify the assemblages of any set of metapelites and by comparing them with the AFM diagrams shown in Fig. 3.3 to deduce the metamorphic zones in which each sample formed. It is frequently the case, however, that the natural assemblages appear to have too many phases, i.e. four rather than three of the phases plotted on the projection. This is too common a circumstance always to result from the fortuitous sampling of rocks that were metamorphosed at the precise P–T conditions of a discontinuous reaction. In general it must result either from persistence of earlier-formed minerals beyond the conditions under which they were stable, or from the presence of additional components in the natural system serving to stabilise a larger number of phases than would ideally coexist. The occurrence of kyanite and sillimanite in the same sample is most likely to result from persistence of the first-formed polymorph outside its own stability field, because both minerals occur as nearly pure $Al_2 SiO_5$ and so their stability is most unlikely to be affected by additional chemical components in the rocks. On the other hand, one of the common assemblages of the kyanite and sillimanite zones is garnet + staurolite + Al-silicate + biotite + muscovite + quartz, while garnet + staurolite + chlorite + biotite + muscovite + quartz may occur in the staurolite zone. In both cases it is notable that the garnet contains Mn and Ca, often in significant amounts ($< 15\%$ grossular component and exceptionally $< 40\%$ spessartine), whereas these elements do not readily substitute into the other KMFASH system phases. Here, it is likely that garnet is being stabilised as an 'extra' phase by the presence of these additional elements.

PLOTTING PELITE REACTION SEQUENCES

An alternative to the AFM projection for the portrayal of the progressive changes in pelites is to plot the X_{Mg} values of individual minerals as a function of temperature, assuming pressure to be either constant or a simple function of temperature (Thompson, 1976). An example of such a T–X_{Mg} diagram is shown in Fig. 3.4 and is based on an example from the Connemara region of Ireland (Yardley, Leake & Farrow, 1980; see also Ch. 7, Fig. 7.9). Univariant reactions are represented as horizontal lines, since they occur at a single temperature for the specified pressure. The composition of the Fe–Mg phases involved in the reaction are represented by points on the line, being uniquely defined in the presence of a univariant assemblage once pressure is fixed. Continuous reactions take place over a temperature range between univariant reactions, and are represented by a series of curves showing the variation in X_{Mg} value for each participating mineral as a function of temperature. The precise reaction path followed by an individual rock depends on both the compositions and the abundances of the participating phases. In the example illustrated in Fig. 3.4, chlorite begins to react at T_1, due to reaction [3.5] and continues to react to T_2, according to reaction [3.6]. At T_2 it is eliminated by reaction [3.7]. Between T_1 and T_2 relatively Fe-rich biotite and staurolite increases in abundance relative to chlorite, so all mineral compositions become more magnesian since the rock composition is constant. From T_2 to T_3 staurolite breaks down according to reaction [3.8] and this causes a reversal in the trend of mineral composition with temperature, since it is now an Fe-rich, rather than an Mg-rich phase which is breaking down. At T_3 reaction ceases due to all the muscovite having been consumed, and so the compositions of staurolite and biotite are now 'frozen-in' apart from small changes due to cation exchange. Even very small variations in the modal abundance of muscovite between samples can lead to significant differences in the temperature, and hence mineral compositions, when reaction [3.8] is terminated.

In the absence of muscovite, reaction [3.10] cannot take place in the specific rock whose history we have been following. Instead, further breakdown of staurolite will commence at a higher temperature, T_4, due to the continuous reaction:

$$\text{staurolite} + \text{quartz} \rightarrow \text{garnet} + \text{sillimanite} + H_2O \qquad [3.11]$$

The use of $T-X_{Mg}$ diagrams to explain reaction sequences in pelitic rocks was first outlined by A.B. Thompson (1976); they can provide a remarkably detailed insight into the reaction history of a group of rocks.

Fig. 3.5   Examples of migmatisation phenomena resulting from anatexis of pelitic schists. Both are from Connemara, Ireland. *a*) Stromatic migmatite composed of thin layers of fine granite leucosome within schist composed of biotite, quartz, plagioclase, K-feldspar, sillimanite and cordierite. Note that the psammite band in the upper left is not migmatised. *b*) Nebulitic migmatite. Part of the outcrop is of undoubted granitic leucosome (G), though containing large relict metamorphic crystals of garnet (xenocrysts). Other parts, at S, are undoubtedly schist, but elsewhere, as at N, the rock has an intermediate or nebulous character, appearing to be schist but containing recrystallised feldspars similar to those of the leucosome. In this outcrop the leucosome is strictly not granite but trondjhemite (i.e. of plagioclase and quartz without K-feldspar).

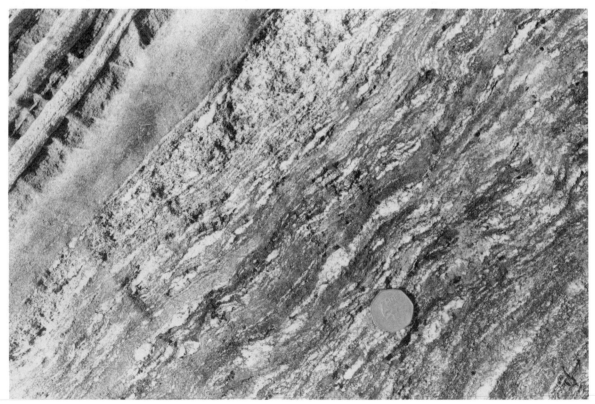

VARIATIONS ON THE BARROVIAN ZONAL PATTERN

Pelitic schists showing similar metamorphic zoning to that found in the Scottish Highlands are widespread and have been described from rocks of many different ages around the world. However, it is not unusual for pelitic rocks to develop different and often equally distinctive mineral assemblages, reflecting different $P-T$ conditions of metamorphism. There are three general ways in which different assemblages and zones may be produced:

1. Metamorphism continues to still higher temperatures so that additional zones are present.
2. Metamorphism takes place at lower pressures.
3. Metamorphism takes place at higher pressures.

HIGH TEMPERATURE METAMORPHISM OF PELITES

Migmatites: an introduction

In some metamorphic belts the sillimanite zone is succeeded by higher grade zones in which the rocks are often **migmatites**. This means that they are literally 'mixed rocks'; usually predominantly schists but with pods, veins or layers of leucocratic material of broadly granitic composition (Fig. 3.5). Although migmatites are best developed in pelitic rocks, they also form in other siliceous metasediments, metabasic rocks, etc. A range of migmatite types may develop in pelitic metasediments, varying principally in the way in which granitic material, known as the **leucosome** from its light colour, occurs. Most commonly the leucosome forms layers more or less parallel to the schistosity of the rest of the rock and usually only a few centimetres at most in thickness; the individual layers are often discontinuous, being more in the nature of elongate lenses. This type of migmatite is known as a **stromatic migmatite** (Fig. 3.5(*a*)). In some instances the leucosomes are not so well oriented and form a more random network of veins; migmatites of this sort have been referred to as **vein-type migmatites**. Another type commonly found in high grade pelites is **nebulite** (Fig. 3.5(*b*)); this has nebulous patches of leucosome passing gradationally through a few centimetres of transitional rock, somewhat darker coloured but with coarse recrystallised feldspar, into undoubted schist. In other cases the leucosomes may form distinct pod-like bodies. A detailed classification of migmatites is given by Mehnert (1968).

Some migmatites may result from injection of material into schists to produce leucosome of unrelated origin, but in many high grade pelitic rocks the leucosomes appear to have segregated out locally from their host rocks over distances of centimetres to metres. For example, leucosomes may be restricted to a particular lithology (Fig. 3.5(*a*)) or the schist at the margin of the leucosome may be unusually dark in colour, having been depleted in light coloured minerals. The schistose portion of the rock is known as the **melanosome** or simply **restite** where it has clearly been depleted in granitic material during the process of migmatisation. Unaffected schist is termed **palaeosome**.

The origin of migmatites has been controversial for many years. Different sides in the 'granite controversy' that developed in the late 1940s saw migmatites as either the result of metasomatic transformation of schist by migrating fluids or the product of partial melting (**anatexis**) of schist due to very high temperatures of metamorphism. Subsequent experimental work, notably by Winkler and co-workers in Göttingen, has shown that metasediments will begin to melt to produce a granitic liquid at temperatures a little above those of the sillimanite zone (see Winkler, 1976) and this process was undoubtedly responsible for the formation of many migmatites. On the other hand it has also been found that many migmatite leucosomes have compositions that do not correspond at all closely with those of the melts produced by experiment; for example, natural leucosomes are quite commonly trondjhemitic, i.e. lacking K-feldspar, rather than truly granitic. It is therefore quite likely that other processes such as metamorphic segregation, which involves solution and reprecipitation of minerals via a fluid phase, may also be found to play a part in the generation of some migmatites (e.g. Olsen, 1984). An important reason why there are still many unanswered problems in migmatite studies is that for many years in the 1960s and 1970s they were rather unfashionable rocks to study. Partly this was because of the stigma remaining from the sometimes less than scientific debates of the days of the granite controversy, partly because their simple mineralogy means that they are not very amenable to the type of phase equilibrium study that came into vogue with the advent of the electron microprobe.

One final, and very important, point to bear in mind about migmatites is that they do vary very greatly in their nature and mode of occurrence. The bulk of the earth's migmatites occur in extensive terranes, often of Precambrian age, that are of fairly uniform high grade and frequently do not exhibit a gradational transition from unmigmatised to migmatised rocks. There are other areas where such transitions do occur and these are central to our understanding of the development of migmatites, however these migmatites are often of much more restricted extent, closely associated with the intrusion of deeper-derived hot magmas, and often differ mineralogically from the migmatites found in the more extensive Precambrian terranes, perhaps because of different pressures of formation. They do not necessarily provide a good model for the formation of all migmatites, but it is on this second type of migmatite, occurring in distinct higher grade zones above the sillimanite zone, that the following section is largely based.

HIGH TEMPERATURE PHASE EQUILIBRIA IN PELITES

The study of high temperature reactions in pelites is made more complex by the development of a melt phase whose composition is not fixed, and indeed the possible range of compositions is not well known. A secondary consequence of melting is that the rock may cease to coexist with an aqueous fluid phase, as has been assumed hitherto, because H_2O is quite soluble in silicate melts and if only a small amount of pore water is present initially it may all be dissolved. Nevertheless, several theoretical studies have been made that predict the sequence of zones that may develop, e.g. by Grant (1973) and A.B. Thompson (1982), and these seem to be in agreement with field observations (e.g. Tracy & Robinson, 1983).

Upper sillimanite zone:
The best known isograd above the first incoming of sillimanite is the so-called 'second sillimanite isograd' (Evans & Guidotti, 1966) which represents the growth of additional sillimanite from the breakdown of muscovite:

$$\text{muscovite} + \text{quartz} \rightarrow Al_2SiO_5 + \text{K-feldspar} + H_2O \qquad [3.12]$$

Hence the upper sillimanite zone is characterised by the coexistence of sillimanite and K-feldspar, rather than by only one mineral.

Reaction [3.12] involves five phases which can be made up of only four components, and is therefore univariant in the ideal case. However, it is commonly the case that the reactant and product phases coexist across a distinct zone, and this may in part result from phengite and paragonite substitution in muscovite or from the occurrence of other volatile species with H_2O in the fluid phase. An alternative explanation of the zone of coexistence is that it is a consequence of the finite time required for reactions to take place (Lasaga, 1986). This is discussed in Chapter 6 (page 186).

The second sillimanite isograd is a particularly useful indicator of metamorphic grade because it develops in a very wide range of rock types, not just 'true pelites' and may even be traced in quartzite in some instances. It was noted above that some true pelites no longer have muscovite in the sillimanite zone, and may therefore retain staurolite. In the case of these pelites the second sillimanite isograd cannot develop, however staurolite disappears from these rocks through reaction [3.11] (above) which takes place at very similar temperatures.

One of the most intensively studied field areas for high grade metamorphism is in the north-east United States, including the recent work of Hess (1969), Tracy (1978) and Tracy & Robinson (1983) in parts of Massachusetts (Fig. 3.6). In these areas migmatite features are first developed close to the second sillimanite isograd and therefore muscovite breakdown may involve a melt phase also. Thompson (1982) has shown that at the moderate pressures of Barrovian metamorphism, reaction [3.12] may be replaced by:

$$\text{muscovite} + \text{quartz} + H_2O \rightarrow \text{sillimanite} + \text{melt} \qquad [3.13]$$

or in most natural rocks, since they contain biotite:

$$\text{muscovite} + \text{biotite} + \text{quartz} + H_2O \rightarrow \text{sillimanite} + \text{melt} \qquad [3.14]$$

Both these reactions involve an aqueous fluid phase which dissolves in the melt produced. In most natural situations it is difficult to imagine sufficient pore fluid being available to permit much H_2O-saturated melt to be produced, since granitic melts may dissolve around 10 per cent H_2O (Clemens, 1984). The melting reaction will therefore cease when all the pore water has been assimilated in the melt, until further heating has occurred. (It should be noted that this natural situation contrasts with the design of many experimental studies, which take place with excess H_2O available.) Further melting can occur through the breakdown of hydrous minerals, which liberate water that is then dissolved in the melt:

$$\text{muscovite} + \text{quartz} \rightarrow \text{K-feldspar} + \text{sillimanite} + \text{melt} \qquad [3.15]$$

The sequence of reaction [3.13] or [3.14] followed by [3.15] would give rise to the development of migmatites a little below the second sillimanite isograd, which is defined by the appearance of sillimanite + K-feldspar and would therefore result from reaction [3.15] in this instance. A very detailed study of mineral assemblages close to this isograd in Massachusetts (Fig. 3.6) by Tracy (1978) has confirmed that this is in fact the sequence that is observed there; i.e. the appearance of sillimanite + K-feldspar results from reaction [3.15] rather than from reaction [3.12].

Cordierite–garnet–K-feldspar zone:

At still higher grades, pelitic rocks develop assemblages with cordierite, garnet, K-feldspar and sillimanite, though not all these necessarily occur together. The assemblages result from continuous reactions such as:

$$\text{biotite} + \text{sillimanite} + \text{quartz} \rightarrow \text{K-feldspar} + \text{cordierite} + \text{melt} \qquad [3.16]$$
$$\text{biotite} + \text{sillimanite} + \text{quartz} \rightarrow \text{K-feldspar} + \text{garnet} + \text{melt} \qquad [3.17]$$

Whether cordierite or garnet develops depends partly on pressure (cordierite is favoured by lower pressures, garnet by higher pressures) and partly on the Mg/Fe ratio of the rock (garnet will form in Fe-rich rocks, cordierite in Mg-rich ones). Reactions [3.16] and [3.17] lead to melting in Mg-rich and Fe-rich compositions respectively due to

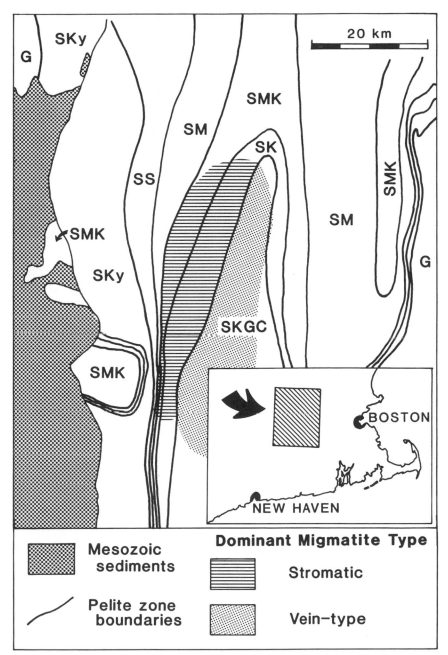

Fig. 3.6 Metamorphic map of Massachusetts, USA, illustrating the occurrence of migmatite types in relation to metamorphic isograds. Simplified from Tracy & Robinson (1983). The metamorphic zones are labelled as follows: G = greenschist facies; SKY = staurolite–kyanite zone; SS = staurolite–sillimanite zone; SM = sillimanite–muscovite zone; SMK = sillimanite–muscovite–K-feldspar zone; SK = sillimanite–K-feldspar zone; SKGC = sillimanite–K-feldspar–garnet–cordierite zone.

dehydration breakdown of biotite, but melting only takes place in the full range of biotite–sillimanite schists when the temperature for the discontinuous reaction is attained:

$$\text{biotite} + \text{sillimanite} + \text{quartz} \rightarrow \text{cordierite} + \text{garnet} + \text{K-feldspar} + \text{melt} \qquad [3.18]$$

The garnet–cordierite–K-feldspar assemblage is typical of high grade pelitic migmatites, and is often taken to mark the beginning of the granulite facies; however, even higher grade assemblages are sometimes found.

Ultra-high grade zones:
High grade granulite facies rocks in central Norway (Touret, 1971a, b) and many other areas contain orthopyroxene. Grant (1973, 1981) has explored the conditions in which such assemblages may be stable compared to more common migmatite assemblages. At medium pressures, very high temperatures are necessary to permit orthopyroxene to form, but the formation of orthopyroxene is also pressure dependent and at higher pressures this phase is stable to lower temperatures. Orthopyroxene assemblages can be related to the more common cordierite–garnet assemblages through the equilibrium:

$$Al_2 SiO_5 + \text{orthopyroxene} = \text{cordierite} + \text{garnet}$$

with sillimanite + orthopyroxene stable at higher temperatures than cordierite + garnet. Orthoamphiboles are also sometimes found in high grade pelitic rocks, though their stability does not extend to such a high temperature as orthopyroxenes.

In a few parts of the world, such as the Archaean granulites of Enderby Land, Antarctica (Grew, 1980; Harley, 1983) the stable association sapphirine + quartz has been found. This assemblage is stable at even higher temperatures than sillimanite + orthopyroxene and is found in rocks that totally lack hydrous minerals, except as products of retrogression. It may be formed on heating through the reaction:

$$\text{sillimanite} + \text{orthopyroxene} \rightarrow \text{sapphirine} + \text{quartz} \qquad [3.19]$$

which will normally give rise to the assemblage sapphirine + quartz + orthopyroxene. This is probably the highest grade assemblage found developed on a regional scale in metasediments, requiring temperatures of at least 850 °C (Droop & Bucher-Nurminen, 1984), and possibly forming under even hotter conditions, up to 1000 °C (Harley, 1983).

METAMORPHISM OF PELITES AT LOW PRESSURES

Pelitic rocks may be heated over a comparable range of temperatures to the Barrovian zones but at lower pressures (i.e. nearer the surface) both in contact metamorphism and in some types of regional metamorphism involving high heat flow.

The principal differences from the Barrovian zonal scheme may be summarised as follows:

1. Kyanite does not occur, but andalusite may be present.
2. Cordierite is more common and forms at lower temperatures.
3. Garnet is less abundant or absent, and staurolite may also be lacking.
4. Migmatites are not developed until above the second sillimanite isograd.

A variety of different zonal patterns have been described from relatively low pressure metamorphism, reflecting the range of pressure–temperature ratios that is possible.

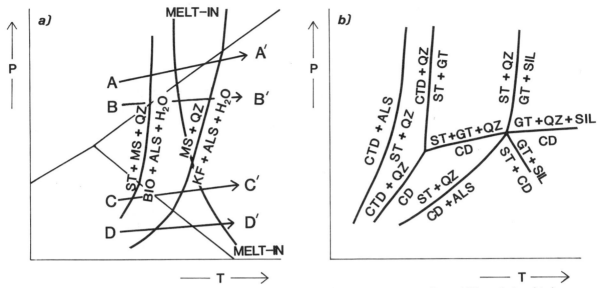

Fig. 3.7 Schematic *P–T* diagrams to illustrate the effect of lowered pressure on metapelites. *a)* The relationship between Al-silicate stability field, first appearance of Al-silicate in normal pelites, muscovite breakdown and the onset of melting. A–A′ represents a Barrovian heating path, B–B′, C–C′ and D–D′ illustrate the different sequences of reactions to be expected at successively lower pressures. *b)* Illustration of the relationships between assemblages involving chloritoid, staurolite, garnet, cordierite and Al-silicate as a function of pressure and temperature. Based on Richardson, 1968.

Figure 3.7 represents schematically some of the differences that result from heating under different pressure conditions.

In Fig. 3.7(*a*) the Al-silicate polymorph stability fields are shown together with curves for some dehydration reactions and for the onset of melting. The path of Barrovian metamorphism (A–A′) results in reaction to produce an Al_2SiO_5 phase while kyanite is the stable polymorph, with subsequent reaction to sillimanite. Only slightly lower pressures are needed to give rise to path B–B′ which will produce a similar zonal sequence to the Barrovian zones except for the absence of a kyanite zone. Regional metamorphism of this type has been described from north-west Maine, USA, by Guidotti (1974). At rather lower pressures, along path C–C′, it is notable that the upper sillimanite zone assemblage K-feldspar + sillimanite will develop before the onset of melting so that the second sillimanite isograd occurs at lower grades than the first occurrence of migmatites. This sequence of high grade zones has been found in Connemara, Ireland, by Barber & Yardley (1985). Another feature of path C–C′ is that the first-formed Al-silicate polymorph is andalusite, which is replaced by sillimanite with further heating in much the same way that sillimanite replaces kyanite in the Barrovian zones.

At extremely low pressures (path D–D′) andalusite may remain stable up to the temperatures of muscovite breakdown, which will result in the stable association andalusite + K-feldspar through reaction [3.12]. This distinctive low pressure, high temperature assemblage has been reported from some thermal aureoles.

Relationships between cordierite, garnet and staurolite are more complex since they depend on rock composition as well as on pressure and temperature. This is because Mg is strongly partitioned into cordierite whereas staurolite and garnet fractionate Fe. As a result there is a broad range of pressures over which the phases overlap, with cordierite being restricted to rocks with high X_{Mg} at higher pressures, but developing in rocks with a wide range of X_{Mg} values at very low pressures.* Relationships between cordierite-bearing assemblages and those found in the Barrovian zones are illustrated schematically in Fig. 3.7(*b*). A simplification is that reactions involving muscovite are not shown, but the general relationship of the different assemblages possible, corresponds well with natural zoning patterns. Figure 3.7(*b*) accounts for the absence of cordierite under Barrovian conditions, the occurrence of coexisting cordierite, staurolite and garnet at intermediate pressures, the absence of garnet and staurolite at very low pressures, and the fact that cordierite only coexists with an Al-silicate phase at very high grades in most compositions of pelitic rock.

Low Pressure Metamorphic Zones in north-east Scotland

Harte and Hudson (1979) have described several separate sequences of metamorphic zones in north-east Scotland, each corresponding to heating over a different range of pressures. Their Buchan (D) zonal sequence provides a good example of distinctly low pressure metamorphism fairly typical of many thermal aureoles. Although the metamorphism here is notionally regional, magmatic heat input was undoubtedly important.

Biotite zone:

The lowest grade rocks in this area are similar to those of the Barrovian biotite zone, having the assemblage biotite + chlorite + muscovite + quartz.

Cordierite zone:

Cordierite appears as the first distinctive index mineral through the continuous reaction:

$$\text{chlorite} + \text{muscovite} \rightarrow \text{cordierite} + \text{biotite} + \text{quartz} + H_2O \qquad [3.20]$$

This reaction, which is probably also responsible for the production of cordierite 'spots' in many 'spotted slates' is apparently very similar to reaction [3.3] that gives rise to the garnet isograd. However, the X_{Mg} value of cordierite is greater than that of coexisting chlorite (Thompson, 1976) whereas in the case of garnet X_{Mg} is much less than for chlorite. This means that whereas reaction [3.3] leads to the first production of garnet in Mg-poor rocks (Fig. 3.2), reaction [3.20] has the opposite effect and cordierite first develops in rocks with high X_{Mg} values, becoming more common with increasing temperature as the continuous reaction progresses, expanding the field of cordierite + chlorite + biotite on the AFM projection (Fig. 3.8 (*a*), (*b*)). Andalusite can occur in this zone, in relatively Fe-rich rocks, and coexists with biotite but *not* cordierite.

Andalusite zone:

Andalusite becomes a possible phase in most pelite compositions at low pressures as a result of the discontinuous reaction:

$$\text{chlorite} + \text{muscovite} + \text{quartz} \rightarrow \text{cordierite} + \text{andalusite} + \text{biotite}$$
$$+ H_2O \qquad [3.21]$$

---

* Most pelitic metasediments have ferrous iron predominating over ferric iron, and yield X_{Mg} values that are seldom significantly greater than 0.5. Hence few rocks have compositions that can be considered really favourable to the production of cordierite, except at very low pressure where even Fe-cordierite is stable (see Fig. 3.11). The sedimentary precursor typically dictates the total amount of Fe in the metamorphic rock, but not its oxidation state. In low to medium pressure metapelites cordierite is often particularly abundant in local zones of magnetite or Fe-sulphide-rich schist. It appears that such rocks can form by local oxidation or sulphidation which makes use of Fe already present in the rock. Hence the reservoir of Fe is depleted in making the opaque phase, leaving less ferrous iron to form aluminosilicates, while the Mg-content is unchanged. This gives rise to rocks with much higher X_{Mg} values than normal, as a result of which they tend to be cordierite rich.

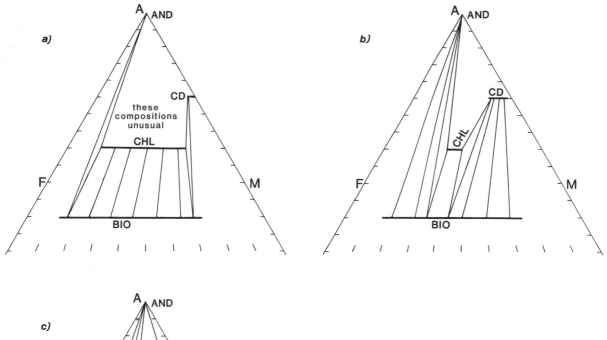

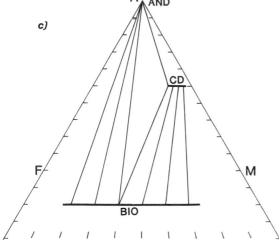

Fig. 3.8 AFM projections showing successive zones of low pressure metamorphism. Based on the Buchan D zonal sequence of Harte & Hudson (1979). *a*) First appearance of cordierite in Mg-rich rocks. *b*) Higher grade cordierite zone with more restricted field of occurrence of chlorite, more widespread cordierite. *c*) Andalusite zone, characterised by coexisting andalusite and cordierite. All assemblages contain muscovite and quartz.

which represents the upper stability limit of chlorite in muscovite–quartz schists and results in its disappearance from the AFM diagram (Fig. 3.8(*c*)). The diagnostic feature of this zone is the *coexistence* of cordierite and andalusite in common pelite compositions that plot below chlorite on the AFM diagram.

Harte & Hudson (1979) found that in their rocks all the chlorite had been consumed by reaction [3.20] before the conditions for reaction [3.21] were attained. As a result reaction [3.21] could not take place, but andalusite was produced at slightly higher temperatures by the continuous reaction:

$$\text{cordierite} + \text{muscovite} + \text{quartz} \rightarrow \text{biotite} + \text{andalusite} + \text{H}_2\text{O} \qquad [3.22]$$

Sillimanite zone:

This zone can be ascribed to the occurrence of the polymorphic reaction:

$$\text{andalusite} \rightarrow \text{sillimanite} \qquad [3.23]$$

However, as in the case of the Barrovian sillimanite zones it seems probable that new sillimanite was also produced by a continuous reaction analogous to [3.22], but proceeding under sillimanite field conditions, since muscovite, cordierite and quartz are all present in this zone.

Upper sillimanite zone:

An upper sillimanite zone with the assemblage sillimanite + cordierite + biotite + K-feldspar + quartz + muscovite is also present in this area, evidently resulting from reaction [3.12].

METAMORPHISM OF PELITES AT HIGH PRESSURES

Until recently there have been relatively few studies of high pressure metamorphism of pelites, and what results were available suggested that, as at low grades, no very clear zonal scheme could be distinguished. More recently a number of distinctive high-P, low-T assemblages have been identified, although identification of these by optical microscopy is often difficult.

The question of the composition of pelitic rocks is particularly important in high-P metamorphism, because many high-P metamorphic belts are characterised by immature greywacke sediments and 'true pelites' of the type described in the earlier part of this chapter are very rare. Most of the high-P belts of the circum-Pacific region fall into this category. Nevertheless, examples of high-P metamorphism of mature 'true pelites' are known, especially from the French and Italian Alps, Greece and Indonesia, and these develop assemblages quite distinct from those of most circum-Pacific regions. The difference results from the very low Ca-content of the mature pelites; if significant Ca is present it combines with Al to form Al-rich minerals such as plagioclase, epidote and lawsonite, thereby inhibiting the formation of the Fe–Mg–Al silicates typical of pelite metamorphism. In this section the high-P metamorphism of true pelites is considered first, but the assemblages found in more immature 'pelites' are also described.

Mineralogy of high pressure pelites

The best known characteristic of pelites in high-P terranes is the absence of biotite and the occurrence instead of phengite-rich muscovite. Garnet, chloritoid, kyanite and chlorite are also common (Koons & Thompson, 1985) but detailed analytical work has shown that in some instances the chloritoid approximates the Mg end-member, whereas only Fe-chloritoid is found at medium and low pressures. Garnet may also be relatively rich in Mg. Furthermore, Abraham & Schreyer (1976) reported an example of talc coexisting with phengite; this assemblage is almost impossible to identify optically but has been found to be widespread in the western Alps by Chopin (1981) and has been shown to be stable only at high pressures. A distinctive index mineral of low-T, high-P metamorphism appears to be Mg–Fe-carpholite. Carpholite is otherwise known as a rare Mn mineral in medium pressure metamorphism, with the formula $(\text{Mn, Mg, Fe}^{2+})\,(\text{Al, Fe}^{3+})_2\,\text{Si}_2\text{O}_6\,(\text{OH})_4$; Mn-poor varieties have been found in pelites only from high-P terranes. Carpholite can be considered as a lower grade equivalent of chloritoid, to which it may be converted on heating by the continuous reaction:

$$\text{carpholite} \rightarrow \text{chloritoid} + \text{quartz} + \text{H}_2\text{O} \qquad [3.24]$$

Zoning in high pressure pelites Chopin & Schreyer (1983) have identified a sequence of four metamorphic zones in high-*P* rocks, and their assemblages are illustrated in Fig. 3.9, projected from phengite + quartz + H_2O. The lowest grade zone has pyrophyllite coexisting with chlorite and in some instances Fe-carpholite; it is represented in Fig. 3.9(*a*). At somewhat higher grades the association Fe-chloritoid + carpholite is found (Fig. 3.9(*b*)), but pyrophyllite is still possible, although it may not coexist with chlorite, thereby distinguishing these rocks from lower pressure terranes. With further heating (Fig. 3.9(*c*), (*d*)) pyrophyllite is replaced by kyanite and chlorite progressively breaks down leading to the formation of distinctive Mg-rich chloritoid and garnet, and permitting the formation of talc in a wide range of pelitic rock types through reactions such as

$$\text{Fe-chlorite} + \text{quartz} \rightarrow \text{garnet} + \text{talc} + H_2O \qquad [3.25]$$
$$\text{Fe-Mg-chlorite} + \text{quartz} \rightarrow \text{Fe-Mg-chloritoid} + \text{talc} + H_2O \qquad [3.26]$$
$$\text{Mg-chlorite} + \text{quartz} \rightarrow \text{kyanite} + \text{talc} + H_2O \qquad [3.27]$$

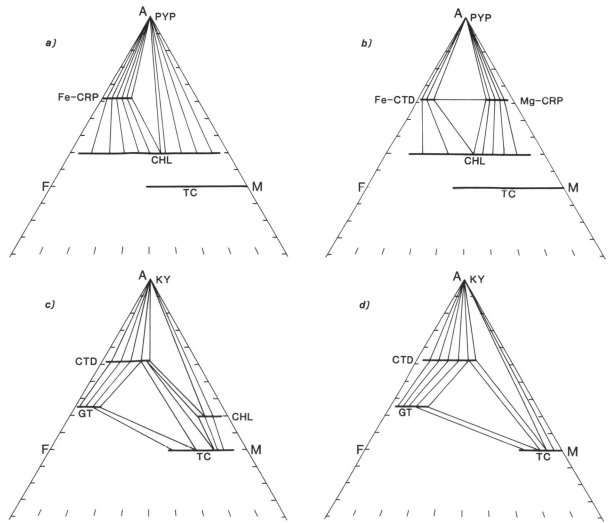

Fig. 3.9 AFM projections for high pressure metamorphism of pelites, based on Chopin & Schreyer (1983): *a*) pyrophyllite–chlorite zone; *b*) chloritoid–carpholite zone; *c*) kyanite–chlorite zone; *d*) talc-kyanite zone (these rocks are sometimes termed **whiteschists**). All assemblages contain phengitic muscovite and quartz.

High pressure migmatite terranes

Some high temperature rocks appear also to have been metamorphosed at rather higher pressures than the migmatites described previously, although no examples have been described in which the distinctive high-P pelite zones can be traced into migmatites. Typical features of high-P, high-T rocks are the presence of garnet without cordierite, kyanite without sillimanite, or the association of kyanite + K-feldspar which must rcsult from breakdown of muscovite + quartz (reaction [3.12]) in the kyanite stability field (see Fig. 3.7(*a*)).

HIGH PRESSURE METAMORPHISM OF 'CIRCUM-PACIFIC PELITES'

The high-P pelites described by Black (1977) from New Caledonia or by Enami (1983) from Japan develop Ca-amphiboles or epidotes in significant quantities and are therefore distinct from the 'true pelites' with which this chapter is principally concerned. Nevertheless, some important metamorphic zonations do occur.

Metamorphism in the Ouegoa district of New Caledonia (see Ch. 4, Fig. 4.3) has undoubtedly taken place at very high pressures and a range of distinctive, dense, high pressure assemblages are developed in the various rock types present. Figure 3.10 illustrates the minerals found in the various zones, though these do not all coexist. Phengite, paragonite, chlorite, albite and quartz occur throughout and biotite is absent as in high pressure 'true pelites'. Garnet is an important constituent at higher grades, but these rocks may also contain lawsonite, replaced by epidote with increasing grade, Na-amphibole (crossite or glaucophane) and at the highest grades omphacitic pyroxene (i.e. an intermediate Ca–Na pyroxene) and hornblende.

In the Sanbagawa belt of Japan (Fig. 3.10, see also Fig. 7.1) metamorphic conditions are

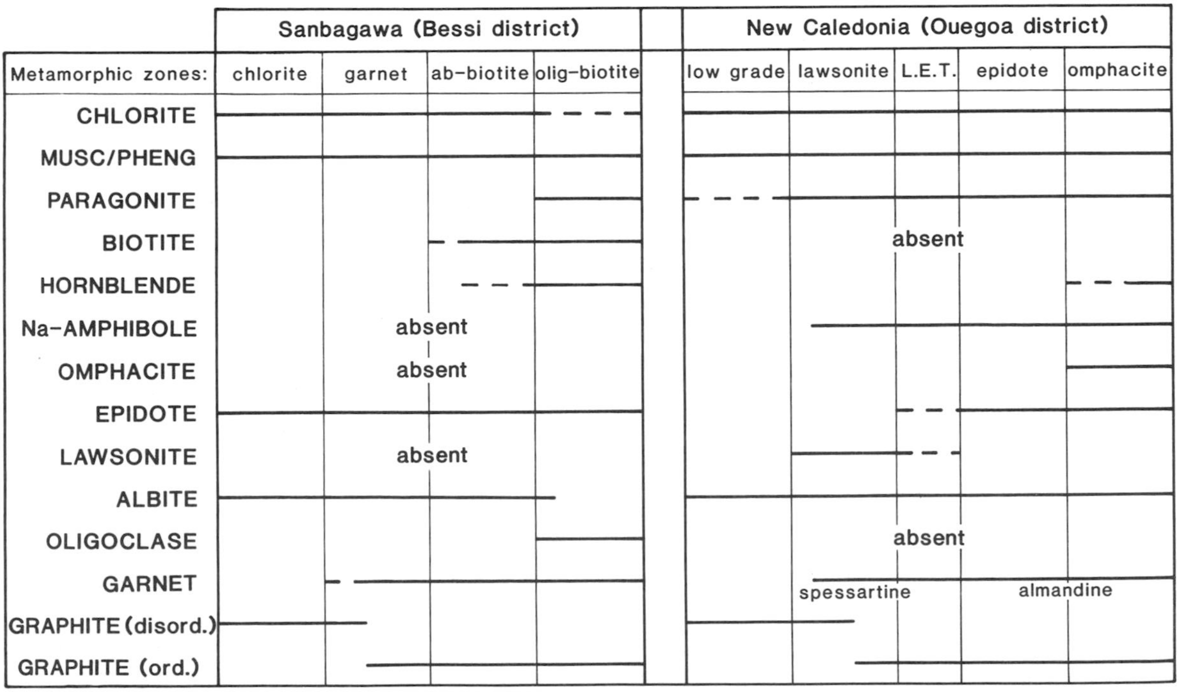

Metamorphic zones:	Sanbagawa (Bessi district)					New Caledonia (Ouegoa district)				
	chlorite	garnet	ab–biotite	olig–biotite		low grade	lawsonite	L.E.T.	epidote	omphacite
CHLORITE										
MUSC/PHENG										
PARAGONITE										
BIOTITE								absent		
HORNBLENDE										
Na–AMPHIBOLE		absent								
OMPHACITE		absent								
EPIDOTE										
LAWSONITE		absent								
ALBITE										
OLIGOCLASE								absent		
GARNET							spessartine		almandine	
GRAPHITE (disord.)										
GRAPHITE (ord.)										

Fig. 3.10   Mineral assemblages of pelites metamorphosed under relatively high pressure conditions in New Caledonia (Brothers & Yokoyama, 1982) and the Sanbagawa belt, Japan (Enami, 1983).

believed to have been intermediate between those of New Caledonia and those of the Barrovian zones of Scotland. Biotite is found, but only appears at the higher grades, after the development of garnet. Again, the rocks are not true pelites for epidote group minerals and hornblende are important.

PRESSURES AND TEMPERATURES OF METAMORPHISM OF PELITIC ROCKS

In this chapter, the relative $P-T$ conditions of formation of many possible pelite assemblages have been established, primarily on the basis of field occurrences with additional constraints provided by the Al-silicate polymorphs present. Over the past 30 years many experimental studies have been carried out on pelitic mineral assemblages and these permit quantitative estimation of the conditions represented by many isograds. In addition, studies using a variety of geothermometers and geobarometers (Ch. 2) have provided independent information about the conditions of formation of natural rocks.

THE PETROGENETIC GRID FOR PELITIC ROCKS

Most experimental studies have been performed using a much simpler chemical system than is found in nature. In the case of reactions between phases of simple chemistry that do not contain significant amounts of the additional components available naturally (such as the Al_2SiO_5 polymorphs) there will be little difference between the stability fields of natural and synthetic assemblages. On the other hand, most experiments on ferromagnesian minerals have been performed in the Fe or Mg end-member systems, so that with one fewer component, reactions that are continuous in nature will be discontinuous in the synthetic system. In such instances the experiments provide only limits to the stability of natural assemblages, unless further calculations are performed, but if Fe and Mg are not strongly partitioned between reactants and products the difference between the equilibrium conditions of natural and synthetic systems may be quite small.

Fig. 3.11 is a petrogenetic grid that has been compiled to illustrate the approximate stability limits of some of the key assemblages discussed in this chapter. It is evident that there are relatively few indicators of pressure among all the reactions shown, although there are a number of good temperature indicators.

The stability of the Al_2SiO_5 minerals is obviously of great importance for determining the depths at which metamorphism has taken place, but unfortunately considerable uncertainty remains. The limits to the kyanite stability field are probably known quite accurately, since there is good agreement between different studies, but the location of the andalusite–sillimanite phase boundary is less certain and there is considerable discrepancy between the experimental studies of Richardson, Gilbert & Bell (1969) and Holdaway (1971) as well as with earlier studies (Fig. 3.12).

More recently, Salje (1986) has shown that there is a sufficient difference in free energy between prismatic and fibrolitic sillimanite to account for the different positions proposed for the andalusite–sillimanite boundary. Prismatic sillimanite was used by Holdaway and has a larger stability field than the finely-ground fibrolite used in the experiments by Richardson and co-workers (Fig. 3.12). Even this may not be the last word on the subject, since in nature fibrolite appears to occur over a wider $P-T$ range than prismatic sillimanite, not the narrower range implied by Fig. 3.12. Possibly the characteristic growth of natural fibrolite in preferred orientations on particular substrates is significant, extending

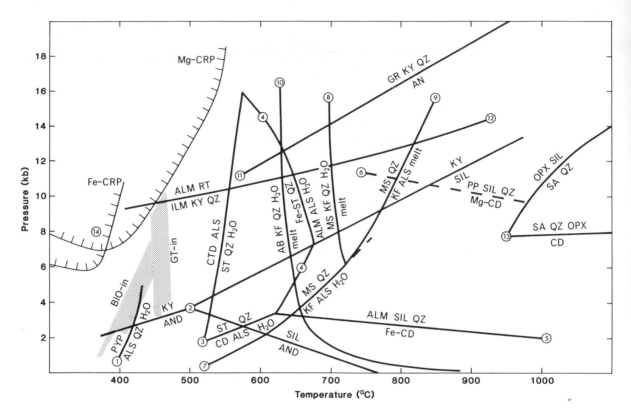

Fig. 3.11 Petrogenetic grid for pelitic metasediments with $P = P_{H_2O}$ (except curve (9)). Abbreviations used are: AB = albite; ALM = almandine; ALS = Al-silicate; AN = anorthite; AND = andalusite; BIO = biotite; CD = cordierite; CRP = carpholite; CTD = chloritoid; GR = grossular; GT = garnet; ILM = ilmenite; KF = K-feldspar; KY = kyanite; MS = muscovite; OPX = orthopyroxene; PP = pyrope; PYP = pyrophyllite; QZ = quartz; RT = rutile; SA = sapphirine; SIL = sillimanite; ST = staurolite. Data sources for the curves are as follows: (1) Kerrick (1968); (2) Holdaway (1971) (see also Fig. 3.12); (3) lower *P-T* limits of Fe-staurolite + quartz fitted to data of Richardson (1968) and Rao & Johannes (1979); (4) Yardley (1981b), compiled from Richardson (1968), Ganguly (1972) & Rao & Johannes (1979); (5) & (6) Holdaway & Lee (1977); (7) Chatterjee & Johannes (1974); (8) & (9) Thompson (1982) (calculated), note that curve (9) is for H_2O-absent conditions; (10) Luth, Jahns & Tuttle (1964); (11) Goldsmith (1980); (12) Bohlen, Wall & Boettcher (1983a); (13) limits to sapphirine + quartz in the Mg end-member system, Grew (1980); (14) inferred limits to carpholite from Chopin & Schreyer (1983). Stippled bands are approximate conditions of the biotite and garnet isograds. Dashed lines are metastable. N.B. Experimental uncertainties are invariably much greater than the thicknesses of the lines drawn.

the stability field of fibrolite towards that of prismatic sillimanite, but it also seems likely that kinetic factors may favour fibrolite growth.

A further pressure indicator for kyanite zone pelites is the nature of the Ti-oxide phase present. The equilibrium curve for:

$$\text{almandine} + \text{rutile} = \text{ilmenite} + \text{kyanite} + \text{quartz} \qquad [3.28]$$

lies within the kyanite stability field (Bohlen, Wall & Boettcher 1983a) and serves to

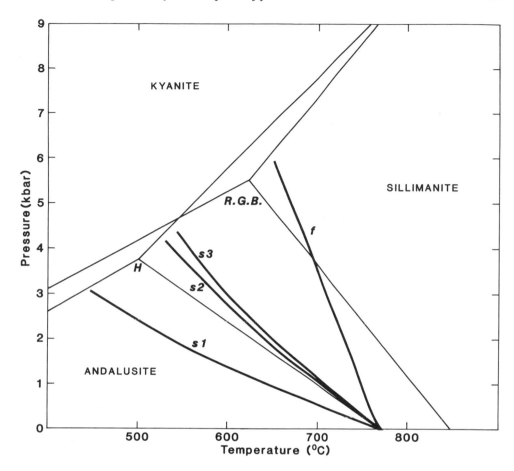

Fig. 3.12 *P–T* diagram illustrating the possible stability fields of the Al_2SiO_5 polymorphs. Light lines are experimental determinations by Richardson, Gilbert & Bell (1969) (*RGB*) who used fibrolite and Holdaway (1971) (*H*) who used prismatic sillimanite. Bold lines are andalusite–sillimanite boundaries calculated by Salje (1986) for specific natural sillimanites. Curve *f* is for fibrolite–andalusite, *s1*, *s2* and *s3* are for equilibrium between andalusite and specific natural prismatic sillimanites. Salje's curves are constrained to meet at the 1 atmosphere equilibrium temperature used by Holdaway. Note how Salje's results account for the discrepancy between the experimental studies but do not explain why the stability field of fibrolite is apparently larger than that of prismatic sillimanite in nature.

separate higher pressure kyanite–garnet schists containing rutile from lower pressure varieties with ilmenite. This is a particularly useful pressure indicator because the end-members involved in reaction [3.28] are the dominant constituents of the natural minerals, so that the equilibrium conditions for natural rocks are close to those in the experimental system.

It is evident from Fig. 3.11 that staurolite provides a good indicator of metamorphic temperature since it is stable only over a narrow temperature range which does not change much with pressure. Another reaction that is quite well known is the muscovite breakdown reaction (reaction [3.12]) that usually gives rise to the second sillimanite isograd. On the other hand many other equilibria are markedly continuous and it is not possible to estimate the conditions of the rocks concerned without accurate mineral analyses, except within very broad limits.

Carmichael (1978) has presented a simple model for interpreting the $P-T$ conditions of formation of pelitic rocks approximately, based on the intersection of the Al_2SiO_5 phase boundaries with curves for reactions [3.8] and [3.12]. His scheme is particularly concerned with defining the pressure range over which heating has occurred and distinguishes six 'bathozones' requiring successively higher pressures:

Zone 1   Andalusite coexists with K-feldspar due to muscovite breakdown in the andalusite stability field ($P < 2.2$ kbar).
Zone 2   Sillimanite is stable by the second sillimanite isograd, but andalusite may have been produced previously through reaction [3.8] ($P = 2.2 - 3.5$ kbar).
Zone 3   Sillimanite is produced by reaction [3.8] and reaction [3.12], however if an Al_2SiO_5 phase is present at lower grades (in veins or rocks of unusual composition) it will be andalusite ($P = 3.5 - 3.8$ kbar).
Zone 4   Similar to Zone 3 except that any early-formed Al_2SiO_5 will be kyanite ($P = 3.8 - 5.5$ kbar).
Zone 5   Kyanite is produced by reaction [3.8] but sillimanite is the stable phase by the temperature at which muscovite breakdown occurs ($P = 5.5 - 7.1$ kbar).
Zone 6   Kyanite is stable with K-feldspar, implying that reaction [3.12] occurs in the kyanite stability field ($P > 7.1$ kbar).

These zones are illustrated in Fig. 3.13. The apparently very precise pressures quoted for the zone boundaries should not be taken too seriously; they assume no errors in the Al_2SiO_5 phase diagram or other equilibrium curves used (a highly controversial assumption), and neglect the fact that reactions [3.8] and [3.12] are both continuous in natural rocks. In fact Carmichael's choice of locations for the key univariant curves differs from that used here in Fig. 3.11. Nevertheless, Carmichael's scheme provides an extremely elegant and simple way of distinguishing areas metamorphosed at different pressures, requiring only thin-section observations. It is notable that the pelite reactions do not provide good estimates of the higher pressure limits that may be attained in metamorphism; for this it is better to turn to the basic rocks described in Chapter 4.

CONTINUOUS AND CATION EXCHANGE REACTIONS AS GEOTHERMOMETERS AND GEOBAROMETERS FOR PELITES

A number of reactions taking place over a range of conditions have been used to try to provide continuous information about changes in temperature and pressure of metamorphism, as opposed to the more limited information available from the petrogenetic grid. Such geothermometers and geobarometers may be (a) calibrated empirically by determining the compositions of coexisting phases in rocks whose $P-T$ conditions of formation are presumed to be independently known (e.g. Thompson, 1976; Perchuk, Podlesskii & Aranovich, 1981) or (b) calibrated experimentally (e.g. Ferry & Spear, 1978; Newton, 1983).

The most widely used geothermometer for pelites is the garnet–biotite cation exchange

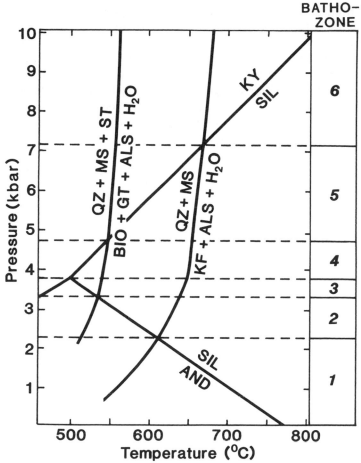

Fig. 3.13 Petrogenetic grid for pelitic rocks divided into bathozones according to the scheme of Carmichael (1978). Note that although the curves shown do not correspond with the better determined ones given in Fig. 3.11, this probably has very little effect on the pressures assigned to the zones, except to raise the lower pressure limit of bathozone *6*.

reaction discussed in Chapter 2; at relatively low grades the muscovite–paragonite solvus may be a useful temperature indicator (Eugster *et al.*, 1972), while the partitioning of Mn between ilmenite and garnet has considerable promise (Pownceby, Wall & O'Neill, 1987). In addition, many studies have made precise calculations based on the breakdown of muscovite or staurolite (reactions [3.12] and [3.11]), thereby treating them as continuous reactions rather than fixed curves on the petrogenetic grid.

The breakdown of anorthite to grossular + Al-silicate + quartz (Ch. 2) is one of the most widely applied geobarometers for metapelites (Newton & Hasleton, 1981), although since anorthite and grossular are usually very minor components of plagioclase and garnet in pelitic rocks, the results are sensitive to the activity–composition models used for these phases. At high grades, the equilibrium

$$\text{quartz} + \text{garnet} + \text{sillimanite} = \text{cordierite} \qquad [3.29]$$

is strongly pressure sensitive. Newton (1983) summarises different approaches to this geobarometer, the main problem with which appears to be the influence of independent variations in P_{H_2O}, since cordierites can contain appreciable water. Martignole & Sisi (1981) provided a useful calibration. Many recent studies have treated reaction [3.28] as a continuous reaction, involving equilibrium between garnet, kyanite, rutile, ilmenite, and quartz, to obtain a precise pressure estimate, however rocks containing ilmenite and rutile coexisting in equilibrium are not common; probably other TiO_2 phases such as anatase, occurring as late-stage alteration products, are sometimes misidentified as rutile.

4 METAMORPHISM OF BASIC IGNEOUS ROCKS

Lava flows, and their related minor intrusions and volcanogenic sediments, are frequently found in metamorphic sequences. In many orogenic belts, basic types of igneous rocks predominate, but elsewhere tuffs and greywackes of intermediate compositions are important. This chapter will be predominantly concerned with rocks of basic composition, known generally as metabasites; however, the mineral assemblages developed in intermediate metavolcanics are often very similar.

Over a wide range of metamorphic conditions, metabasic rocks are dominated by amphibole and are known as **amphibolites**. It is also possible for amphibolites to be formed by metamorphism of marly sediments containing a mixture of clays and carbonate material, or by metasomatic interactions between contrasting sedimentary layers during metamorphism (Orville, 1969). The problem of distinction between meta-igneous or **ortho-amphibolites** and metasedimentary or **para-amphibolites** has been widely discussed in the past, however Leake (1964) showed that although the chemical composition of an individual sample might not permit an unequivocal identification of its parentage, the chemical trends obtained when a suite of analyses from an amphibolite unit are plotted on chemical variation diagrams, *can* permit the distinction between rocks of sedimentary and igneous origin.

The basic rocks show two important differences in their metamorphism when compared with pelitic metasediments. In the first place, the initial igneous assemblage is made up of mostly anhydrous minerals stable at high temperatures, in marked contrast to clay. As a result the first changes that take place when igneous rocks are buried and heated in the midst of a sedimentary sequence, will involve the formation of hydrous minerals stable at low temperatures. The extent of such retrograde reaction depends on the amount of water that is able to penetrate, and while permeable tuffs may retain none of their original minerals, nearby massive lava flows and dykes may retain extensive relics of igneous minerals and textures. Deformation also facilitates the influx of water.

A second contrast with the metapelites concerns the type of mineral assemblage and reactions that occur. Metabasic rocks at most grades contain a relatively small number of minerals, many of which show extensive solid solution, and fewer isograds can be detected in the field. Most reactions are continuous and involve progressive changes in mineral compositions over broad intervals of pressure and temperature. Only at very low grades of metamorphism are there significant changes in mineralogy over a sufficiently narrow temperature interval that zones can be defined which are comparable to the pelite zones. The lack of variation in their mineralogy was for a long time a major obstacle to the study of metabasic rocks, however the development of the electron microprobe, permitting rapid analysis of minerals in thin section, has made it possible for the progressive variation in the composition of solid solution minerals to be studied in large numbers of samples spanning a range of metamorphic grades. The most important mineral in this respect is amphibole, common in metabasites at all grades except the very lowest. Laird & Albee (1981) have documented the variation in amphibole chemistry over a wide range of types of metamorphism from many different areas.

THE FACIES CLASSIFICATION

The mineral assemblages of metabasites are less sensitive to changes in pressure and temperature than those of pelites except at very low grades of metamorphism and the zones that they define therefore represent a broader range of possible conditions of formation. In this sense metabasites are less useful than pelites as metamorphic indicators, but on the other hand, 'true pelites' are not abundant in many metamorphic terranes, and the detailed zones identified in pelites can seldom be traced accurately through other rock types. However, metabasites are found in most metamorphic belts and the zones identified in most other rock types can be correlated with metabasite zones. For this reason Eskola based his scheme of metamorphic facies, introduced in Chapter 2, on the assemblages of metabasites.

MINERALOGICAL CHANGES DEFINING THE FACIES

The main mineralogical changes that occur can be grouped as follows:

1. *Changes in the composition of amphibole*   Broadly speaking the lower temperature facies are characterised by actinolite; hornblende occurs at higher temperatures and glaucophane at higher pressures. Although some intermediate amphibole compositions are found, the breaks between these three main types are often quite abrupt, and there has been considerable discussion of the extent to which this may result from miscibility gaps in the amphibole solid solution series, succinctly summarised by Robinson *et al.* (1982).

   The reason for suspecting that actinolite and hornblende are immiscible, at least under greenschist facies conditions, is the frequency with which they occur in the same rock as discrete phases. On the other hand, most textures of rocks with both calcic amphiboles present, suggest that one has grown as a replacement of the other. Furthermore, exsolution textures of actinolite in hornblende, or vice versa, have never been found. Robinson *et al.* (1982) conclude that there probably is a solvus between actinolite and hornblende under greenschist facies conditions, although a continuous range of compositions appears to be present at higher grades, above about 600 °C (Misch & Rice, 1975). The difficulty in resolving the problem stems in part from the relatively constant composition of metabasic rocks; in part from a possible correlation between the width of a solvus gap (which is in terms of $Al^{IV} \rightleftharpoons Si$ substitution) and other variables such as X_{Mg}; and in part from the fact that irrespective of any solvus, many occurrences of coexisting actinolite and hornblende clearly have disequilibrium textures.

   The frequent occurrence of replacement textures is also a problem in investigations of immiscibility between sodic and calcic amphiboles. Nevertheless, the balance of evidence again suggests that a miscibility gap does exist, however the solvus appears to be strongly asymmetrical since glaucophane coexisting with a calcic amphibole never shows appreciable Ca substitution, whereas calcic amphiboles range up to appreciable Na-contents (barroisite, intermediate between hornblende and glaucophane, is not uncommon in high pressure rocks).

In addition to the more common Ca- or Na-bearing monoclinic amphiboles, Ca-poor amphiboles (cummingtonite and the orthorhombic amphiboles, anthophyllite and gedrite) are also found in some metabasites, especially at low pressures.

2. *Formation of pyroxene under extreme conditions* Both clinopyroxene (diopside–augite) and orthopyroxene (typically a pleochroic hypersthene) may develop at very high temperatures, and are characteristic of the **granulite facies**. At high pressures and low temperatures, albite is replaced by a jadeite-rich clinopyroxene, while at both high temperatures and high pressures in the **eclogite facies** an omphacitic pyroxene, intermediate between jadeite and diopside, is found.

3. *Changes in feldspar composition* Albite occurs at low temperatures but is replaced by an intermediate plagioclase at higher grades in much the same way as was described for pelitic schists on page 67. There is a tendency for higher anorthite contents with increase in temperature. Plagioclase feldspar is entirely absent at very high pressures.

4. *Formation of hydrous Ca–Al silicates at low grades* Zeolites, prehnite and pumpellyite are characteristic of very low grade metamorphism, while lawsonite requires high pressures and low to medium temperatures. However, epidote minerals are stable over a wide range of P–T conditions, although progressively replaced by plagioclase at high temperature.

Both orthorhombic zoisite and monoclinic clinozoisite occur in metabasites, and there is continuous solid solution between clinozoisite ($Ca_2Al_3Si_3O_{12}.OH$) and epidote ($Ca_2AlFe^{3+}_2Si_3O_{12}.OH$). Zoisite and clinozoisite are polymorphs, and might each be expected to characterise distinct P–T conditions, as in the case of the Al-silicate polymorphs. Unfortunately this is not the case. Clinozoisite shows extensive solid solution by replacement of Al by Fe^{3+}, whereas zoisite accepts only small amounts of Fe. The effect of the contrasting behaviour of this additional component is to permit clinozoisite–epidote to occur in Fe-bearing rocks over a wide range of conditions where zoisite is also stable, and indeed the two can occur together. For a long time it has even been controversial which is the higher temperature polymorph, but recent work by Jenkins, Newton & Goldsmith (1984) has established that in Fe-free conditions zoisite has the higher entropy and is therefore the higher temperature form. In metamorphic rocks of basaltic composition it is generally true that zoisite and Fe-poor clinozoisite are typical of relatively high pressure or high temperature metamorphism, while epidote is more widespread at low grades.

5. *Presence or absence of garnet* Garnet is indicative of medium to high pressure and is absent from metabasites metamorphosed under low pressure conditions, as in most thermal aureoles.

CHARACTERISTICS OF THE METAMORPHIC FACIES

The scheme of metamorphic facies adopted in this book was illustrated in Fig. 2.8. It is based on that of Turner (1981) but differs principally in that fewer subdivisions are made at low grade; the **prehnite–pumpellyite facies** of this book includes the prehnite–pumpellyite, pumpellyite–actinolite and lawsonite–albite facies of other authors. The reason for this difference of opinion is that these low grade mineralogical changes in metabasites define a series of zones that each represent only a narrow range of P–T conditions, comparable to the range represented by the high grade pelite zones. Such zones do not really merit the status of facies, because they cannot be recognised in other rock types. However, because other metamorphic facies have come to be equated with mineralogical zones in metabasites, many authors have designated each metabasite zone as a separate facies. The critical assemblages of each facies are given in Table 4.1.

Table 4.1   Mineral assemblages or sub-assemblages diagnostic of the metamorphic facies

Facies	Metabasic rocks	Pelitic rocks (with quartz)
Zeolite	Laumonite (most typical), analcite, heulandite, wairakite. Incompletely reacted relics widespread	Mixed-layer clays
Albite–Epidote Hornfels	Albite + epidote + actinolite + chlorite Actinolite + oligoclase	Muscovite + biotite + chlorite
Hornblende Hornfels	Hornblende + plagioclase ± cummingtonite	Cordierite + chlorite + biotite + muscovite Andalusite + biotite + muscovite Cordierite + andalusite + muscovite (higher temperature zone)
Pyroxene Hornfels	Clinopyroxene + orthopyroxene + plagioclase ± olivine ± hornblende	Cordierite + andalusite + K-feldspar
Sanidinite	Not well defined	Corundum + magnetite + anorthite (no quartz), glass
Prehnite–Pumpellyite	Prehnite + pumpellyite ± chlorite ± albite ± epidote (lower temperature zone) Pumpellyite + actinolite (higher temperature zone) Lawsonite + albite (higher pressure zone)	Illite/muscovite + chlorite + albite + quartz Stilpnomelane, pyrophyllite
Greenschist	Actinolite + epidote ± albite ± chlorite ± stilpnomelane (lower temperature zone) Hornblende ± actinoline + albite + chlorite + epidote ± garnet (higher temperature zone)	Chlorite + muscovite + albite (lowest temperature zone) Chlorite + muscovite + biotite + albite Garnet + chlorite + muscovite + biotite + albite (highest temperature zone) Chloritoid, paragonite + muscovite + albite
Amphibolite	Hornblende + plagioclase ± epidote ± garnet	Staurolite, kyanite *or* illimanite + muscovite (lower temperature zone) Sillimanite + K-feldspar ± muscovite + cordierite *or* garnet Sillimanite + garnet + cordierite, *no* K-feldspar (higher temperature zone)

Facies	Metabasic rocks	Pelitic rocks (with quartz)
Granulite	Orthopyroxene + clinopyroxene + plagioclase ± olivine ± hornblende (low pressure) Garnet + clinopyroxene + orthopyroxene + plagioclase ± hornblende (medium pressure) Garnet + clinopyroxene + quartz + plagioclase ± hornblende (high pressure)	Cordierite + garnet + K-feldspar + sillimanite (moderate pressure) Kyanite + K-feldspar (high pressure) Hypersthene, sapphirine + quartz (high temperature)
Blueschist	Glaucophane + lawsonite	Phengite + chlorite *or* talc + garnet, *no* biotite Mg-chloritoid, carpholite
Eclogite	Omphacite + garnet, *no* plagioclase, *no* lawsonite	Talc + kyanite ± garnet ± muscovite (phengitic)

METAMORPHISM OF BASIC ROCKS AT LOW GRADES: ZEOLITE AND PREHNITE–PUMPELLYITE FACIES

ZEOLITE FACIES

Until quite recently, the low temperature hydration and alteration of igneous rocks was considered to be an unfortunate affliction that inhibited the work of igneous petrologists, rather than a branch of metamorphic petrology. In the 1950s, D.S. Coombs first demonstrated that systematic metamorphic zoning patterns could none the less be present in such rocks and as a result of his work Fyfe, Turner & Verhoogen (1958) erected a zeolite facies. Coomb's original results from southern New Zealand are summarised, with comparisons with other areas, in Coombs *et al.* (1959).

The type area for the zeolite facies is the Taringatura Hills of Southland, New Zealand (Figs 4.1 and 4.2(*a*)). The rocks here are predominantly volcanogenic greywackes and tuffs, and are therefore particularly susceptible to metamorphism because they contained unstable glass shards as well as the high temperature igneous minerals, and also had the porosity and permeability of sandstone. In the Taringatura Hills the volcanic material is of andesitic, dacitic and even rhyolitic composition, and therefore not strictly basic igneous material at all. However, subsequent work on the nearby Otama complex, which does include basic rocks, shows that the changes are essentially the same for both basic and intermediate compositions, except that the precise species of zeolite minerals formed are different. The section in the Taringatura Hills comprises 10 km of Triassic greywackes, from which a further 3–5 km of overlying sediment may have been removed by erosion. The rocks have not been very much deformed and the series of progressive mineralogical changes in the greywackes is simply related to depth of burial. Hence Coombs coined the term **'burial metamorphism'** for this sort of metamorphism that is of regional extent but unrelated to orogenic deformation.

Mineralogical variation in the Taringatura Hills sequence is illustrated in Fig. 4.2(*a*). In the upper part of the succession volcanic glass has been altered to the zeolite heulandite, or

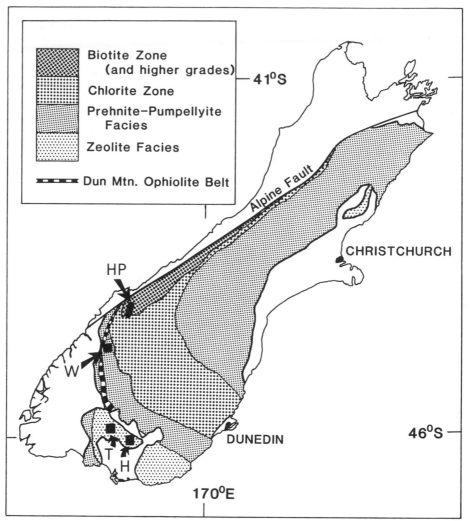

Fig. 4.1 Map of the metamorphic zonation in the Wakatipu metamorphic belt, South Island of New Zealand, compiled from Landis & Coombs (1967) and New Zealand Geological Survey (1972). Locations discussed in this chapter are indicated as follows: H, Hokonui Hills; HP, Haast Pass; T, Taringatura Hills; W, upper Wakatipu district.

more rarely to analcite, and some secondary quartz and fine phyllosilicates (essentially montmorillonite or celadonite) are present; however, the crystalline igneous phases are preserved more or less intact.

With increasing depth a second zone is reached in which the original calcic plagioclases are replaced by pseudomorphs of albite, and laumontite replaces the original zeolites. Chlorite also appears in this zone, and adularia may occur in pseudomorphs after analcite.

In the third and highest grade zone, zeolites become scarce and laumontite is replaced by the hydrous Ca–Al silicates prehnite and pumpellyite; minor epidote also makes its appearance.

The principal difference found in the basic rocks of the Otama complex nearby is that thomsonite is the dominant zeolite mineral.

In a subsequent study of zeolite facies metamorphism in the Hokonui Hills (Fig. 4.1), along strike from Coombs's type area, Boles and Coombs (1975) were able to show that it

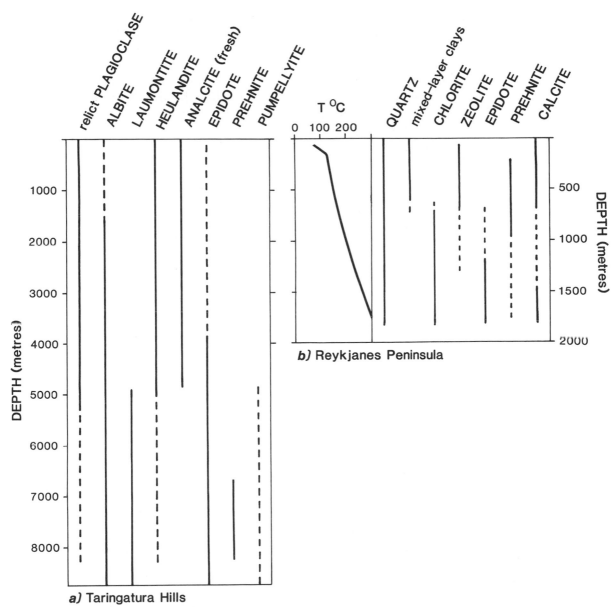

Fig. 4.2 Variation in low grade mineral assemblages with depth in *a*) the burial metamorphic zones of the Taringatura Hills (Coombs, 1954) and *b*) a representative well in the Reykjanes geothermal field (Tomasson & Kristmansdottir, 1972). The temperature profile of the geothermal well is also illustrated. Note the difference in depth scales between the two parts of the figure. Broken lines denote sparse occurrences.

was accompanied by appreciable metasomatic changes in the chemical composition of the original tuffs, and demonstrated that the chemistry of the circulating fluids was actually a critical factor in determining which metamorphic minerals were stable.

PREHNITE–PUMPELLYITE FACIES

The low grade metamorphic effects described by Coombs have since been found in many parts of the world and are recognised as resulting from distinct conditions of metamorphism. Most of the Taringatura Hills sequence can be assigned to the zeolite facies of metamorphism, but the appearance of prehnite and pumpellyite marks the onset of the **prehnite–pumpellyite facies**. In some areas a transition is seen from prehnite–pumpellyite-bearing rocks to progressively higher grade zones not found in the Taringatura Hills. For example, Kawachi (1975) described the assemblages in the Wakatipu district of southern New Zealand (Fig. 4.1), in which the prehnite–pumpellyite zone is succeeded by a pumpellyite–actinolite zone. Both these zones are included here in the prehnite–pumpellyite facies.

Typical minerals of prehnite–pumpellyite facies metabasites include prehnite, pumpellyite, actinolite, chlorite, epidote, albite, quartz, sericite, lawsonite, sphene and stilpnomelane and the rocks often have a distinct blue–green colour imparted by Fe-pumpellyite. Relic igneous phases are also frequently present, and it appears that an original igneous rock may react directly to zeolite, prehnite–pumpellyite or even greenschist or blueschist facies assemblages according to the conditions at which fluid is first able to penetrate. In other words, metamorphism is not strictly progressive in originally igneous rocks at these low grades; a common example is the growth of actinolite as fringes around relic augite rather than from pre-existing metamorphic minerals.

LOW GRADE METAMORPHISM IN GEOTHERMAL FIELDS

Rather similar metamorphic assemblages to those described by Coombs have been found forming today in active geothermal systems, where these are developed in basic or intermediate rocks. Samples are obtained from drilling, and because temperature can be measured down the drill hole, the conditions of formation of the different zones are well known.

One of the best-studied active geothermal fields is the Reykjanes field of south-west Iceland, and it is of particular interest because the fluid circulating within it is dominated by sea water, rather than the meteoric water found in most sub-aerial geothermal systems. This means that the Reykjanes field may provide a closer analogue for hydrothermal metamorphism on the sea floor.

Tomasson and Kristmansdottir (1972) have described the alteration sequence at depth in the Reykjanes system, revealed by drilling for commercial exploitation, and the zones observed are summarised in Fig. 4.2(*b*), with the zones from the Taringatura Hills for comparison. The rock types affected include hyaloclastic tuffs and breccias as well as lava flows, and alteration principally affects glassy or very fine grained material. Of the magmatic minerals, olivine is usually completely altered but pyroxene and to some extent plagioclase feldspar are relatively resistant.

The principal alteration products in the upper part of the system are clay minerals, dominated by montmorillonite near the surface but giving way to mixed-layer clays and chlorite at depth. A variety of zeolite minerals is also found in the upper part of the system, and includes mordenite, stilbite, mesolite, analcite and wairakite. Secondary calcite and quartz are found throughout. In the deeper parts of the system, corresponding to temperatures in excess of 230 °C, prehnite, epidote and chlorite are found, and plagioclase shows sporadic alteration to albite or, rarely, K-feldspar. Elsewhere in Iceland, where drilling has penetrated still hotter rocks at temperatures in excess of 300 °C, actinolite has been found. Note the very small vertical extent of the zones compared with burial metamorphism, because of the steep geothermal gradient.

Geothermal fields are of considerable importance for metamorphic petrology, both because they allow us to study active metamorphism and because they provide an analogue for the much more extensive hydrothermal metamorphism that is believed to affect newly formed oceanic crust. This is described more fully in the last part of this chapter.

ON THE LACK OF LOW TEMPERATURE MINERALS IN SOME METABASITES

No discussion of low grade metamorphism would be complete without noting that it sometimes does not appear to have taken place! The distinctive metamorphic zones described by Coombs were looked for in other parts of the world where low grade (e.g. slaty) rocks occur, but were not always found. It was pointed out by Zen (1961) that metamorphism sometimes seems to proceed direct to greenschist facies assemblages from the original minerals, and phases such as zeolites, prehnite or pumpellyite are not developed. He suggested that this would be the case if the fluid phase that entered the rock contained appreciable CO_2, since minerals such as laumontite, prehnite and pumpellyite are hydrous Ca–Al silicates that can react with CO_2 to produce calcite, together with epidote, quartz and chlorite which are also widespread in the greenschist facies. In other words, in the presence of CO_2, the minerals distinctive of the lower grade facies may never form. This appears to be the case in the Salton Sea geothermal field (California) where zeolites are absent but epidote and chlorite are common and fluid contains CO_2 (Muffler and White, 1969). Thompson (1971a) has shown that only very small amounts of CO_2 are necessary to suppress the formation of laumontite and prehnite.

METABASITES FROM THE BARROVIAN ZONES: GREENSCHIST AND AMPHIBOLITE FACIES

In the regions of the Scottish Highlands from which Barrow and Tilley described the pelite zonal scheme outlined in Chapter 3, there are also quite widespread occurrences of metabasite. In the lowest grade regions of the south-west Highlands, pillow lavas, sills, etc., can sometimes still be identified but for the most part the rocks are too deformed and recrystallised for their original character to be determined. Metabasites are also common elsewhere in the world in association with pelites that have been subjected to similar 'Barrovian' metamorphism.

The first systematic study of Scottish metabasites was made by Wiseman (1934) who attempted to correlate their assemblages with the pelite zones. Wiseman's observations were as follows:

Chlorite and biotite zones:
Original igneous *textures* are sometimes still present, but the mineral assemblage is entirely metamorphic (e.g. ophitic feldspar proves to be albite in composition). The dominant minerals are chlorite, epidote, albite, pale-green actinolite and quartz, often with calcite and biotite. Stilpnomelane has subsequently been recognised in these rocks.

Garnet zone:
Garnet in fact first appears in metabasites at somewhat lower grades than in pelites. At the same time chlorite and calcite become scarcer, actinolite is replaced by a blue–green hornblende and a more calcic plagioclase feldspar may appear.

Staurolite and kyanite zones:
Biotite and chlorite are absent, and often so is calcite. Green hornblende and plagioclase predominate and epidote is scarce.

Sillimanite zone:
The rocks are dominated by green to brownish-green hornblende and an intermediate plagioclase, and no epidote remains.

The progressive change in amphibole type from actinolite through blue-green hornblende to brownish-green hornblende has been documented from other parts in the world, however a detailed reinvestigation of the Scottish zones by Graham (1974) reveals that it is an oversimplification. Instead, blue–green hornblende occurs, albeit rarely, in rocks down to the chlorite zone, and both hornblende and actinolite persist through the pelitic chlorite, biotite and garnet zones. Nevertheless, hornblende is very rare at the lower grades and restricted to rocks of unusual composition, while actinolite is similarly of restricted occurrence in the garnet zone. Where both amphiboles occur together, hornblende is usually seen to enclose actinolite, and therefore grew later. The only unequivocal phase changes that Graham reported were the disappearance of stilpnomelane and growth of garnet within the pelitic biotite zone, although neither of these phases is invariably present in the metabasites.

FACIES OF BARROVIAN METAMORPHISM

The lowest grade, chlorite zone, rocks in Scotland can probably be assigned to the greenschist facies, for pumpellyite has never been reported and epidote coexists with actinolite. Furthermore, around Lake Wakatipu in New Zealand (Fig. 4.1) there is a transition from pumpellyite-bearing, sub-greenschist facies rocks to a chlorite zone with very similar assemblages to those found in Scotland (Landis and Coombs, 1967). The higher grade parts of the Scottish zonal sequence belong to the amphibolite facies, and the precise boundary between the facies is usually linked to the composition of plagioclase.

It was pointed out in Chapter 3 (page 67) that within the garnet zone there is a gap in the plagioclase feldspar solid solution series known as the peristerite gap. At lower grades only albite is found, at higher grades oligoclase occurs, and in some intermediate types both albite and oligoclase are present. This transition has been studied in metabasites from the Haast Pass, New Zealand, by Cooper (1972) and provides a convenient basis for defining the boundary between the greenschist and amphibolite facies. Cooper found that where garnet first appears, albite is the only plagioclase present, but at about this grade oligoclase also appears, occurring as a distinct phase coexisting with albite. With increasing grade in the garnet zone, albite becomes more calcic but oligoclase becomes more sodic. In other words, although they persist as separate phases, the difference between their compositions becomes less. Finally, albite disappears and there is only one plagioclase present which is oligoclase. This change marks the onset of the oligoclase zone which succeeds the garnet zone and is approximately equivalent to the staurolite zone of pelites. The beginning of the amphibolite facies corresponds to the onset of the oligoclase or staurolite zones.

REACTIONS DURING BARROVIAN METAMORPHISM OF BASIC COMPOSITIONS

Harte and Graham (1975) have summarised the changes that take place in going from a greenschist facies metabasite to one of the amphibolite facies as follows:

1. *Decrease in abundance*: actinolite, stilpnomelane (vanishes), chlorite, epidote, albite (vanishes).
2. *Increase in abundance*: hornblende, garnet, Ca-plagioclase.

It is however very difficult to write specific reactions to account for these changes because the rocks contain relatively few phases (often only four to five) and these are made up of a large number of components, e.g. Na_2O, K_2O, CaO, MgO, FeO, Al_2O_3, SiO_2, H_2O and often CO_2 and Fe_2O_3 may all be important. Hence reactions are likely to be continuous, and the reacting assemblage may have several degrees of freedom.

An important reaction at the lowest grades is:

$$\text{chlorite} + \text{calcite} \rightarrow \text{epidote} + \text{actinolite} + CO_2–H_2O \text{ fluid} \qquad [4.1]$$

which accounts for the fact that the association chlorite + calcite becomes scarce with increasing grade.

Cooper (1972) has suggested a number of possible reactions that may lead to the transition from the greenschist to the amphibolite facies. These are written as reactions between end-members of the naturally occurring solid solutions, and if these compositions could be synthesised artificially and reacted together, the reactions would be univariant. In nature, many of the constituents of the reactions occur as only one component of a solid solution, and so the reactions are continuous and lead to an increase or decrease in the concentration of the various components in the solid solutions present. For further simplicity, Mg^{2+} and Fe^{2+} may be treated as a single component, denoted MF. One of the reactions proposed by Cooper is:

$$3 \text{ MF}_{10}\text{Al}_4\text{Si}_6\text{O}_{20}(\text{OH})_{16} + 12 \text{ Ca}_2\text{Al}_3\text{Si}_3\text{O}_{12}(\text{OH})_2 + 4 \text{ SiO}_2 \rightarrow$$
$$\text{chlorite} \qquad\qquad \text{Al-epidote} \qquad\qquad \text{quartz}$$

$$10 \text{ Ca}_2\text{MF}_3\text{Al}_4\text{Si}_6\text{O}_{22}(\text{OH})_2 + 4 \text{ CaAl}_2\text{Si}_2\text{O}_8 + 2 \text{ H}_2\text{O}$$
$$\text{tschermakite hornblende} \qquad \text{anorthite} \qquad \text{fluid} \qquad [4.2]$$

This leads to the production of a calcic plagioclase component which may combine with albite to produce the oligoclase that is actually observed, and also contributes to the growth of hornblende. Other reactions that contribute to the overall changes in mineral abundance and composition may include:

$$\text{NaAlSi}_3\text{O}_8 + \text{Ca}_2\text{MF}_5\text{Si}_8\text{O}_{22}(\text{OH})_2 \rightarrow$$
$$\text{albite} \qquad\qquad \text{actinolite}$$

$$\text{NaCa}_2\text{MF}_5\text{AlSi}_7\text{O}_{22}(\text{OH})_2 + 4 \text{ SiO}_2$$
$$\text{edenite hornblende} \qquad \text{quartz} \qquad [4.3]$$

$$\text{Ca}_2\text{MF}_5\text{Si}_8\text{O}_{22}(\text{OH})_2 + 7 \text{ MF}_{10}\text{Al}_4\text{Si}_6\text{O}_{20}(\text{OH})_{16} + 28 \text{ SiO}_2$$
$$\text{actinolite} \qquad\qquad \text{chlorite} \qquad\qquad \text{quartz}$$

$$+ 24 \text{ Ca}_2\text{Al}_3\text{Si}_3\text{O}_{12}(\text{OH}) \rightarrow 25 \text{ Ca}_2\text{MF}_3\text{Al}_4\text{Si}_6\text{O}_{22}(\text{OH})_2 + 44 \text{ H}_2\text{O}$$
$$\text{Al-epidote} \qquad\qquad \text{tschermakite hornblende} \qquad \text{fluid} \qquad [4.4]$$

EFFECTS OF LOWERED PRESSURE: HORNFELS FACIES

In areas where contact or regional metamorphism at low pressure has led to the production of andalusite- or cordicrite-bearing assemblages in pelites, many metabasites show few differences from the assemblages found in the Barrovian zones. However, they do not normally develop garnet in such terranes, and the relationship between changes in amphibole type and changes in plagioclase type may be rather different. In some low pressure metabasites an association of actinolite with intermediate plagioclase (rather than albite) has been reported. In other words, whereas in the Barrovian sequence actinolite gives way to hornblende before albite is replaced by oligoclase, this sequence appears to be reversed at low pressures (Miyashiro, 1973).

An additional feature of low pressure metabasites is that Ca-poor amphiboles, notably cummingtonite, are more widespread, appearing in the hornblende hornfels facies or the low pressure part of the amphibolite facies. The reason for this is that at low pressures there is little substitution of Al for Fe and Mg in the octahedral sites of hornblende. For a given rock composition that might have crystallised as essentially hornblende + plagioclase at higher pressures, the modal amount of plagioclase must be greater at lower pressure, since this is the alternative phase in which the Al can occur. The consequence of this is that there is insufficient Ca available to combine with Mg and Fe as hornblende, and so some Mg–Fe amphibole forms instead. The reaction can be represented in terms of breakdown of the idealised aluminous hornblende end-member tschermakite as:

$$7Ca_2Mg_3Al_4Si_6O_{22}(OH)_2 + 10SiO_2 = 3Mg_7Si_8O_{22}(OH)_2 + 14CaAl_2Si_2O_8 + 4H_2O$$

    tschermakite   quartz   cummingtonite   anorthite   fluid

[4.5]

where the assemblage on the left is favoured by increased pressure.

BASIC IGNEOUS ROCKS METAMORPHOSED AT HIGH PRESSURES: BLUESCHIST AND ECLOGITE FACIES

Metabasites are especially important to the understanding of metamorphism at relatively high pressures and low temperatures because they undergo a number of conspicuous mineralogical changes under such conditions whereas pelites, as was noted in the previous chapter, undergo relatively few changes with increased pressure.

THE BLUESCHIST FACIES

The most characteristic effect of high pressure metamorphism on basic rocks is the replacement of the calcic amphiboles found in the Barrovian sequence by the sodic amphibole glaucophane, which imparts a characteristic slaty lilac to blue colour to them, hence the name blueschist. Other minerals diagnostic of high pressure metamorphism are lawsonite, jadeite-rich pyroxene and aragonite.

Ernst, Onuki & Gilbert (1970) have contrasted the high pressure metamorphism of the Sanbagawa terrane of Japan with that of the Franciscan terrane of California. In parts of Sanbagawa, pumpellyite-bearing sub-greenschist facies rocks pass progressively into blueschists with sodic amphiboles (see. Ch. 7, page 190, for details), whereas in the lowest grade parts of the Franciscan high pressure terrane, such as the Stoneyford Quadrangle, mafic volcanic and related rocks with well-preserved igneous textures and relic igneous minerals directly develop assemblages that include distinctive blueschist facies phases. In these rocks albite, chlorite, sphene, quartz, stilphomelane, sericite and pumpellyite occur, but lawsonite, Na-amphibole and aragonite (usually altered in part to calcite) are also found, and all these require relatively high pressure conditions (Fig. 4.7). Indeed the assemblage lawsonite + glaucophane is usually taken to be diagnostic of the **blueschist facies**. In higher grade parts of the Franciscan high pressure metamorphic terrane (e.g. the area around Goat Mountain described by Ghent, 1965), metabasites no longer preserve any original igneous relics except for occasional augite. Instead they are thoroughly recrystallised schists composed predominantly of glaucophane and lawsonite with chlorite, sphene, quartz, muscovite and often pumpellyite, magnetite, pyrite or calcite.

The comparison of the Sanbagawa and Franciscan terranes suggests that there may in fact be a range of different metamorphic types possible at conditions involving higher pressures and/or lower temperatures than the Barrovian zones, and this is a conclusion that is borne out by the study of other areas of high pressure metamorphism. In the case of the Sanbagawa terrane, sub-greenschist facies rocks rather similar to those of New Zealand (where the metamorphism is dominated by the occurrence of greenschist facies assemblages indicative of only moderate pressures) pass into sodic-amphibole-bearing rocks instead, but in the Franciscan terrane the entire metamorphic sequence is different and thereby implies more extreme conditions.

New Caledonia

One of the most complete sets of high pressure metamorphic zones has been described from the Ouega district of New Caledonia and has already been mentioned in Chapter 3. This account is based primarily on that of Black (1977). Metamorphism occurred between 38 and 21 Ma and the resulting zonal sequence is shown in Fig. 4.3. At the lowest grades, in the west, metabasites retain original igneous textures and relic plagioclase and pyroxene; the alteration phases are not particularly distinctive and include phengite, chlorite, albite and sphene. The first distinct isograd marks the appearance of pumpellyite, and stilpnomelane and actinolite appear at about the same grade. The next zone is marked by the appearance of lawsonite and glaucophane or crossite, and therefore belongs to the blueschist facies rather than the prehnite-pumpellyite facies. Igneous relics still remain and influence the way in which the metamorphic phases grow. For example, lawsonite can form directly from original plagioclase through the reaction:

$$\text{plagioclase} + H_2O \rightarrow \text{lawsonite} + \text{albite} \qquad [4.6]$$

The next isograd marks the appearance of epidote and major recrystallisation at this grade has led to the destruction of relic igneous features. Near the epidote isograd, three minerals appear in close succession with increasing grade: omphacite pyroxene (replacing remaining augite relics) then epidote, then garnet. At the same time pumpellyite, stilp-nomelane and then lawsonite disappear. Epidote zone metabasites are therefore predominantly glaucophane schists but can contain a variety of additional minerals, as shown in Fig. 4.3(*b*). In the highest grade parts of the epidote zone, hornblende is sometimes present and demonstrates that high temperatures as well as pressures were attained. Omphacite and garnet are more abundant and may coexist with paragonite as a result of the continuous reaction:

$$\text{albite} + \text{epidote} + \text{glaucophane} \rightarrow \text{omphacite} + \text{paragonite} + \text{hornblende} + H_2O \qquad [4.7]$$

Transitional Blueschists

The transition from the blueschist facies to the eclogite facies

Reaction [4.7] marks the transition from the blueschist facies to the eclogite facies in garnetiferous rocks, since eclogites are characterised by the presence of abundant garnet and omphacite, and absence of plagioclase feldspar of any sort.

An alternative reaction (Ridley, 1984) to mark the upper temperature limit of blue-schists is:

$$\text{zoisite} + \text{glaucophane} = \text{garnet} + \text{omphacite} + \text{paragonite} + \text{quartz} + H_2O \qquad [4.8]$$

This reaction appears to be more appropriate for the transition from blueschist to eclogite recorded from many Alpine–Tethyan occurrences, where it may take place at higher pressures than in New Caledonia (see Fig. 2.8).

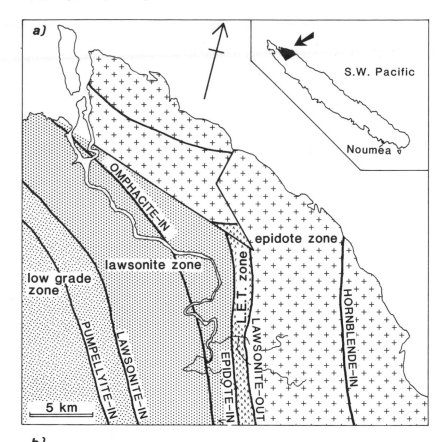

Metamorphic zones:	'low grade'	lawsonite	L.E.T.	epidote
CHLORITE	————	————	————	————
PHENGITE	————	————	————	————
PARAGONITE				———
Na–AMPHIBOLE		————	————	———
ACTINOLITE		———————	————	———
HORNBLENDE				———
OMPHACITE			————	————
EPIDOTE			————	————
LAWSONITE		———————	———	
PUMPELLYITE	————	———		
STILPNOMELANE	————	———		
ALBITE	————	————	————	
GARNET				————

It is clear that blueschists are stable over a very wide range of temperatures (see below, page 115), and it is possible to separate out the highest temperature blueschists, in which lawsonite has broken down due to the reactions:

$$\text{lawsonite} + \text{albite} = \text{zoisite} + \text{paragonite} + \text{quartz} + H_2O \qquad [4.9]$$

or at higher pressure:

$$\text{lawsonite} + \text{jadeite} = \text{zoisite} + \text{paragonite} + H_2O \qquad [4.10]$$

Both these reactions were studied experimentally by Heinrich & Althaus (1980); they may lead to the growth of distinctive pseudomorphs of zoisite and paragonite that retain the shape of the original lawsonite, within blueschist. 'High grade' blueschists of this type are well known from the Sesia–Lanzo and Zermatt–Saas zones of the western Alps, as well as from the Greek Cyclades; typically such blueschists are closely associated with eclogites and are transitional to the eclogite facies.

The transition from the greenschist to the blueschist facies   Greenschist facies metabasites contain albite, chlorite, actinolite and epidote, whereas in the blueschist facies the common phases include glaucophane, lawsonite and zoisite. It is possible to write a variety of reactions relating greenschist facies and blueschist facies assemblages, although these do not indicate reactions that have necessarily taken place, rather they are statements of the chemical equivalence of certain blueschist and greenschist assemblages. For example Brothers and Yokoyama (1982) point out that the diagnostic assemblages of the two facies can be related by the equation

$$\underset{\text{blueschist}}{\text{glaucophane} + \text{lawsonite}} = \underset{\text{greenschist}}{\text{albite} + \text{chlorite} + \text{actinolite}} \qquad [4.11]$$

The precise way in which such a reaction is balanced, and even whether or not additional phases such as quartz or fluid are involved, depends on the compositions of the solid solution phases involved.

Another important equation is:

$$\text{glaucophane} + \text{zoisite} + \text{quartz} = \text{albite} + \text{chlorite} + \text{actinolite} + H_2O \qquad [4.12]$$

This is written to involve only end-members that do not contain Fe^{3+}, however both glaucophane and the monoclinic polymorph of zoisite form solid solutions with Fe^{3+} end-members (riebeckite and epidote respectively). Brown (1974) showed that it was possible to write a comparable equation between oxidised blueschists and greenschists:

$$\text{crossite} + \text{epidote} = \text{albite} + \text{chlorite} + \text{actinolite} + \text{Fe-oxide} + H_2O \qquad [4.13]$$

Equation [4.13] represents a continuous reaction involving Fe^{3+} as well as the components involved in reaction [4.12]. However, Fe^{3+} only substitutes into the solid solution phases on the left-hand side of the equation, and so since this substitution dilutes the concentration of glaucophane and zoisite in their respective solid solutions, but leaves the concentrations of albite, chlorite and actinolite unchanged, the effect will be to extend the stability field of Na-amphibole and epidote (see Ch. 2, page 55). As a result crossite and epidote may develop in oxidised metabasites at lower pressures than those required for glaucophane and zoisite to form in reduced metabasites. This provides an explanation for

---

Fig. 4.3 opposite   *a*) Metamorphic map of the Ouegoa district of northern New Caledonia, showing zones and some additional metabasite isograds. *b*) Variation in mineralogy of the metabasites across the metamorphic zones of the Ouegoa district. In both parts of the figure L.E.T. denotes lawsonite–epidote transition zone. After Black (1977).

the occurrence of interbanded blueschists and greenschists in some parts of the world: both rock types were subjected to the same pressures and temperatures, but the blueschists are crossite schists that developed in oxidised layers, while the greenschists are reduced and do not contain appreciable amounts of Fe^{3+}.

Many of the world's 'blueschist' terranes prove to be composed predominantly of crossite–epidote schists and lack the higher pressure indicators such as glaucophane + lawsonite, jadeite or omphacite, or aragonite. For example, large parts of the Sanbagawa terrane of Japan or the Shuksan blueschists of Washington, USA, are of this type. There is no doubt, however, that the crossite–epidote schists do require higher pressures to form than those experienced in the Barrovian zonal scheme, because as Brown (1974) points out, metabasites in the Haast Schists of New Zealand, that display a Barrovian type of metamorphism, contain the assemblage albite + chlorite + actinolite + magnetite corresponding to the right-hand side of reaction [4.13].

Most authors would restrict the blueschist facies to those rocks in which glaucophane develops in reduced as well as in oxidised rocks, and so the glaucophane + lawsonite association, corresponding to the left-hand side of reaction [4.11], is diagnostic. Crossite–epidote 'blueschists' are often referred to a 'transitional blueschist–greenschist facies'.

The low grade limits of the blueschist facies are also controversial. Liou, Maruyama & Cho (1987) include all lawsonite-bearing assemblages in the blueschist facies, effectively extending it down to pressures as low as 3 kbar, below the limits of glaucophane (Fig. 4.7). Although a good case can be made for this definition on the grounds of phase equilibrium, the rocks concerned cannot readily be distinguished in the field from those of the prehnite–pumpellyite facies, and have therefore been treated as transitional in Fig. 2.8.

ECLOGITES

Under extreme conditions of metamorphism at both high pressure and moderate to high temperatures, rocks of basaltic composition recrystallise to a distinctive red and green, dense rock, dominated by garnet and omphacitic pyroxene and known as **eclogite**. Other minerals that are commonly present in small amounts include quartz, rutile, kyanite, amphibole (usually Na-rich) and pyrite, but plagioclase is never present.

Eclogites may be the product of extreme metamorphism of lower grade metabasites, or may be produced directly from basalt, gabbro or basaltic melt by cooling under high pressure conditions. Yoder & Tilley (1962) and Green & Ringwood (1967) have experimentally investigated these relationships between eclogite and gabbro.

Coleman *et al.* (1965) classified natural eclogite occurrences into three types according to their mode of occurrence:

Group A: occur as zenoliths in kimberlites or basalts (e.g. Oahu, Hawaii).
Group B: occur as bands or lenses in migmatitic gneisses (e.g. west Norway).
Group C: occur as bands or lenses associated with blueschists (e.g. New Caledonia, Franciscan terrane of California, Alpine–Tethyan chain).

There are also mineralogical differences between the three groups, especially in the Mg-content of the garnet. In Group A eclogites, garnet is Mg-rich with ~55 per cent pyrope end-member, garnet from Group B eclogites has 30–55 per cent pyrope, while in most Group C eclogites the garnets have ~30 per cent pyrope only.

In the light of subsequent work on the temperature dependence of Fe–Mg exchange between garnet and clinopyroxene (Ellis & Green, 1979, see below), the three groups of eclogites can be seen to result from crystallisation at different temperatures as well as in different geological settings. Group C eclogites form at the lowest temperatures; Group A at the highest temperatures.

Since Yoder and Tilley found that basalt magma would crystallise to eclogite under upper mantle pressures, it has been widely believed that eclogites may originate in the mantle. This is most probably the case for Group A eclogites which often occur in association with mantle-derived xenoliths in kimberlites and are not considered in detail here. In contrast, Group C eclogites have strong affinities with crustal rocks, as do most (perhaps all) Group B eclogites.

Group C eclogites   The development of Group C eclogites by progressive metamorphism of blueschist in New Caledonia has already been touched on. In the Zermatt–Saas zone of the Swiss Alps, Bearth (1959) has described pillow lavas in which the pillow core is now composed of eclogite while the rim is glaucophane schist. One significant conclusion from this occurrence is that there can be no doubt that the parental material of the eclogite was a lava erupted at the surface and subsequently metamorphosed to high pressures, rather than an upper mantle rock or melt.

This type of eclogite typically contains amphibole as well as garnet and omphacite; and zoisite, phengite, paragonite and quartz also occur commonly. Chloritoid is sometimes present.

One of the areas in which Group C eclogites were first studied in detail was in the Franciscan terrane of California, where they occur as isolated 'knockers' or large blocks in a matrix of black argillaceous material in a chaotic mélange. Blueschists also occur as 'knockers' in the same matrix, but it is impossible to study field relations between the two types. In New Caledonia, however, the field relationships between blueschists and Group C eclogites can be studied, and here they appear to be intercalated.

Classic Group B eclogites of western Norway   Eclogites, occurring as small bodies within migmatitic gneiss, are widespread in the Western Gneiss Region of Norway (Fig. 4.4) where Eskola carried out some of the first detailed studies of eclogites. Quartz and rutile are generally present with garnet and omphacite, and kyanite, zoisite and paragonite are also widespread. Garnets with relic inclusions of glaucophane have been described. Retrograde effects are often very important, and in parts of some bodies the later amphibolite facies recrystallisation has destroyed most of the original eclogite minerals.

The origin of the Norwegian eclogites is still controversial. One school of thought points to the very high pressures of metamorphism (c.40 kbar) that can be obtained from thermodynamic calculations based on the composition of the naturally occurring phases, and argues that such pressures are so great that they can only be realised in the mantle. Hence the eclogite pods must have originated in the mantle and subsequently been tectonically explaced into the crust (e.g. Lappin and Smith, 1978). On the other hand other workers point out that the field relationships of many of the eclogite bodies suggest that they were original minor basic intrusions such as dykes, emplaced into the host gneisses at a relatively high level in the crust because in some cases the original igneous plagioclase and augite are partially preserved, and display normal basaltic textures (e.g. Bryhni *et al.*, 1970; Bryhni, Krogh & Griffin, 1977). These observations appear to preclude an origin by tectonic emplacement from the mantle for many of the eclogites, and this school of workers would argue that there are large uncertainties in the calculations that appear to require mantle pressures for eclogite formation (Krogh, 1977). On balance, the evidence appears to favour the 'crustal eclogite' school in this writer's opinion, but very thick crust is evidently required for such eclogites to form.

Transition from Group C to Group B eclogites   The distinction between Group C and Group B eclogites has not really survived more modern work, because many eclogites which display the field associations of Group C have mineral compositions typical of Group B. The eclogites of the Tauern Window (Miller, 1974; Holland, 1979b) are of this type, and Newton (1986) reviews a number of

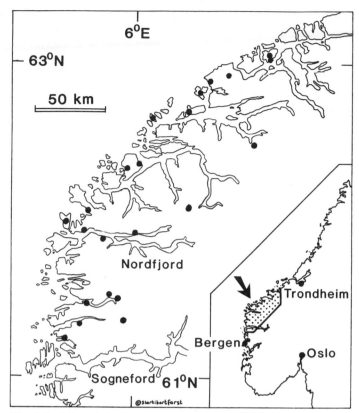

Fig. 4.4 Map of the distribution of the major eclogite bodies within the acid gneisses of the Western Gneiss Region of Norway after Krogh (1977). Individual bodies have dimensions that are typically of the order of tens of metres, and many additional bodies are undoubtedly present in this inhospitable terrain.

other occurrences. It is probably better to think of them as higher temperature crustal eclogites, rather than keep rigidly to the Coleman *et al.* classification. Kyanite is a common phase of higher temperature eclogites and the association of talc + kyanite is particularly distinctive because it results from the breakdown of chlorite + quartz with increased temperature,* and hence provides a temperature indicator.

Origin of the blueschist–eclogite association

The association of blueschists with intercalated eclogites has been reported worldwide, and although one rock type is often seen to replace the other, it seems clear that they can also be **isofacial**, i.e. formed at the same conditions of pressure and temperature due to differences in rock or fluid composition. Many workers have suggested that eclogites form in relatively dry conditions, while blueschists require high water pressure (e.g. Bearth, 1959; Black, 1977). Fry & Fyfe (1969) calculated that eclogites would not be stable under any possible crustal conditions of pressure and temperature in the presence of an H_2O phase. On the other hand Essene, Hensen & Green (1970) synthesised eclogite at high water pressures at 700 °C, while Holland (1979b) reported evidence for a H_2O–fluid in equilibrium with kyanite eclogite from the Tauern Window, Austria. Ridley (1984) showed that bulk composition was an important factor in determining which rocks became eclogites and which became blueschists on Syros (Cyclades); implying that P_{H_2O} was not

---

*Schreyer (1973) termed talc–kyanite rocks 'whiteschists' and inferred that they formed at elevated *P* and *T*. This is confirmed by the occurrence of talc + kyanite in eclogites.

the dominant factor. Fe-rich metabasites and silica-deficient varieties recrystallised to eclogite while Mg-rich metabasites remained as glaucophane schists.

The evidence to date seems to suggest that eclogites can form in water-rich environments at temperatures in excess of perhaps 500–550 °C, but at lower temperatures it is more likely that low water pressures are required to stabilise eclogite. In a number of areas, gabbro retaining primary igneous assemblages in part is recrystallised patchily to eclogite but reaction is often incomplete, and the local disequilibrium is strongly suggestive of water-deficient conditions. Pognante & Kienast (1987) report temperatures below 500 °C for eclogites formed in this way in the western Alps, while comparable omphacite-bearing rocks in the same region which, however, lack garnet and so are not truly eclogites, yield temperatures as low as 350 °C and almost certainly formed similarly by incomplete breakdown of gabbro in water-deficient conditions.

HIGH TEMPERATURE METAMORPHISM: GRANULITE FACIES

At the highest grades of metamorphism, where temperatures exceed those of the Barrovian zones, hornblende in metabasites begins to break down and pyroxenes appear. At the same time, partial melting may occur to produce migmatitic metabasites with leucosomes of plagioclase and quartz.

Pyroxene-bearing metabasites, other than those high pressure rocks in which the pyroxene is sodic, are typical of the granulite facies and the transition from the amphibolite to the granulite facies. Many granulites occur in distinct terranes, and indeed they are typical of Precambrian Shield terranes, although much younger granulites do occur. The boundaries of these terranes are often unconformities or faults, and so it is relatively rare to be able to study the progressive metamorphism of amphibolites to granulites. In most cases where amphibolite facies rocks occur with those of the granulites facies, the amphibolites are of retrograde origin due to later reworking and infiltration of fluids. A well-known example of this is the Lewisian terrane of north-west Scotland, in which granulites formed during the Scourian event around 2700 Ma have locally been reworked and recrystallised to amphibolites in the Laxfordian event at 1700 Ma.

One area in which a prograde transition to the granulite facies is seen, is the Willyama Complex of the Broken Hill district of New South Wales, Australia. According to Binns (1962, 1965a, b) the zonal sequence in metabasites is as follows:

A. Bluish-green hornblende + plagioclase ± garnet ± epidote or clinozoisite + ilmenite.
B. Bluish-green or green–brown hornblende + plagioclase + clinopyroxene ± garnet + ilmenite.
C. Green–brown hornblende + plagioclase + clinopyroxene + orthopyroxene + ilmenite.

This sequence of appearance of clino- and orthopyroxene has also been found in experimental studies by Spear (1981) described below (page 118). Zone A is typical of the amphibolite facies while Zone C belongs to the granulite facies, for which the occurrence of coexisting orthopyroxene and clinopyroxene in metabasite is diagnostic. Zone B is of an intermediate or transitional character.

The granulite facies embraces a wide pressure range of high temperature metamorphism, and is distinguished from the eclogite facies at high pressures by the presence of plagioclase in granulite facies metabasites. However, there are a number of differences between different granulite occurrences that can be ascribed to pressure differences. Green and Ringwood (1967) have therefore proposed a three-fold subdivision of the granulite facies based on metabasite assemblages:

1. *Low pressure granulites* contain orthopyroxene + clinopyroxene + plagioclase, and in the more basic varieties olivine + plagioclase may occur. It is the latter association that is diagnostic of low pressure. Associated pelites have abundant cordierite although garnet and hypersthene may occur.
2. *Medium pressure granulites* are characterised by the association of garnet + clino-pyroxene + orthopyroxene + plagioclase, and hornblende is usually present also. Quartz is a possible accessory but does not normally coexist with both garnet and clinopyroxene. A well-known example of medium pressure granulites is the Scourian granulites of north-west Scotland described by O'Hara (1961). Pelitic rocks at this grade are typically migmatites with coexisting garnet and cordierite.

Medium pressure granulites may be related to those found at low pressure by the equations:

$$Fe_2SiO_4 + CaAl_2Si_2O_8 = CaFe_2Al_2Si_3O_{12} \qquad [4.14]$$
$$\text{olivine} \qquad \text{plagioclase} \qquad \text{garnet}$$

$$3\,Mg_2SiO_4 + 4\,CaAl_2Si_2O_8 = 2\,CaMgSi_2O_6 \cdot CaAl_2SiO_6 + \qquad [4.15]$$
$$\text{olivine} \qquad \text{plagioclase} \qquad \text{Al-diopside}$$
$$2\,MgSiO_3 \cdot MgAl_2SiO_6 + SiO_2$$
$$\text{Al-enstatite} \qquad \text{quartz}$$

All the compounds involved in these equilibria, apart from quartz, occur as components in complex solid solutions.

3. *High pressure granulites* are associated with pelites that contain K-feldspar + kyanite and are also distinguished by the lack of orthopyroxene in plagioclase-bearing clinopy-roxene metabasites. Their assemblages can be related to those found at medium pressures by equations such as:

$$4\,MFSiO_3 + CaAl_2Si_2O_8 = MF_3Al_2Si_3O_{12} + CaMFSi_2O_6 + SiO_2$$
$$\text{orthopyroxene} \quad \text{plagioclase} \qquad \text{garnet} \qquad \text{clinopyroxene} \quad \text{quartz}$$
$$[4.16]$$

$$2\,MFSiO_3 + CaAl_2Si_2O_8 = CaMF_2Al_2Si_3O_{12} + SiO_2 \qquad [4.17]$$
$$\text{orthopyroxene} \quad \text{plagioclase} \qquad \text{garnet} \qquad \text{quartz}$$

where in both cases 'MF' denotes $Mg^{2+} + Fe^{2+}$. Note that quartz coexists with garnet and clinopyroxene in high pressure granulites.

Some granulite terranes, such as that of Madras, India, include rocks of the **charnock-ite series** which contain plagioclase, hypersthene and augite, together with K-feldspar, and may also have garnet, hornblende and quartz.

GRANULITES AND PROGRESSIVE METAMORPHISM

Many granulites, such as those of the Willyama complex, appear to be the end-products of progressive metamorphism of basic rocks up to very high temperatures. However, there are two alternative origins for basic granulites that may be important in some instances.

Not all granulites are of unequivocally metamorphic origin and it is possible that some may be the products of direct crystallisation of basic magmas under high pressure conditions where garnet is stable on the basalt solidus. For example, Howie (1955), after a painstaking review of the textural evidence for the origin of charnockites, concludes that there was evidence for both metamorphic and igneous origins in different specimens. A second possibility is that some granulites may owe their dehydrated character not to extreme temperatures but to flushing of amphibolites by mantle-derived CO_2. This is outlined further below (page 119).

REACTION TEXTURES IN GRANULITE

A feature of some granulite terranes is the development of **corona textures**. These are rims of metamorphic minerals that form around the original igneous mineral grains in rocks such as gabbro or anorthosite, and develop as a result of reaction between the enclosed grain and other minerals in the rock matrix. Often there is more than one rim, and concentric shells each composed of a distinct mineral or mineral assemblage may develop. Examples of mineral zones found in coronas by Griffin and Heier (1973) are illustrated schematically in Fig. 4.5.

In many cases the mineral zones found in corona textures can be closely related to reactions analogous to eqns [4.14] to [4.17]. For example, one common corona texture (Fig. 4.5) develops around original olivine in contact with plagioclase and consists of an inner rim of pyroxene with an outer rim of garnet. Clearly this results from the combined

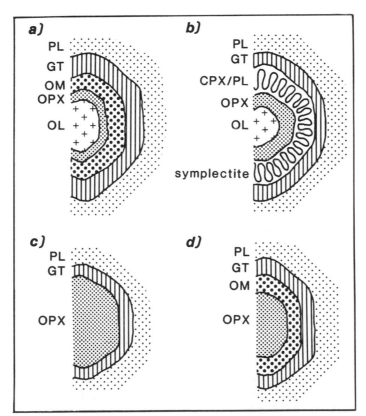

Fig. 4.5 Schematic representations of some types of corona texture described by Griffin & Heier (1973). *a*) and *b*) develop between olivine and plagioclase and are similar except that in *a*) omphacite is stable whereas in *b*) omphacite has broken down and exsolved into a fine intergrowth or **symplectite** of calcic clinopyroxene and plagioclase. *c*) and *d*) are examples of coronas between orthopyroxene and plagioclase.

effects of reactions [4.14] and [4.15], with the reaction products developing at the interface between the reactants and thereby tending to inhibit further reaction.

The development of many corona textures implies a direct transition from an unaltered igneous assemblage to an assemblage typical of high temperature, and usually high pressure, metamorphism, with no intervening stage of lower grade metamorphism. There are two ways in which this might be achieved. The effect of increased temperature is to stabilise the low pressure granulite mineralogy (also typical of basic igneous rocks) to higher pressures (Fig. 4.6). Thus a magma crystallising at moderate depth in the crust might develop olivine + plagioclase initially, but as it cooled below the solidus would enter the medium pressure granulite field so that garnet is produced during the original cooling of the igneous rock (path A, Fig. 4.6). The second way of developing coronas would be for large basic igneous bodies to cool initially near the surface without metamorphism and also to survive subsequent regional burial and reheating without metamorphism taking place until very high temperatures were attained (path B, Fig. 4.6). This might occur if the intrusion were emplaced into dry country rocks, such as acid gneisses, and was buried without being deformed, since both these circumstances would inhibit fluid entering the intrusion to produce the hydrous minerals typical of lower grades of metamorphism. Griffin and Heier (1973) favoured the first type of model for the development of the Norwegian corona textures, but subsequent work on the geological evolution of the region has made the second model look more realistic (Mørk, 1986).

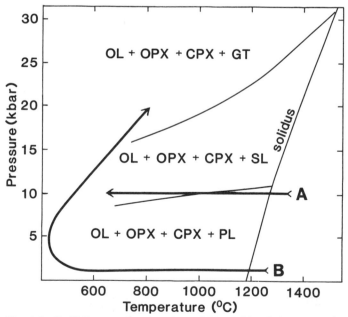

Fig. 4.6 *P–T* diagram showing the stability fields of plagioclase lherzolite, spinel lherzolite and garnet lherzolite, based on Green & Ringwood (1967). Paths A and B illustrate different possible origins for corona textures between plagioclase and olivine. Path A denotes cooling of basaltic magma under lower crustal conditions to produce coronas in a single stage of cooling (at slightly higher pressures, corona formation could commence above the solidus). Path B illustrates initial high level cooling to produce gabbro followed by regional burial and heating to high pressure conditions.

THE *P–T* CONDITIONS OF FORMATION OF METABASIC ROCK TYPES

This final section of the chapter is concerned with defining the conditions of pressure and temperature that are required to form the metabasite assemblages characteristic of the different facies. In the case of some rock types, especially some eclogites and granulites, it is possible that their formation is dependent not only on the P and T prevailing, but also on the presence or absence of particular fluids. A general petrogenetic grid showing the most relevant experimentally determined reactions is given in Fig. 4.7, and Fig. 2.8 effectively shows the probable range of conditions of formation of the major metabasic rock types.

THE LOW GRADE FACIES

Information on the conditions of formation of these zones comes from three main sources: experimental studies of mineral stabilities, oxygen isotope geothermometry and direct measurement in geothermal fields. There is reasonably good agreement between all these independent methods to suggest that zeolite facies metamorphism takes place over a range of temperatures from as low as around 100 °C up to 250–300 °C. The range of stability of prehnite in natural rocks is not very well known but there have been a number of experimental studies of prehnite and pumpellyite. Very Fe-rich pumpellyites, such as occur in some zeolite facies rocks, are stable below 200 °C and break down between 200 °C and 250 °C (Schiffman & Liou, 1983), but the more magnesian varieties typical of the prehnite–pumpellyite facies are stable to temperatures between 350 and 400 °C, and this provides a temperature for the transition from prehnite–pumpellyite to greenschist facies that is in agreement with the occurrence of sub-greenschist facies assemblages in deep geothermal wells at temperatures in excess of 300 °C. Additional confirmation that the greenschist facies has a lower temperature limit around 400 °C comes from the application of oxygen isotope geothermometers (page 56) to chlorite zone metamorphic rocks, e.g. by Graham *et al.* (1983), Matthews & Schliestedt (1984) and Yardley (1982).

The upper pressure limit for the low grade facies is provided by the appearance of blueschist facies assemblages. Laumontite is replaced by lawsonite at pressures of around 3 kbar (Fig. 4.7), resulting in lawsonite–albite-bearing rocks that are transitional between the low grade facies and the blueschist facies, and have in fact been given the status of a separate **lawsonite–albite facies** by some authors.

THE BLUESCHIST FACIES

The greatest obstacle to understanding the conditions of blueschist formation has been uncertainties about the conditions under which glaucophane is stable. Although several experimental studies have been made, it has been pointed out by Maresch (1977) that experimental products to that date had never corresponded to true glaucophane, while natural glaucophane used in experiments with H_2O present breaks down to Na-micas. These problems have usually been considered to be a consequence of the short time available for experiments, so that equilibrium is not attained, however Koons (1982) found that a synthetic amphibole close to glaucophane in composition was stable in experiments where there was no excess of water. This led him to suggest that in nature glaucophane schists may develop only where no free fluid phase is present. This suggestion certainly goes against the established prejudices of most petrologists, (not least when relations to

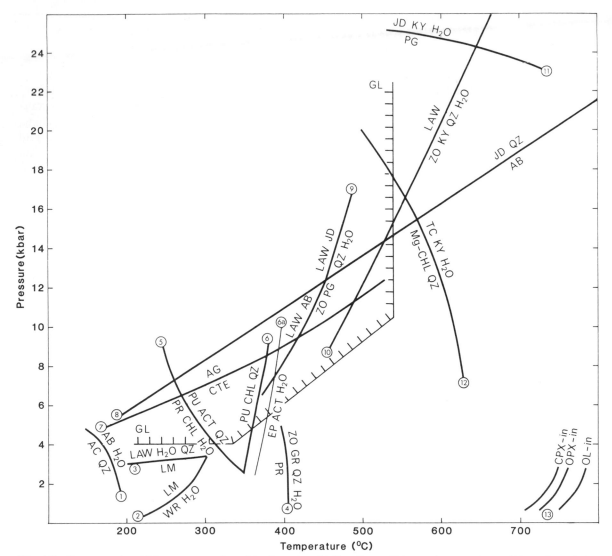

Fig. 4.7 Petrogenetic grid for metamorphosed basic rocks with P = P$_{H_2O}$. Abbreviations used are: AB = albite; AC = analcite; ACT = actinolite; AG = aragonite; CHL = chlorite; CPX = clinopyroxene; CTE = calcite; EP = epidote; GL = glaucophane; GR = grossular; JD = jadeite; KY = kyanite; LAW = lawsonite; LM = laumontite; OL = olivine; OPX = orthopyroxene; PG = paragonite; PR = prehnite; PU = pumpellyite; QZ = quartz, TC = talc; WR = wairakite; ZO = zoisite. Data sources for the curves are as follows: (**1**) Thompson (1971b); (**2**) & (**3**) Liou (1971b); (**4**) Liou (1971a); (**5**) & (**6**) Nitsch (1971), using natural minerals; **6a** Schiffman & Liou (1980), for Mg–Al pumpellyite; (**7**) Johannes & Puhan (1971); (**8**) Holland (1980); (**9**) Heinrich & Althaus (1980); (**10**) Newton & Kennedy (1963); (**11**) Holland (1979a); (**12**) Massonne, Mirwald & Schreyer (1981); (**13**) mineral appearances in natural basalt compositions, Spear (1981); Hachured line is the supposed limit to glaucophane stability according to Maresch (1977).

eclogite are considered) but Thompson (1983) has also presented a case for fluid being, in general, absent from metamorphic rocks as a distinct phase, and it remains an intriguing and important possibility. The glaucophane stability field shown in Fig. 4.7 is from the review by Maresch, and it indicates only possible limits; in particular the low pressure limit may be too low, while the upper temperature limit is almost certainly not high enough (c.f. Fig. 4.8).

It is apparent from the field descriptions of glaucophane-bearing rocks already given, that they are stable over a wide range of temperatures. In the Stoneyford quadrangle (California) the blueschist assemblages appear directly at very low grade whereas in New Caledonia they overlap with hornblende-bearing assemblages more comparable to the amphibolite facies. Taylor and Coleman (1968), in one of the first extensive studies applying oxygen isotope geothermometry to metamorphic rocks, obtained temperatures of around 300 °C for low grade Franciscan blueschists, while Black (1974) applied the same technique to obtain temperatures for the New Caledonia high pressure sequence that ranged from 250 °C for the lowest grade part to over 500 °C for the highest grade omphacite- and hornblende-bearing blueschists. Matthews and Schliestedt (1984) have

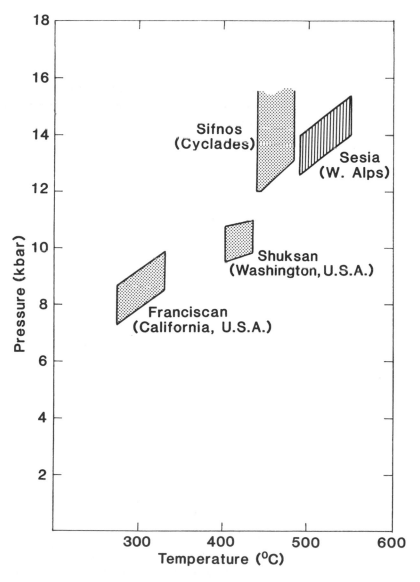

Fig. 4.8 P–T conditions of formation of a range of different blueschist types. After Matthews & Schliestedt (1984). The ruled symbol denotes blueschists coexisting with eclogite.

compiled temperature estimates for a variety of blueschists, and confirm that values can range from 300 to 550 °C; their results are illustrated in Fig. 4.8.

Because of the uncertainty about glaucophane stability, many estimates of the pressure of blueschist metamorphism have been based on the reaction of albite to jadeitic pyroxene, because the conditions of the equilibrium:

$$\text{albite} = \text{jadeite} + \text{quartz} \qquad [4.18]$$

are quite well known. In addition, the activity–composition relationships of jadeite–diopside solid solutions have been investigated experimentally by Holland (1983).

This equilibrium represents the upper stability limit of albite, and some blueschists do indeed lack albite, and contain a jadeite-rich pyroxene. The equilibrium curve therefore provides only a lower pressure limit for such rocks. In other blueschists, albite is present and coexists with a jadeite-bearing pyroxene that also contains significant amounts of other components. In such cases it is possible to calculate a P–T curve along which the assemblage equilibrated, if the actual mineral compositions are known (see Ch. 2, Fig. 2.9(*a*)).

It appears that some blueschists have formed at pressures of around 9 kbar and temperatures around 400 °C, but others have formed at significantly higher pressures. Okrusch, Seidel & Davis (1978) calculate a minimum pressure of 14 kbar for glaucophane schists from Sifnos in the Cyclades Archipeligo, Greece (cf. Fig. 4.8). There are few reactions to provide upper pressure limits to blueschists because those reactions that have been studied experimentally, such as the breakdown of paragonite to jadeite + kyanite (below, page 117), lie at too high pressures to take place normally at low to medium temperatures. However, paragonite breakdown does provide useful information on the conditions of eclogite formation and thereby helps to constrain the conditions of the blueschist facies.

Some low grade blueschists contain aragonite (page 102), and this provides a further useful indicator for their minimum pressure of formation. It is apparent from Fig. 4.7 that aragonite may be the stable $CaCO_3$ polymorph when most blueschists form, and the reasons why it is not found in higher grade blueschists are probably kinetic. They are considered on page 182.

The conditions of the transition zone between the blueschist and greenschist facies are not very well defined, but Brown (1977) has estimated pressures of around 7–8 kbar at temperatures around 400 °C, for the formation of oxidised crossite–epidote schists interbedded with reduced greenschists, and these are in accord with most estimates of conditions for both greenschists and blueschists.

THE ECLOGITE FACIES

The eclogite facies is distinguished by the lack of plagioclase, so reactions that define the stability limits of albite and anorthite are obviously important in defining a minimum stability field for eclogites. Albite is limited at high pressures by reaction [4.18], while anorthite breaks down according to:

$$3 \text{ anorthite} = \text{grossular} + 2 \text{ kyanite} + \text{quartz} \qquad [4.19]$$

Since both grossular and jadeite occur as components of complex solid solutions in natural eclogites, these end-member reactions do not in fact provide very precise constraints.

Temperatures for eclogites may be as low as 500 °C as noted above, for example in the case of the 'low grade' Group C eclogites of New Caledonia, the Western Alps, etc., where they are associated with blueschists. Phase equilibria do not provide very good temperature indicators for eclogites, but geothermometry based on Fe–Mg cation exchange between garnet and clinopyroxene (Ellis and Green, 1979) indicates temperatures of 600–750 °C for Norwegian Group B eclogites. Newton (1986) has used this geothermometer to divide Group B and Group C eclogites into high, medium and low temperature types.

The precise pressures at which eclogites form have been the subject of controversy, although it is universally agreed that higher pressures are required than for other metabasites. Group B eclogites from Norway and similar eclogites from the European Alps include varieties that contain paragonite and kyanite, and in these the amount of jadeite in the pyroxene is controlled by equilibrium with paragonite and kyanite:

$$NaAl_3Si_3O_{10}(OH)_2 = NaAlSi_2O_6 + Al_2SiO_5 + H_2O \qquad [4.20]$$
$$\text{paragonite} \quad \text{jadeite component} \quad \text{kyanite} \quad \text{fluid}$$
$$\text{of omphacite}$$

Reaction [4.20] proceeds to the right with increasing pressures (Fig. 4.7) and so the jadeite content of the omphacite can be analysed and used to calculate a pressure of formation for any particular kyanite–paragonite eclogite. Holland (1979a, b) used this approach, together with his own experimental results, to obtain a pressure of 19.5 kbar at a temperature of 620 °C for an eclogite from Austria.

Some eclogites contain orthopyroxene, and for these the amount of Al dissolved in the orthopyroxene can be used as a pressure indicator. Harley (1984 a, b) has estimated pressures of 19 kbar for Norwegian orthopyroxene eclogites using this method, but there is a large uncertainty of ± 9 kbar. It seems unlikely that eclogites would form at pressures below 12 kbar, because this would imply conditions not much different from those for Barrovian metamorphism, whereas eclogites are never found in Barrovian zonal sequences. There is some evidence that very much higher pressures may have been attained by some eclogites, because coesite, the high pressure polymorph of SiO_2, has been found as inclusions in pyroxene from eclogite, and this requires pressures of about 30 kbar (Smith, 1984).

THE GREENSCHIST AND AMPHIBOLITE FACIES

Although the lower temperature limit of the greenschist facies is quite well defined by the breakdown of pumpellyite (Schiffman & Liou, 1980, 1983), changes in metabasites within the greenschist and amphibolite facies are the result of markedly continuous reactions and no simple experimental system is of direct relevance. Even the changes found at lower pressures, in the hornfels facies, are not clearly defined. As a result metamorphic conditions within these facies are best defined by pelite assemblages rather than metabasites. The continuous reactions result in changes in amphibole chemistry and Laird and Albee (1981) have presented a series of plots of amphibole composition that allow the approximate pressure and temperature regime of formation of metabasites to be deduced by comparison with areas where the conditions of metamorphism are known independently from pelitic rocks. An example is illustrated in Fig. 4.9. Additional geothermometers include the exchange of Fe and Mg between garnet and hornblende (Graham and Powell, 1984), and exchange of NaSi for CaAl between plagioclase and hornblende (Spear, 1980).

THE GRANULITE FACIES

There can be no doubt that basic granulites form at very high temperatures because they are characterised by the dehydration of amphibole to pyroxene and associated pelitic rocks

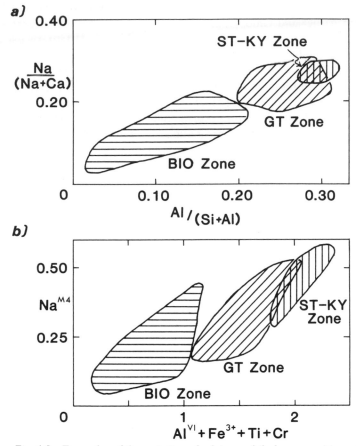

Fig. 4.9 Examples of the variation of calcic amphibole composition with grade during a Barrovian type of metamorphism (data from Vermont). After Laird & Albee (1981). Plot *a*) is based on analysed atomic ratios alone, whereas for plot *b*) these have been recalculated to allow for the oxidation of Fe, and assigned to sites in the amphibole structure. Na^{M4} denotes Na replacing Ca on the M4 site (some additional Na may occur in the A site) while the other axis is a summation of trivalent ions, plus Ti occurring on octahedral sites. Each ruled field denotes the range of amphibole compositions found in metabasites from a specific pelite zone.

have usually been partially melted. Direct experimental evidence for the temperatures required for the transition from amphibolite to granulite is available only at low pressures (Fig. 4.7). Spear (1981) found that clinopyroxene appeared in amphibolite at 770–790 °C in the 1–2 kbar pressure range, orthopyroxene appeared at a temperature 10–30 °C higher, and olivine after a further 10–30 °C heating. This is in accordance with Binn's field observations of the sequence in which the phases appeared in the Willyama complex.

    Most estimates of the conditions of formation of granulites are based on the extent of solid solution and cation exchange between coexisting garnet, orthopyroxene and clinopyroxene, using the same thermometers and barometers that are applied to the orthopyroxene eclogites. Since the calibrations of these geothermometers and geobarometers have changed over the years in the light of new experimental studies or theoretical treatments, considerable care is needed in comparing results from studies carried out at different times or in different laboratories. Despite this uncertainty it appears clear that basic granulites form over a temperature range of at least 700–1000 °C and a pressure range from about 15 kbar down to very low values around 5 kbar, and these results are in good agreement with the values estimated for related pelitic migmatites in the previous chapter (see compilations in Tarney & Windley, 1977, p. 164; Turner, 1981, p. 440; Newton, 1983; and Harley, 1984b).

Some workers have suggested more extreme temperatures of formation for granulites, up to 1250 °C (O'Hara and Yarwood, 1978). They point to a fundamental uncertainty in the calculation of pressure and temperature for high grade rocks using geothermometers and geobarometers. The basis for the calculations is that it is possible to measure today the compositions that the minerals had when they were in equilibrium with one another at the peak of metamorphism. However, at high temperatures minerals can change their composition by diffusion (below, page 150) and so the compositions measured today may have been changed from the original values as the rocks slowly cooled. As a result they may indicate a lower temperature than that at which the rock originally crystallised (i.e. a 'blocking temperature'). As an example, Barnicoat (1983) calculated temperatures of 820–920 °C for most samples of Archaean granulites from north-west Scotland, but a few, much coarser grained specimens gave temperatures of around 1000 °C, because diffusion is less effective in changing mineral compositions where the grains are larger, requiring material to move over longer distances through the lattice. Nevertheless, it appears unlikely that temperatures as high as 1000 °C have been very widely attained because of the scarcity of metamorphic belts that contain the high temperature pelite assemblage sapphirine + quartz (page 78).

The role of fluids in the formation of granulites   The fact that granulites have a lower water content than amphibolites has led many workers to suggest that they form under relatively 'dry' metamorphic conditions (e.g. Buddington, 1952), and that it may be this, rather than differences in temperature that distinguishes granulites from amphibolites. More recently, this question has been reformulated and controversy has centred on whether the dry conditions for granulites are due to removal of water by dissolution in melts formed at extreme temperature (Powell, 1983), or to infiltration of deep, possibly mantle-derived, CO_2 which promoted dehydration at temperatures not much greater than those of the amphibolite facies.

The evidence for the CO_2- infiltration model came initially from fluid inclusion studies, notably the work of Touret (1971a, b) who studied the transition from amphibolites to granulites in south Norway, and in particular examined the fluid inclusions present within minerals. He found that whereas the relic fluid preserved in inclusions in the amphibolites was H_2O, that in the granulites was dominantly CO_2; what H_2O was present appeared to have been introduced at a late stage during retrogression. Touret argued that the transition from amphibolite to granulite did not reflect increased temperature so much as infiltration of a CO_2 fluid, possibly derived from the mantle. Many subsequent studies have confirmed that CO_2 dominates the fluid inclusion composition of granulites.

On the other hand Phillips (1980) has re-examined the amphibolite to granulite facies transition in the Willyama complex at Broken Hill, which was described above, and also found evidence for a deficiency of water in the basic granulites. However, he attributes this to the extensive melting in adjacent rocks having taken up all available water. Furthermore, Valley & O'Neill (1978), used stable isotopes to show that granulite samples from nearby layers in the Adirondack suite, New York, were not in isotopic equilibrium with one another during metamorphism, which appears to rule out the possibility that they were all infiltrated by a uniform, deep-derived fluid.

There seems little doubt that many granulites have been metamorphosed at temperatures sufficiently high for melting to have taken place were there even small traces of H_2O present, and furthermore many granulite facies rocks are migmatitic. In addition, many granulites are depleted in those elements, such as K, that would most readily enter melt, and this also supports an anatectic origin for granulites (Fyfe, 1973). Nevertheless, the geochemistry of certain other granulites is apparently more likely to result from metasomatism by CO_2-rich fluids than by melting (Glassley, 1983; Weaver & Tarney, 1983). The solution to this problem is not clear at present because although there is no doubt that many granulites have undergone melting, this does not rule out the possibility that influx of CO_2 has been the dominant process in some instances.

HYDROTHERMAL METAMORPHISM OF BASALTIC ROCKS

Although much metamorphism is essentially isochemical (apart from gain or loss of volatiles), metabasalts actually show a wider range of chemical composition than known fresh basalts. This is now believed to result primarily from metasomatism accompanying hydrothermal metamorphism.

Spilites are basalts in which the plagioclase is albite, and which invariably contain some secondary hydrous phases replacing glass or primary minerals. For many years the origin of these rocks was contentious; the albite pseudomorphs primary plagioclase so perfectly that many workers considered it was itself of igneous origin and supposed that spilites formed from a sodic, and perhaps also hydrous, type of basaltic magma that is not represented by contemporary volcanoes. The case for a metamorphic origin of spilites was made convincingly by Vallance (1965) who showed that some, though not all, had undergone metasomatism. Subsequently Cann (1969) found spilitic rocks dredged from the ocean floor, demonstrating that spilites had formed more recently than hitherto had been supposed. Other extreme rock compositions also occur in association with spilites, for example **epidosites**, that is rocks composed almost exclusively of epidote and found as thin layers with thicknesses of millimetres to centimetres, up to pods with dimensions of metres. Spilites and epidosites are typical of low grade metamorphism, however in some higher grade (amphibolite facies) regions, rocks rich in Ca-poor amphiboles, such as the cordierite–anthophyllite rocks originally described by Eskola (1914) may also represent hydrothermally metamorphosed basalts. They are discussed further below.

The origin and significance of hydrothermally altered basalts has become much clearer in recent years with the prediction and discovery of active convection cells circulating sea water through young oceanic crust in the deep oceans, giving rise to metamorphism on the sea floor.

SEA-FLOOR METAMORPHISM

It has been known for many years that dredge-haul collections of samples from the sea bed in the vicinity of the mid-ocean ridges include both fresh and metamorphosed basalt, and more rarely metagabbro. Cann (1969) studied samples dredged from the Carlsberg ridge in the Indian Ocean and found a range from fresh to thoroughly metamophosed greenschist facies basalt. Even in the intensely metamorphosed basalts igneous textures are still apparent, but plagioclase is largely albitised and olivine and glass are replaced by chlorite. Other metamorphic minerals include quartz, sphene, fine needles of actinolite and epidote and most of these metamorphic minerals occur in veins as well as in the bulk of the rock. Cann pointed out that the metamorphism was not strictly isochemical; for example glassy pillow rims may convert to pure chlorite.

The wide range of minerals and facies types found in ocean floor rocks is illustrated in Table 4.2, prepared from a compilation by Humphris and Thompson (1978). Their

Table 4.2 Mineralogy of sea-floor metabasalts. (After Humphris & Thompson, 1978)

Metamorphic facies	Metamorphic minerals reported
Zeolite	Zeolites (e.g. analcite, heulandite, stilbite, natrolite, mesolite, scolecite), mixed-layer clays
Prehnite–Pumpellyite	Prehnite, chlorite, calcite, epidote
Greenschist	Actinolite, tremolite, hornblende, albite, chlorite, talc, epidote, nontronite, quartz, sphene, magnetite
Amphibolite	Hornblende, actinolite, plagioclase, chlorite, biotite, epidote, quartz, sphene, magnetite, epidote

temperatures of formation probably range from 100 to 500 °C but pressures are low, as indicated by, for example, the low Na-contents of the actinolitic amphiboles. The highest grade rocks are relatively rare, and are often metagabbros, zeolite facies rocks are also scarce, and greenschist facies rocks are the most abundant. This section is only concerned with the relatively high temperature phenomena (> 250 °C) related to circulation of sea water along cracks in newly created crust, not the more widespread hydration effects that can affect sea-floor rocks, notably glasses, at lower temperatures.

Appreciable metasomatic changes often accompany sea-floor metamorphism and Humphris and Thompson (1978) found that there are two distinct types of alteration. Most commonly metabasalts are chlorite-rich, but in some cases there is abundant epidote. The chlorite-enriched rocks have lost Ca and Si and gained Mg by comparison with the precursor basalt, while those rich in epidote have high Ca, low Mg and are somewhat oxidised. Humphris and Thompson pointed out that since sea-floor metamorphism involves removal of Mg from sea water to form chlorite, it provides an explanation for the imbalance between the relatively large amounts of Mg that enter the oceans in river water and the low Mg contents of the oceans themselves. In order to account quantitatively for the Mg-content of the modern ocean, sea-floor metamorphism must be occurring on a worldwide scale today.

Direct evidence for the process came with the discovery of warm springs on the ocean floor at the Galapagos Ridge (Corliss *et al.*, 1979), and shortly afterwards, with the more spectacular findings of high temperature fluid discharges at 22 °N on the East Pacific Rise (Speiss *et al.*, 1980). These occur in areas where sea water is drawn down into young, hot oceanic crust which heats it and reacts with it (Fig. 4.10). The heated sea water discharges at temperatures of around 350 °C and precipitates fine grained sulphides when it re-enters the ocean, giving rise to a characteristic 'black-smoke' of sulphide particles. Hence the name **black smokers**. Subsequently, hot springs have been found in a number of localities elsewhere in the oceans and experimental studies have been made of how sea water reacts with basalt at 350–400 °C (see Seyfried, 1987 for a review).

A significant feature of hydrothermally metamorphosed rocks is that those samples which have been subjected to the most extreme metasomatic change have the fewest minerals. The term **water–rock ratio** is used to denote the volume of water that has flowed through a particular unit volume of rock during metamorphism, and where the water–rock ratio is high, the rock reacts so extensively with the fluid that it begins to take on a mineralogy and chemical composition that is dictated by the chemistry of the fluid as much as by the original nature of the rock. This is explained more fully in the accompanying box. In the case of sea-floor metamorphosed basalts, Mottl (1983), reported a progression from assemblages of chlorite + epidote + albite + actinolite + quartz at relatively low water–rock ratios, though chlorite + albite + quartz to chlorite + quartz at

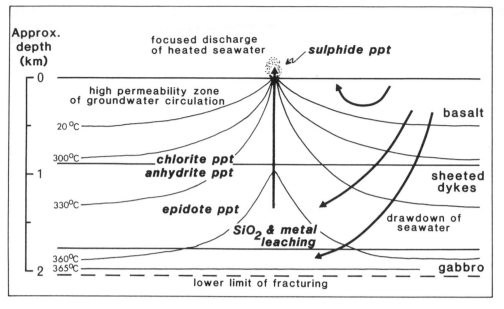

Fig. 4.10   Schematic section showing the lithological layering, thermal structure and hydrothermal circulation beneath an active mid-ocean ridge hydrothermal vent, to illustrate the chemical and mineralogical changes taking place. Based on Mottl (1983), Seyfried (1987) and Richardson et al. (1987). Bold horizontal lines show the lithological stratification of the oceanic crust, while labelled isotherms indicate the thermal structure around the vent. Bold arrows indicate generalised pathways of fluid flow (right half), while locations of major mineralogical changes are shown on the left.

high water–rock ratios. At the same time, chlorite became more Mg-rich the larger the amount of sea water that had passed through the rock. In nature, high water-rock ratios probably occur in the immediate vicinity of fractures, as illustrated in Fig. 1.1(*c*).

CORDIERITE–ANTHOPHYLLITE AND RELATED ROCKS

Cordierite–anthophyllite rocks are a distinctive class of metamorphic rock characterised by the presence of Ca-poor amphiboles coexisting with aluminous phases, notably cordierite. In addition to cordierite and anthophyllite, other minerals found include cummingtonite, plagioclase, chlorite, quartz *or* corundum, spinel, biotite, talc and Fe–Ti oxides. Related rocks include assemblages with staurolite, gedrite or hornblende. Despite their unusual mineralogy, this particular suite of assemblages has been found in many regions of amphibolite facies metamorphism since it was first reported by Eskola (1914). In particular, J.B. Thompson Jr, P. Robinson and their students have enthusiastically sought out every conceivable variation in amphibole composition and paragenesis in rocks of this suite from Massachusetts and New Hampshire, USA.

Fig. 4.11 *opposite*   Graphical representation of the mineral assemblages of cordierite–anthophyllite and related rocks. *a*) Demonstration of the 'plagioclase projection' on to the AMF face of the Al_2O_3–FeO–MgO–CaO tetrahedron. *b*) Mineral assemblages of cordierite–anthophyllite rocks and related parageneses from the sillimanite zone, south-west New Hampshire. After Robinson *et al.* (1982).

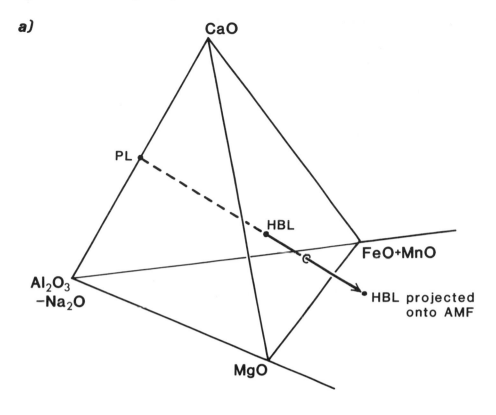

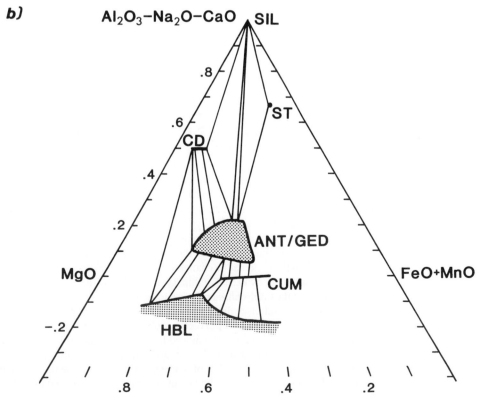

The mineral assemblages of quartz-bearing cordierite–anthophyllite rocks can be represented graphically in an **AMF projection** (also known as a **plagioclase projection**), comparable to the AFM projection used for pelites, developed by Robinson & Jaffe (1969). In these rocks, K_2O is not an important component whereas CaO is, and allowance must also be made for Na_2O in plagioclase. Assemblages coexisting with quartz and water are shown in Fig. 4.11(*a*), a tetrahedron whose corners are CaO, MgO, FeO and Al_2O_3–Na_2O. Subtraction of Na_2O has removed the effect of albite, so that all plagioclases plot at the anorthite point. This allows us to project rock and mineral compositions that lie within the tetrahedron on to the AMF face from plagioclase. The resulting projection for assemblages from south-west New Hampshire, corresponding to the sillimanite zone as defined by the interbedded pelites, is illustrated in Fig. 4.11(*b*).

The problem presented by this class of metamorphic rocks is that there is no igneous or sedimentary rock of comparable bulk composition to have been their precursor. As a result, most authors have accepted a metasomatic origin, but the causes and timing of the metasomatism have been controversial. We have seen above (and in the accompanying box) that in open systems the number of mineral phases is reduced according to the number of components that are mobile. Where appreciable metasomatism has taken place, as must be the case here, we would expect the rocks produced by the open system hydrothermal metamorphic event to contain only a few minerals, which appears to contradict the complexity of the assemblages of some cordierite–anthophyllite rocks. However, subsequent metamorphism of metasomatic rocks can increase the number of phases present; for example a pure serpentine rock produced by open system metamorphism of peridotite will recrystallise to olivine + talc when heated above the upper stability limit of serpentine (Ch. 2, page 44). Vallance (1967) found that cordierite–anthophyllite rocks had the same composition as chlorite + quartz-rich regions of hydrothermally altered basalts, and proposed that they resulted from subsequent high temperature metamorphism of the altered basalts. By comparison with basalts, cordierite–anthophyllite rocks are depleted in Ca and enriched in Mg, and to a lesser extent Fe, while retaining the same Al-content. These are precisely the types of chemical change that Humphris & Thompson (1978) found were characteristic of sea-floor metamorphism.

APPLICATION OF THE PHASE RULE TO ROCKS UNDERGOING HYDROTHERMAL METAMORPHISM

Rocks undergoing metamorphism can behave either as **closed systems**, in which case no material is added or removed, or as **open systems** involving loss or gain of some components. Most metamorphic rocks behave as open systems to some extent, because they lose volatiles during progressive heating. However, the amounts involved are relatively small and so there is no appreciable loss of other components in solution. In hydrothermal metamorphism where large volumes of fluid may pass through the rock, the concentrations of a number of constituents of the rock may be changed appreciably, giving rise to significant metasomatism.

It was pointed out by D.S. Korzhinsky in the 1930s (see Korzhinsky (1959) for an English summary) that the Gibbs phase rule as outlined in Chapter 2 can be modified to be applied to open systems. Korzhinsky distinguished between **inert components**, i.e. those that neither enter nor leave the rock, and **perfectly mobile components**, which are present in the fluid phase and may be added to, or removed from, the rock according to whether their chemical potential in the fluid is greater or less than in the mineral matter of the rock. The important point is that chemical potentials of perfectly mobile components are controlled by external factors; for example that of a solute component may be controlled by mineral solubility in the source region from which the fluid was derived. If we consider that metamorphic P–T conditions will normally lie within divariant fields on a P–T diagram for the system concerned (i.e. F = 2), the normal situation is that the *total* number of phases will equal the number of components during isochemical metamorphism, i.e. P = C. Korzhinsky showed that if we designate the number of inert components as C_i and the number of perfectly mobile components as C_m, then the number of *solid* phases $P_s = C_i = C - C_m$.

As an example consider the mineral assemblage talc + forsterite + tremolite in the aureole of the Bergell tonalite (Figs 2.4 and 2.7). The chemical system is MgO–CaO–SiO_2–H_2O, and if a fluid phase is present also, from the Gibbs phase rule P = C = 4. On the other hand, we can treat water as a perfectly mobile component because its pressure, and hence chemical potential, is independently controlled by the mechanical strength of the rock which prevents P_f exceeding P_{load} (see page 18). In this case $P_s = C_i = 3$ from Korzhinsky's phase rule, effectively the same result as from the Gibbs phase rule because only water was considered to be mobile.

Korzhinsky's phase rule is really of use when it comes to hydrothermal processes where large fluxes of fluid through rock volumes permit a range of normally inert components, including most of the major element constituents of rocks, to become mobile. Clearly, as more components become mobile there must be fewer inert components, and hence the number of phases in the rock will be smaller. As a result, *hydrothermally metamorphosed rocks and related veins contain only a small number of coexisting minerals, and may be monomineralic.* It could be argued that many amphibolites in any case have a very restricted number of phases, even though their metamorphism was isochemical. However, in that case the minerals are solid solutions, such as amphibole and plagioclase, that can display considerable compositional variability between nearby samples. Metasomatic rocks formed in open hydrothermal systems not only have few minerals, those minerals are of very uniform composition even if they have potential for solid solution. The reason is that element ratios in the solid solution are being fixed at a constant value by the composition of the infiltrating fluid.

Keeping to the example of metamorphosed ultrabasic rocks, ultrabasic pods emplaced along shear zones into siliceous rocks are not uncommon in orogenic belts, and have often recrystallised to monomineralic talc rocks with addition of SiO_2 and removal of CaO. In this case only MgO can be considered inert, and so the final rock has only one mineral. Where less water gained access, CaO does not behave as a perfectly mobile component however, and bimineralic talc–tremolite rocks result.

5 METAMORPHISM OF MARBLES AND CALC-SILICATE ROCKS

The third group of metamorphic rocks to be considered in detail are those derived from calcareous sediments: limestones and marls. Pelitic rocks provided relatively clear-cut isograds that could be related to specific discontinuous reactions, while metabasites introduced the additional complexity of a zonal sequence dominated by continuous reactions. In this chapter we will see the effects of a further variable on the stability of metamorphic mineral assemblages: the composition of the fluid phase.

In the sedimentary record limestones are sometimes very pure and because calcite itself is stable under most crustal conditions these rocks may not develop new minerals during metamorphism. However, many limestones contain other constituents such as detrital grains or diagenetic dolomite, and these may react extensively with calcite during metamorphism. Marly sediments containing a mixture of carbonate and silicate components are also common, and there is a complete spectrum possible between purely carbonate and purely silicate sediments.

Metamorphic rocks reflect the variability of the sedimentary record, and so also include both pure marbles and a range of metasediments with variable proportions of carbonate. However, it is also not unusual, especially at medium to high grades, to find metasediments that are rich in Ca- or Ca–Mg-silicates (such as zoisite, grossular, amphibole or diopside) but which contain little or no carbonate. These rocks are known as **calc-silicates**, and in many cases are probably the products of metamorphism of originally carbonate-bearing sediments. We infer this because calcite and dolomite are the major Ca- and Mg-bearing constituents of sediments, and the reactions in which they participate typically involve breakdown of carbonates with loss of CO_2 in the production of silicates. **Skarns** are a variety of calc-silicate rock formed by metasomatic interaction between marble and silicate rock. The most spectacular examples result from intrusion of granite into marble.

In practice, therefore, it becomes convenient for the description of metamorphosed calcareous sediments to divide them into two categories: marbles in which carbonates are abundant; and calc-silicates with little or no carbonate. The possible range in mineralogy of calc-silicates is very large, since it depends on the precise mixture of sedimentary components in the original layer as well as being susceptible to metasomatic interactions with adjacent layers. For this reason, no attempt will be made here to provide a comprehensive guide to the mineralogy of calc-silicates, even though they are often sensitive indicators of metamorphic grade. However, some examples of the types of mineralogical zoning most commonly found are outlined later in the chapter. The compositions of the phases discussed in this chapter are listed in the Glossary.

MARBLES

CALCITE MARBLES

The term marble is used for metamorphosed calcareous rocks in which carbonate minerals dominate. Many marbles are composed only of calcite with minor quartz and phyllosilicates, originally of detrital origin. There is sometimes graphite derived from organic debris, and pyrite is also a common accessory. The mineral assemblage in a marble of this type provides few clues as to the conditions of formation, since calcite is stable at all but the highest pressures (Fig. 5.1), and even where aragonite does form during burial, it is likely in most cases to change back completely to calcite during uplift, except at very low temperatures (Ch. 6, page 182). At very high temperatures and low pressures, calcite may react with any quartz present to produce calcium silicate, wollastonite. Despite the lack of mineralogical reaction in calcite marbles, they are susceptible to extensive textural changes due to recrystallisation of calcite to produce a coarser grain size and often a preferred orientation.

The reaction to form wollastonite provides a simple example of one of the most common types of reaction to occur in carbonate rocks, i.e. a decarbonation reaction:

$$CaCO_3 + SiO_2 \rightarrow CaSiO_3 + CO_2$$
$$\text{calcite} \quad \text{quartz} \quad \text{wollastonite} \quad \text{fluid} \qquad [5.1]$$

Like H_2O, CO_2 forms a supercritical fluid under metamorphic conditions, with a density that is broadly similar to that of supercritical water, though slightly greater under most metamorphic conditions (e.g. see Touret, 1977).

Reaction [5.1] was studied experimentally by Harker and Tuttle (1956). In their work the pressure of the fluid in the experimental capsule (CO_2) was equal to the total

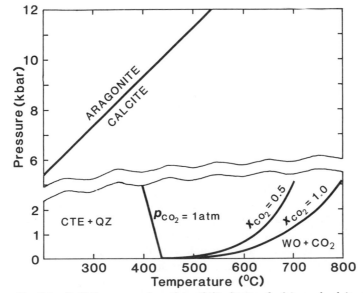

Fig. 5.1 *P–T* diagram to show the stability limits of calcite and calcite + quartz. Curves for calcite + quartz breakdown are given for various values of X_{CO_2} and for $P_{CO_2} = 1$ bar. Data of Johannes & Puhan (1971), Greenwood (1962, 1967) and Harker & Tuttle (1956).

pressure applied. This is usually the case in high pressure experiments because the noble metal capsules containing the reacting minerals (or **experimental charge**) are weak and collapse on to the grains in the charge forcing them together until the pressure on the fluid in the remaining interstices is equal to the applied pressure.

Harker and Tuttle's results, shown in Fig. 5.1, demonstrated that at pressures of more than a couple of kilobars the temperature required to form wollastonite is beyond the normal range of regional metamorphism. This is consistent with the fact that most wollastonite occurrences are in thermal aureoles formed by contact metamorphism at relatively low pressures. Nevertheless, wollastonite is sometimes found in situations where it apparently formed at significantly higher pressures but without excessive temperatures (e.g. Misch, 1964). An explanation for these occurrences necessitates considering the possibility of metamorphic fluids intermediate in composition between H_2O and CO_2.

MIXED-VOLATILE FLUIDS

The principle introduced in Chapter 2 that higher pressures of CO_2 inhibit decarbonation reactions has been understood at least since the time of James Hutton and Sir James Hall in the eighteenth century. Nevertheless, if it were possible to apply a high pressure to the solid phases in a marble while allowing CO_2 to escape at low pressure, then, since the molar volume of wollastonite is less than that of 1 mol quartz + 1 mol calcite, we might expect the reaction to take place at *lower* temperatures than when the total pressure on the system is also low. This effect was discussed in the early 1950s by H. Ramberg and T.F.W. Barth. It is difficult to envisage a natural situation that would correspond to such an experiment but in 1962 H.J. Greenwood and P.J. Wyllie independently pointed out that a very similar effect would be produced if the fluid phase in contact with the calcite and quartz were rich in H_2O. At the temperatures of the greenschist facies and above, H_2O and CO_2 supercritical fluids are completely miscible (except where the aqueous fluid contains large amounts of dissolved salts). Hence the partial pressure due to the CO_2 in a mixed H_2O-CO_2 fluid may be very much less than the total fluid pressure, even if $P_{fluid} = P_{lithostatic}$. Fluid composition is conveniently expressed in terms of the mole fraction of CO_2 or X_{CO_2}:

$$X_{CO_2} = \frac{n_{CO_2}}{n_{CO_2} + n_{H_2O}}$$

where n denotes the number of molecules of the subscripted species in the system.

Partial pressures are then given by:

$$P_{CO_2} = P_{fluid} \times X_{CO_2} \text{ and } P_{H_2O} = P_{fluid} \times X_{H_2O}$$

A major innovation by Greenwood (1962) was to carry out experiments at controlled values of X_{CO_2} in the fluid phase, and hence with $P_{CO_2} < P_{total}$. His work is included in Fig. 5.1 and demonstrates the lowering of the temperature for the appearance of wollastonite in the presence of an H_2O-rich fluid. These results provide a possible explanation of the occasional occurrences of wollastonite in regionally metamorphosed marbles: i.e. that it occurs where water was able to infiltrate the marble from adjacent schists and, by flushing away CO_2 as fast as it was produced, provided an environment with low P_{CO_2} favourable to wollastonite growth.

The observed effect of adding H_2O to experiments on the equilibrium between calcite, quartz, wollastonite and fluid accords with the phase rule. In the H_2O-absent system there are four phases and three components (CaO, SiO_2, CO_2), and hence one

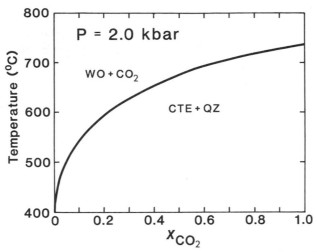

Fig. 5.2 $T-X_{CO_2}$ diagram for $P = 2$ kbar to show the effect of fluid composition on the stability of calcite + quartz. Based on Greenwood (1962, 1967).

degree of freedom, i.e. the full assemblage can occur stably only along a univariant curve on a $P-T$ diagram. Adding H_2O increases the number of components by one, but does not change the number of phases if it is miscible with CO_2. Hence there are now two degrees of freedom when calcite, quartz, wollastonite and fluid coexist. Here, fluid composition is a variable in addition to T and P, and by specifying one of these three variables, the equilibrium conditions can be represented by a univariant curve on a plot with the other two variables as axes. For example on Fig. 5.1, a plot of P versus T, there is a series of univariant curves, each valid for a specific value of X_{CO_2} only. Another widely used way of representing this sort of equilibrium is on a plot of T versus X_{CO_2}, constructed for some specified constant value of total pressure. An example of such a plot for reaction [5.1], constructed for a total pressure of 2 kbar, is shown in Fig. 5.2. (As an exercise, construct a similar curve for a pressure of 1 kbar from the curves in Fig. 5.1.) Plots of this type are known as **isobaric $T-X_{CO_2}$ diagrams** (or simply $T-X_{CO_2}$ diagrams). Divariant equilibria, such as reaction [5.1] taking place in the presence of H_2O, plot on such a diagram as a line known as an **isobaric univariant curve**, i.e. when P is fixed, one degree of freedom remains.

DOLOMITIC MARBLES

The number of phases that can form from limestones composed only of $CaCO_3$ + quartz is clearly limited, as shown by Fig. 5.1. Only at exceptionally high temperatures and low pressures do other phases such as spurrite and larnite appear, and since the classic locality, Scawt Hill in Northern Ireland, where C.E. Tilley described this extreme type of metamorphism, results from the chance heating of chalk by basalt lava in the immediate vicinity of the surface, they cannot be considered as of widespread geological importance. For this reason the relationships of these, and other associated phases, are not considered further here, even though the paper in which they were interpreted by Bowen (1940) still stands as one of the classics of the metamorphic literature.

Limestones that contain dolomite provide much more useful indicators of metamorphic grade because a range of Ca–Mg-silicates can form in the more usual $P-T$ conditions of metamorphism, for example talc, tremolite and diopside. The general sequence of mineral zonation in dolomitic marbles was first described by Eskola (1922) and subsequently refined by Bowen (1940) and Tilley (1951), who first recognised the importance of talc at the lowest grades. The sequence of mineral-appearance isograds in regionally metamorphosed dolomitic limestones appears to be:

– talc (not always present)
– tremolite
– diopside *or* forsterite
– diopside + forsterite.

Most earlier studies reported the appearance of forsterite before diopside, but the precise conditions for the growth of either mineral are dependent on rock composition, and since both appear at very similar temperatures, chance variation in lithology can dictate the relative order of appearance. For this reason these two minerals have been grouped together here, although it is clear that higher temperatures are needed for them to *coexist* than for one *or* the other (according to rock composition) to occur.

The mineral assemblages of impure dolomitic marbles can be conveniently represented in a triangular diagram with CaO, SiO_2 and MgO at the apices. CO_2 and H_2O are treated as being available in excess to produce carbonate or hydrous phases. The locations of the common phases of metamorphosed marbles, plotted on such a diagram, are shown on Fig. 5.4(*h*).

In addition to the Ca–Mg silicates and carbonates, and quartz, impure marbles can contain additional phases such as mica, feldspar, garnet, etc., which involve further components, but many marbles nevertheless have compositions that can be modelled very closely in the system $CaO–MgO–SiO_2–CO_2–H_2O$. Minor amounts of other phases do not substantially change the reactions among the Ca–Mg-silicates.

Dolomitic marbles of the central Alps

One of the most extensive studies of the regional metamorphism of marbles under medium pressure conditions is the classic work by Trommsdorf in the central Alps (Trommsdorf, 1966, 1972). Fig. 5.3 is a map showing the metamorphic zoning of dolomitic marbles in the region from Trommsdorf's study. The metamorphic grade increases southwards from low grade rocks with talc through tremolite marbles to diopside and forsterite-bearing rocks. The diagnostic three-phase assemblages found by Trommsdorf are represented graphically in Fig. 5.4, and from this diagram it is possible to suggest reactions to describe the changes between the zones, on the basis of the shift of the pattern of tie-lines.

Figure 5.4(*a*) represents the original sedimentary assemblages of dolomite + calcite + quartz. The change between this diagram and Fig. 5.4(*b*), representing the low grade rocks shown in the northern part of the region on Fig. 5.3, is the replacement of the dolomite–quartz tie-line by a talc–calcite tie-line, and this change can be represented by the reaction:

$$3 \text{ dolomite} + 4 \text{ quartz} + 1 \text{ } H_2O \rightarrow 1 \text{ talc} + 3 \text{ calcite} + 3 \text{ } CO_2 \qquad [5.2]$$

Note that rocks containing all four of these solid phases are not uncommon. They might be expected where insufficient water was added to the marble to convert all the available reactants to talc. The appearance of tremolite leads to a more complex situation. In some rocks, talc persists with either dolomite or quartz and calcite, but in others tremolite occurs without talc. Figure 5.4(*c*) represents the main paragenesis found when tremolite first reacts in. The composition of tremolite plots within the talc–calcite–quartz triangle of Fig. 5.4(*b*), from which we can deduce that the reaction is:

$$5 \text{ talc} + 6 \text{ calcite} + 4 \text{ quartz} \rightarrow 3 \text{ tremolite} + 2 \text{ } H_2O + 6 \text{ } CO_2 \qquad [5.3]$$

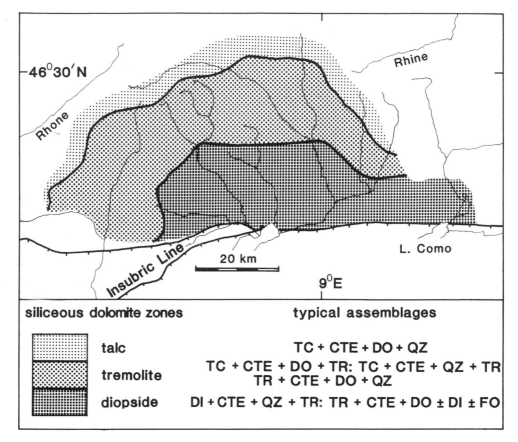

Fig. 5.3 Metamorphic zonation of siliceous dolomitic marbles in the Lepontine Alps. Based on Trommsdorff (1966). Abbreviations as in glossary.

In the case of quartz-poor rocks, quartz may be completely consumed by this reaction to give the assemblage talc + calcite + tremolite, but in more siliceous rocks the talc is consumed to give tremolite + calcite + quartz. Final disappearance of talc in these rocks can be ascribed to the reaction:

$$2 \text{ talc} + 3 \text{ calcite} \rightarrow 1 \text{ tremolite} + 1 \text{ dolomite} + 1 \text{ CO}_2 + 1 \text{ H}_2\text{O} \qquad [5.4]$$

This gives rise to the phase relations illustrated in Fig. 5.4(*d*). At higher grades than the disappearance of talc + calcite, diopside + calcite or forsterite + calcite appear, although tremolite commonly persists. Figure 5.4(*e*) shows the appearance of the diopside within the tremolite–calcite–quartz field, through the reaction:

$$1 \text{ tremolite} + 3 \text{ calcite} + 2 \text{ quartz} \rightarrow 5 \text{ diopside} + 1 \text{ H}_2\text{O} + 3 \text{ CO}_2 \qquad [5.5]$$

In contrast, the association of forsterite + calcite implies the replacement of the tremolite–dolomite tie-line (Fig. 5.4(*f*)) due to the reaction:

$$1 \text{ tremolite} + 11 \text{ dolomite} \rightarrow 8 \text{ forsterite} + 13 \text{ calcite} + 1 \text{ H}_2\text{O} + 9 \text{ CO}_2 \qquad [5.6]$$

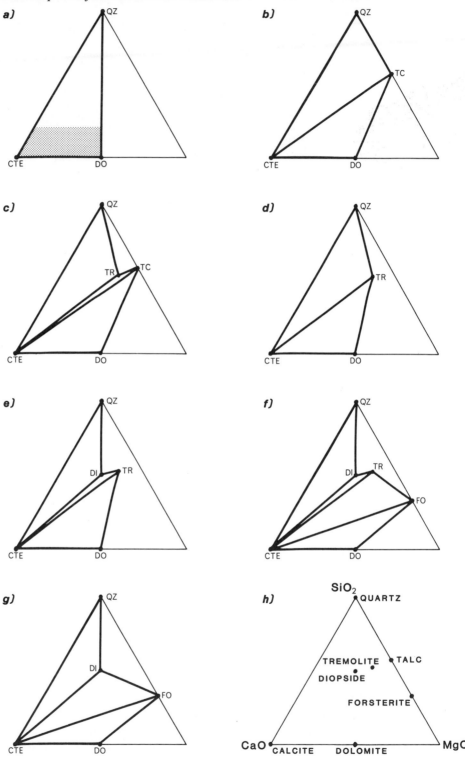

Fig. 5.4   Phase compatibilities in the system $CaO-SiO_2-MgO + CO_2-H_2O$ fluid, for siliceous dolomites of the Lepontine Alps: *a*) to *g*) illustrate progressive changes with increasing metamorphic grade (see text); *h*) is a key to mineral compositions in the system $CaO-SiO_2-MgO$ ($+CO_2+H_2O$). The stippled region in *a*) represents the common compositions of siliceous marbles.

It is clear from Fig. 5.4 that forsterite first appears in rocks whose compositions are silica-poor and lie near the base of the triangle, while diopside appears in relatively silica-rich or dolomite-poor rocks. This may account for the discrepancies in their order of appearance between different studies. Diopside and forsterite can coexist only when the tremolite–calcite tie-line has been removed, due to the reaction:

$$3 \text{ tremolite } + 5 \text{ calcite } \rightarrow 11 \text{ diopside } + 2 \text{ forsterite } + 3 \text{ H}_2\text{O} + 5 \text{ CO}_2 \quad [5.7]$$

This gives rise to the phase relations shown in Fig. 5.4(*g*), and results in the assemblages reported from the southern part of the central Alps (Fig. 5.3).

Despite the apparent simplicity of the reaction sequence outlined above and illustrated in Fig. 5.4, the natural sequence of assemblages, although conforming in a general way with the idealised sequence, is actually far more complex. Assemblages involving both reactants and products of these reactions may occur over large areas of countryside, and apparently low grade assemblages may persist alongside higher grade assemblages. Why this complexity should be present is apparent if we analyse the assemblages found using the phase rule. Our simplified system has five components: CaO, MgO, SiO_2, H_2O and CO_2. No assemblages have more than five phases, i.e. four solid phases plus a single mixed H_2O–CO_2 fluid phase. Hence even these assemblages have two degrees of freedom and can occur over a range of pressures and temperatures according to the composition of the fluid phase. Much of the diversity in the mineral assemblages within the broadly defined zones of Fig. 5.3 can therefore be explained if the fluid composition varied between samples. Why such variations in fluid composition might occur will be discussed in the following section.

CONTROLS ON THE FLUID COMPOSITION IN MARBLES

We have seen in this chapter that the *P–T* conditions at which many reactions in carbonate rocks take place are dependent on the composition of the fluid phase. However, the self-same reactions involve components of the fluid and so change its composition. A marble with calcite + quartz may have a nearly pure H_2O fluid phase present as it is heated until wollastonite begins to form, but at this point the fluid must become enriched in CO_2 from the decarbonation reaction (reaction [5.1]). We can imagine two possible scenarios representing extreme cases, natural processes may be intermediate between the two in many instances.

In the first case, no more water gains access to the marble after wollastonite starts to form. If the initial fluid composition is represented by X_A on Fig. 5.5(*a*), wollastonite will first appear at a temperature above T_A,* say T_B, and will continue to grow until the fluid composition has shifted to X_B before there is any further rise in temperature. Further heating is accompanied by further increments of reaction which cause the fluid to become progressively enriched in CO_2 so that the fluid composition and temperature evolve together along the path of the isobaric univariant curve as long as calcite, quartz, wollastonite and fluid are all present in the rock. In terms of the phase rule, the assemblage has two degrees of freedom, and so if *P* and *T* are independently determined, the fluid composition is fixed by the presence of the particular mineral assemblage. In this type of behaviour, fluid composition is controlled by the mineral assemblage of the rock itself, and

---

* Reactions that produce new minerals can only take place after the equilibrium conditions (in this case, temperature T_A) have been overstepped by a finite amount (see Ch. 6).

is described as being **internally buffered** (see Ch. 2, page 37). In the example illustrated, suppose reaction proceeds until a temperature T_C is reached, at which point one of the reactants (calcite or quartz) is completely consumed; fluid composition then remains constant at X_C with further heating. If, however, the CO_2 released by reaction is able effectively to flush away all the H_2O present initially, reaction will proceed to point T_D and the system will then become effectively a three-component system only (no H_2O remains); reaction will go to completion at T_D. Greenwood (1975) has pointed out that natural rocks probably have such a low porosity, and hence contain so little H_2O initially, that this alternative is more probable (below). Trommsdorf (1972) showed that the complex pattern of mineral assemblages in the central Alps could be explained by internal buffering, with different rock layers undergoing different reactions initially, which then led to the fluid composition being buffered to different values of X_{CO_2} in each layer.

The second way in which fluid composition can be controlled in marbles is by **external buffering**. In this process, the fluid is derived from adjacent rock units or from magmatic processes, for example, and it is the source region that defines the fluid's composition in exactly the same way as in the open systems discussed in Chapter 4. If such a fluid passes through the marble in sufficiently large quantities it may be able to flush away the fluid being generated by reactions going on within the marble without having its composition substantially modified. In this case the fluid composition is externally buffered. This type of reaction is represented in two ways on Fig. 5.5(*b*). If influx of fluid of composition X_A continued steadily as the rock was heated over a range of temperature, then the heating of the rock is represented by the vertical dotted line, fluid composition remains constant at X_A and the reaction will go to completion at a temperature close to T_A. On the other hand, fluid infiltration may take place in a single event at a fixed temperature, say T_E. If the initial fluid composition is X_E and the rock is composed of calcite and quartz, infiltration of fluid X_A will shift the fluid composition at constant temperature, indicated by the horizontal dotted line at $T = T_E$, and again this causes reaction to produce wollastonite. External

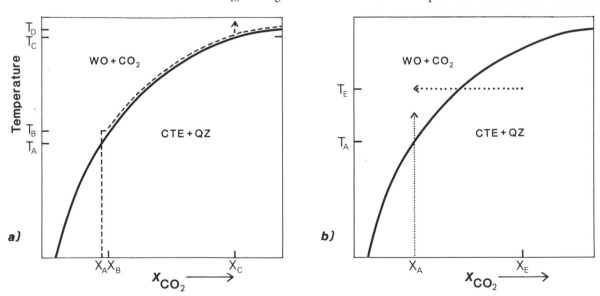

Fig. 5.5 Influence of internal versus external buffering of fluid composition on the breakdown of calcite + quartz. a) Fluid composition is internally buffered; for an initial fluid X_A, reaction takes place over a large temperature interval from T_B to approximately T_D. b) Externally buffered fluid. If fluid is buffered at X_A during heating, reaction goes to completion at T_A; alternatively if the initial fluid is X_E, an influx of fluid X_A at temperature T_E will cause the reaction to go to completion.

buffering is probably responsible for examples of regional wollastonite, noted above. Tracy *et al.* (1983) have documented the development of progressively decarbonated calc-silicate assemblages adjacent to quartz veins near the edges of marble lenses, and interpret them as resulting from flow of water derived from adjacent schists through original cracks, now represented by the quartz veins. The efficiency of the external buffering decreased away from the vein, because the amount of water penetrating the rock declined rapidly, away from the fissure, so that in the bulk of the marble the fluid was internally buffered.

THE EFFECT OF REACTION ON FLUID COMPOSITION

The reactions that take place in marbles and other carbonate-bearing rocks can be grouped into six types, each of which has a distinctive shape to the equilibrium curve on an isobaric *T–X* diagram.

1. *Decarbonation reactions* We have already seen that reactions such as [5.1] can be represented by equilibrium curves on the *T–X* diagram which reach a temperature maximum at $X_{CO_2} = 1.0$, and fall to lower temperatures at small values of X_{CO_2}. This is shown by curve i on Fig. 5.6.

2. *Dehydration reactions* The equilibrium curve for a dehydration reaction is the mirror image of that for a decarbonation reaction (Fig. 5.6, curve ii).

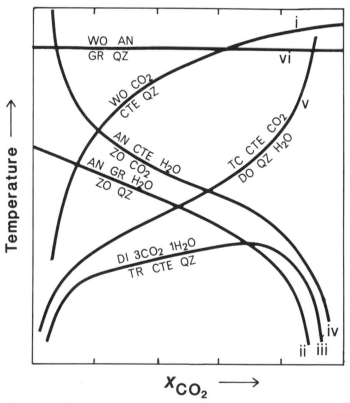

Fig. 5.6 Schematic representation of the form of $T–X_{CO_2}$ curves for different types of mixed-volatile reactions. After Kerrick (1974). Note that although specific examples are used (see Glossary for abbreviations), they are not shown in their correct relative positions along the temperature axis.

3. *Dehydration–decarbonation reactions* Many reactions in marbles give off a mixture of CO_2 and H_2O, for example reactions [5.3] to [5.7] above. In the case of dehydration or decarbonation reactions, the highest temperature on the equilibrium curve on a T–X diagram is attained when the composition of the fluid phase in the system corresponds to that being given off by the reaction, i.e. when it is pure H_2O or pure CO_2 as the case may be. In the same way, an isobaric univariant curve for reactions giving off a mixed-volatile fluid has a temperature maximum on a T–X diagram corresponding to the X_{CO_2} value of the fluid produced. For example, in the case of reaction [5.4] the maximum will be at $X_{CO_2} = 0.5$, for reactions [5.3] and [5.5] it is at $X_{CO_2} = 0.75$ and for reaction [5.6] it is at $X_{CO_2} = 0.9$. The reaction takes place at lower temperatures if the fluid phase present in the rock does not correspond to that given off by the reaction, irrespective of whether it is richer in CO_2 or richer in H_2O (curve iii, Fig. 5.6). This type of reaction can internally buffer the fluid composition in the same way as was demonstrated in Fig. 5.5, until the overall fluid composition corresponds to that of the fluid given off by the reaction itself.

4. *Hydration–decarbonation reactions* Other mixed-volatile reactions, for example reaction [5.2], have H_2O and CO_2 on different sides of the reaction. The equilibrium curve for such reactions (e.g. curve v on Fig. 5.6) displays a particularly large range of equilibrium temperatures according to fluid composition. At low values of X_{CO_2} the equilibrium temperature drops as for other decarbonation reactions, favouring production of CO_2. However, at high X_{CO_2} values it is the behaviour of the aqueous component that dominates the form of the equilibrium curve; as X_{H_2O} diminishes so the hydration aspect of this type of reaction is inhibited by the absence of water, hence the equilibrium temperature approaches infinity as X_{CO_2} approaches 1.

5. *Carbonation–dehydration reactions* These are similar to hydration–decarbonation reactions except that the effect of increased temperature is to cause CO_2 to be consumed and H_2O to be released. An example is the reaction:

$$\text{zoisite} + CO_2 \rightarrow \text{anorthite} + \text{calcite} + H_2O \qquad [5.8]$$

The form of the equilibrium curve is therefore the mirror image of that for hydration–decarbonation reactions (Fig. 5.6, curve iv).

6. *Fluid-absent reactions* Where no fluid is released or consumed, as in the reaction:

$$\text{grossular} + \text{quartz} \rightarrow 2\ \text{wollastonite} + \text{anorthite} \qquad [5.9]$$

the equilibrium conditions will be unaffected by the composition of any fluid phase that may be present. Such reactions may be represented by a horizontal line on an isobaric T–X_{CO_2} diagram, indicating a unique equilibrium temperature at each pressure (Fig. 5.6, curve vi).

INTERNALLY BUFFERED REACTION SEQUENCES IN MARBLE

In 1975, Greenwood published an influential paper in which he cast doubt on the importance of many simple mixed-volatile reactions, including some of those presented above, in the metamorphism of many natural marbles. We have seen already that these reactions are divariant, and the fluid composition changes as reaction proceeds if it is internally buffered. Greenwood pointed out that the composition of the pore fluid can be changed very considerably while only traces of the solid products are produced, because pore volume makes up only a small part of the rock. As an example, consider what happens when wollastonite is produced by reaction [5.1] in a siliceous marble with an initially aqueous pore fluid. Growth of 5 per cent wollastonite (by volume) liberates 5–10 per cent fluid (according to P and T). Since the porosity of the rock is likely to be small (<1 per cent perhaps) the production of even this small amount of CO_2 is clearly sufficient to flush away the original water, shifting the fluid composition to virtually pure CO_2. Hence the reaction will be spread over the entire temperature interval between the two curves for reaction [5.1] at $P_{CO_2} = 1$ bar and $P_{CO_2} = P_{fluid}$ (Fig. 5.1). In other words, only small amounts of wollastonite can be produced at $P_{CO_2} < P_{fluid}$ and so there will be no marked wollastonite isograd until the temperature for the reaction to proceed at $X_{CO_2} = 1$ has been attained.

Greenwood's view of the metamorphism of marbles is that internal buffering is common and therefore divariant reaction over a temperature interval makes only a small or negligible contribution to the development of new minerals. The only reactions that can give rise to distinct isograds in the field are those that allow mineralogical changes to occur without changing the fluid composition.

There are two general ways in which these isograd reactions can take place in rocks with a mixed-volatile fluid. Firstly, if the fluid present in the pore spaces is of the same composition as that being given off in the reaction, then the fluid composition cannot change further, no matter how much reaction takes place. Hence, reaction will proceed at the same temperature until one of the reactants is used up. The growth of wollastonite by reaction [5.1] in the presence of a pure CO_2 fluid is a simple example of this. Dehydration–decarbonation reactions can also behave in this way (curve iii, Fig. 5.6). The fluid composition is buffered by very small amounts of reaction until the temperature rises to the maximum on the curve, which corresponds to the point at which the fluid mixture being given off is of the same composition as the pore fluid. Further reaction proceeds until one of the reactants is used up, without any further rise in temperature, and so there is a sharp isograd although traces of the higher grade minerals may occur in rocks from below the isograd. One of the isograds described by Trommsdorf from the central Alps appears to correspond to reaction of this type (i.e. reaction [5.4], Fig. 5.7(*b*)).

The second way in which an isograd can arise is illustrated in Fig. 5.7(*a*). The equilibrium curve for reaction [5.2] represents the conditions under which talc, calcite, dolomite, quartz and fluid can coexist in the system $CaO–MgO–SiO_2–CO_2–H_2O$. Similarly, the equilibrium curve for reaction [5.3] in the same system gives the conditions for coexistence of talc, calcite, tremolite, quartz and fluid. These two curves cross at the point I at which all five phases involved in the two equilibria must be able to coexist. The assemblage at I has one less degree of freedom and is univariant, which is why on the isobaric T–X diagram it is represented by a unique or **isobaric invariant point**. Since talc, calcite, tremolite, dolomite and fluid are all stable at point I, the curve for reaction [5.4] must also pass through this point, as must the curve for the reaction:

$$\text{dolomite} + \text{quartz} + H_2O = \text{tremolite} + \text{calcite} + CO_2 \qquad [5.10]$$

A dolomite–quartz–calcite marble with an initial fluid composition X_A and temperature T_A, shown on Fig. 5.7(*a*), will begin to react on heating to produce talc by reaction [5.2]; whether or not appreciable amounts of talc can form depends on how much water can enter the rock. If the fluid composition is internally buffered it will evolve along curve [2] until point I is reached at temperature T_I, and here tremolite will appear. Reaction will proceed at this temperature until talc, quartz, calcite or dolomite is consumed. This is not a simple reaction with a unique stoichiometry, rather it varies according to the fluid composition represented by point I, which will vary with pressure; for the fluid composition plotted ($X_{CO_2} = 0.55$) the reaction produces tremolite + dolomite. The tremolite isograd corresponds to reaction at point I.

The temperature and fluid composition of the marble cannot depart from point I as long

a)

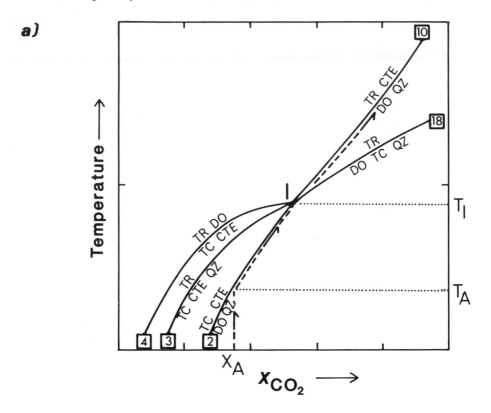

b)

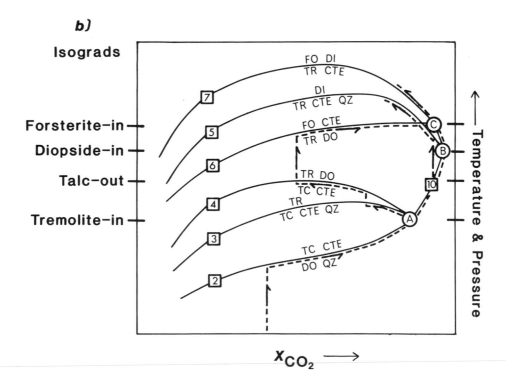

as talc + calcite are present, but supposing all the talc is consumed the remaining assemblage must satisfy the equilibrium conditions for reaction [5.10]. Fluid composition will then evolve further along curve [10], although only a small amount of reaction is likely until the next isobaric invariant point is reached.

Figure 5.7(*b*) portrays Trommsdorf's interpretation of the reaction sequence in the marbles of the central Alps, assuming that the fluid was internally buffered. Another well-documented example of an internally buffered reaction sequence comes from the aureole of the Marysville stock, near Helena, Montana, described by Rice (1977a). Again, Rice was able to show that the isograds identified in the field correspond to the special cases of univariant reactions similar to those outlined above.

A PETROGENETIC GRID FOR REACTIONS IN MARBLES

At first glance the task of relating the assemblages of impure marbles to prevailing temperatures and pressures is a daunting one, because the temperature at which reaction takes place is often strongly dependent on fluid composition. However, if the fluid composition is internally buffered as reaction proceeds, then it becomes possible to assign temperatures to specific isograds, because as we have seen above, they correspond to reaction among a univariant assemblage of phases.

A number of pioneering experimental studies of mixed-volatile reactions were carried out in Göttingen, under the direction of H.G.F. Winkler, in the 1960s. Continued improvement of experimental techniques has led to much better agreement between different studies, and early results have often been greatly improved on. G.B. Skippen (1971) carried out a series of experiments on reactions in dolomitic marble, and used his results to obtain thermodynamic data from which the equilibrium conditions of a number of other reactions were calculated. He presented a series of isobaric T–X_{CO_2} diagrams (Skippen, 1974) of which examples are illustrated in Figs 5.8(*a*) and (*b*). Further experiments by Slaughter, Kerrick and Wall (1975) on some additional reactions are broadly consistent with Skippen's work, although there are some discrepancies. Comparable T–X_{CO_2} diagrams from Slaughter *et al.* are shown in Figs 5.8(*c*) and (*d*); note that Figs 5.8(*b*) and (*c*) are both for 2 kbar and should be the same. The most important discrepancy between the two studies concerns the stability of the assemblage talc + calcite. (Slaughter *et al.* note significant uncertainties for their curves involving forsterite.) According to

Fig. 5.7 opposite Reactions in siliceous dolomites in which the fluid composition is internally buffered. *a*) DO–QZ–CTE marble with initial fluid X_A begins to react at temperature T_A to produce talc. Talc is progressively produced by reaction [5.2] until the isobaric invariant point I is reached at T_I. Here tremolite appears, giving rise to a distinct isograd because reaction continues at this temperature until all talc is consumed. Reaction numbers correspond to those used in the text, except reaction 18 which has the stoichiometry: $4DO + 8QZ + TC = 2TR + 8CO_2$, and is therefore restricted to rock compositions that are more Mg-rich than pure dolomite. It is therefore not important in metamorphosed marbles. For simplicity this reaction is omitted from subsequent figures. *b*) Pathways of changing fluid composition produced by internally buffered reactions in dolomitic marbles of the central Alps. After Trommsdorf (1972). Different dotted lines denote different pathways followed by rocks of different bulk composition. Note that the equilibrium curves on this T–X_{CO_2} diagram were derived before the publication of the experimental studies shown in Fig. 5.8; the diagram is empirically constrained to fit the field observations, and so the vertical axis is one of increasing metamorphic grade, not temperature alone.

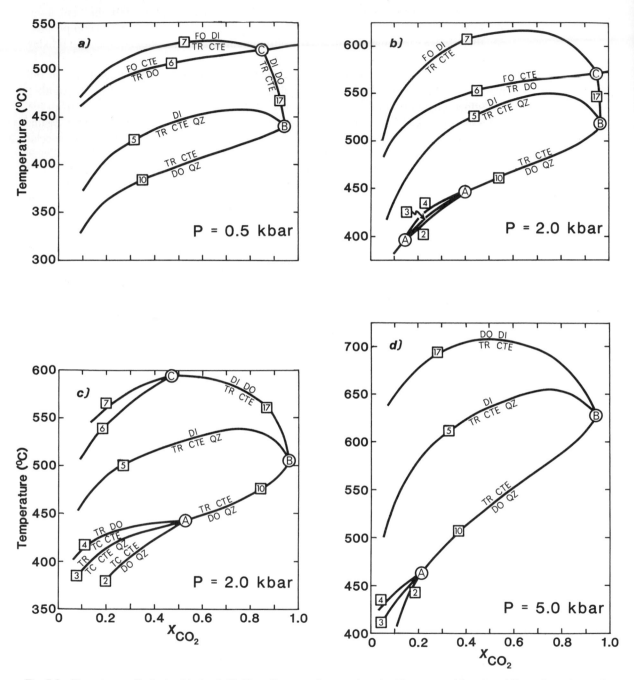

Fig. 5.8 Experimentally derived isobaric T–X_{CO_2} diagrams for reactions in siliceous marbles: *a*) and *b*) are from the work of Skippen (1974); *c*) and *d*) are from Slaughter, Kerrick & Wall (1975), to the same scale. Reaction 17 is not discussed in the text but has the stoichiometry: TR + 3CTE = 4DI + DO + CO_2 + H_2O.

Skippen, talc + calcite should become stable over a wider range of temperatures at higher pressures, while Slaughter *et al.* indicate a smaller stability field at higher pressures. On the whole, Skippen's interpretation appears to accord better with natural examples, e.g. the widespread talc + calcite in the medium pressure Alpine rocks described by

Trommsdorf. However, Tilley originally recognised talc as an important member of the zonal sequence in a thermal aureole.

Apart from the petrogenetic grid, there are few other ways of determining pressures and temperatures of formation of mineral assemblages in marble. The calcite–aragonite transition is one of the only reactions that is strongly pressure sensitive, and this is only useful in very low grade rocks. One important temperature indicator is the calcite–dolomite geothermometer. Calcite and dolomite form a restricted solid solution at metamorphic temperatures. The solvus is asymmetrical, and the most reliable way of determining the temperature of formation of coexisting calcite and dolomite is from the Mg-content of the calcite; the dolomite may be almost stoichiometric within the limits of analytical error. The following formulation for the calculation of the temperature of formation of coexisting calcite and dolomite has been presented by Rice (1977b):

$$\log_{10} X^{Cte}_{MgCO_3} = \frac{-1690}{T} + 0.795$$

where $X^{Cte}_{MgCO_3}$ is the mole fraction of $MgCO_3$ (*not* dolomite) in calcite coexisting with dolomite, and T is in Kelvin. The degree of mutual solid solution between calcite and dolomite can also be influenced by the presence of Fe in the system, and this problem has been addressed recently by Powell, Condliffe & Condliffe (1984) and Anovitz & Essene (1987).

METAMORPHISM OF CALC-SILICATES

At the beginning of this chapter, calc-silicates were defined as being rocks rich in Ca–Mg-silicate minerals, but with only minor amounts of carbonate. The usefulness of calc-silicates as indicators of metamorphic grade was recognised early in this century by Goldschmidt and Eskola, and subsequently Kennedy (1949) developed a zonal scheme for calc-silicates in Scotland which he was able to correlate with the Barrovian zones based on pelite mineralogy (see Ch. 1). Kennedy's work was of profound importance because it allowed metamorphic zones to be mapped across large areas where no pelites were present, and permitted him to produce a zonal map for the entire Scottish Highlands. The diagnostic features of the zones that he identified, and their correlation with the pelite zones, are tabulated in Table 5.1.

Table 5.1 Comparison of calc-silicate zones and pelite zones of the Scottish Highlands. (After Kennedy, 1949). Cf. the more recent work of Ferry, (page 143 and Fig. 5.10)

Pelite zone	Calc-silicate zone
Garnet	{ Zoisite–calcite–biotite { Zoisite–hornblende
Staurolite } Kyanite }	Anorthite–hornblende
Sillimanite	Anorthite–pyroxene

Because calc-silicates contain significant amounts of other chemical components, especially A1, K and Fe, their mineralogy is more complex than that of the dolomitic marbles. Common additional phases include zoisite, garnet, hornblende, calcic plagioclase, margarite, K-feldspar, phlogopite and vesuvianite. Phase relations are correspondingly more complex also, and Kerrick (1974) provides a useful summary. In general, zoisite, margarite and grossular garnet are only stable if the fluid phase is rich in water, while calcic plagioclase is indicative of a fluid richer in CO_2. Of the hydrous minerals, probably only phlogopite can remain stable in the presence of CO_2-rich fluid, although all these relationships are also temperature dependent.

FLUID INFILTRATION IN CALC-SILICATES

A number of studies in recent years have demonstrated that infiltration of fluid from surrounding rocks has played an important role in determining the metamorphic reactions in calc-silicates. This was first shown by Carmichael (1970), in a study of regional metamorphism in the Whetstone Lake area of Ontario. Carmichael mapped the isograds in both pelitic and calc-silicate lithologies, and his map is reproduced in Fig. 5.9.

The pelite isograds indicate a steady increase in metamorphic grade westwards. The

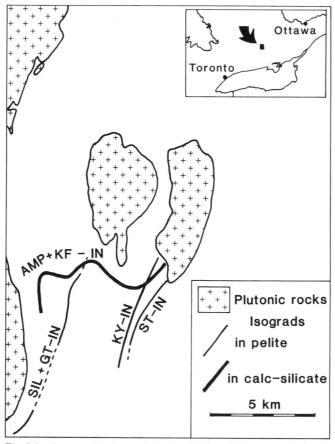

Fig. 5.9   Intersecting isograds in the vicinity of Whetstone Lake, Ontario (after Carmichael, 1969). The isograd marking the appearance of K-feldspar + amphibole in calc-silicates clearly cross-cuts the pelite isograds.

lowest grade rocks have the assemblage garnet + muscovite + chlorite + quartz ± biotite, and these are succeeded at the staurolite isograd by the assemblage staurolite + garnet + biotite + muscovite + quartz. Successive isograds mark the appearance of kyanite, often with staurolite, the appearance of sillimanite replacing kyanite, and the appearance of sillimanite in staurolite–garnet–biotite schists that lack kyanite. This is a typical Barrovian zonal sequence of the type outlined in Chapter 3; note that the disposition of the isograds is not influenced by the presence of two small granite plutons.

The major mineralogical change observed in the calc-silicates is the appearance of calcic amphibole coexisting with K-feldspar, through the reaction:

$$\text{biotite} + \text{calcite} + \text{quartz} \rightarrow \text{Ca-amphibole} + \text{K-feldspar} + CO_2 + H_2O \; [5.11]$$

Although the assemblage Ca-amphibole + K-feldspar can be broadly considered to occur at higher grades than biotite + calcite + quartz, the isograd representing reaction [5.11] clearly cuts across the pelite isograds at a high angle (Fig. 5.9). Carmichael interpreted the intersecting isograds as resulting from variations in the composition of the fluid phase. It is apparent from Fig. 5.6 that if the fluid is water rich, the temperature for the pelite dehydration reactions is at a maximum whereas the calc-silicate dehydration–decarbonation reaction occurs at relatively low temperatures. This might correspond to the situation in the northern part of the map area. On the other hand at higher values of X_{CO_2} in the fluid, reaction [5.11] could occur at higher temperatures relative to the dehydration reactions in the pelites, and this would correspond to the situation in the southern half of the area.

It is implicit in this interpretation of the intersecting isograds that the composition of the fluid phase is the same in both lithologies where they occur nearby, and is not greatly influenced by the fluid being released from reactions. This is an example of external buffering of the fluid phase, and implies that fluid from some external source is continually passing through the sequence of metasediments as the reactions proceed. Carmichael suggests that one of the granite plutons may have been the source of this water.

Calc-Silicates from the Vassalboro Formation, Maine, USA

One of the most thorough studies of metamorphic rocks in recent years has been the work of J.M. Ferry on calcareous metasediments in south-central Maine (e.g. Ferry, 1976, 1983a, b). The calcareous layers studied by Ferry occur in the Silurian Vassalboro Formation, and were metamorphosed during the Upper Palaeozoic Acadian orogeny (part of the Variscan belt). The formation consists of finely interbanded semi-pelites, pelites and argillaceous calcareous rocks, typically layered on a scale of only a few centimetres. Metamorphism occurred at low pressures (2.5 – 3.8 kbar) and is spatially related to syn-metamorphic granite stocks (Fig. 5.10). The distribution of the isograds in the calcareous rocks of the Vassalboro Formation and in the pelites of the adjacent Waterville Formation is illustrated in Fig. 5.10. The zonal sequence described by Ferry is as follows:

Ankerite zone:
The lowest grade rocks contain the assemblage ankerite ($Ca(Mg, Fe)(CO_3)_2$) + quartz + albite + muscovite + calcite ± chlorite. A range of possible accessories include pyrite, graphite and ilmenite.

Biotite zone:
This zone is characterised by the coexistence of biotite and chlorite without amphibole. The biotite isograd can be related to the reaction:

$$\begin{aligned}\text{muscovite} + \text{quartz} + \text{ankerite} + H_2O \rightarrow \\ \text{calcite} + \text{chlorite} + \text{biotite} + CO_2\end{aligned} \qquad [5.12]$$

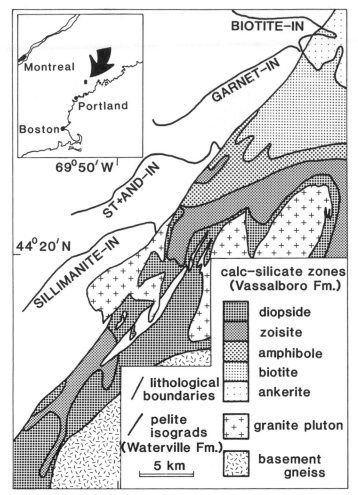

Fig. 5.10   Calc-silicate zones in the Vassalboro Fm. (north-east USA). After Ferry (1983b). Pelite isograds in the adjacent Waterville Fm. are also shown.

Within the biotite zone, albite is replaced by an intermediate plagioclase (oligoclase or labradorite) through the reaction:

$$\text{muscovite + calcite + chlorite + quartz + albite} \rightarrow$$
$$\text{biotite + plagioclase + H}_2\text{O + CO}_2 \qquad [5.13]$$

As a result, muscovite becomes rare and is often absent by the upper biotite zone.

Amphibole zone:
The appearance of calcic amphibole is accompanied by a further shift towards more calcic plagioclase compositions, and the characteristic assemblage is Ca-amphibole + quartz + Ca-plagioclase + calcite + biotite ± chlorite. The isograd reaction is inferred to be:

$$\text{chlorite + calcite + quartz + plagioclase} \rightarrow$$
$$\text{Ca-amphibole + Ca-plagioclase + H}_2\text{O + CO}_2 \qquad [5.14]$$

Zoisite zone:
Zoisite first appears rimming plagioclase at contacts with calcite grains, suggesting growth is due to the reaction:

$$\text{Ca-plagioclase} + \text{calcite} + H_2O \rightarrow \text{zoisite} + CO_2 \qquad [5.15]$$

Within the same zone, the assemblage K-feldspar + Ca-amphibole first appears, due to the same reaction (reaction [5.11]) that was studied by Carmichael. The appearance of K-feldspar and Ca-amphibole here is close to the sillimanite isograd in nearby pelites, a broadly similar situation to that obtaining in the southern half of Carmichael's area (Fig. 5.9). However, the very different pressures of metamorphism in the two areas preclude a closer comparison between them.

Diopside zone:
At the highest grades, the mineral assemblage is diopside + zoisite + Ca-amphibole + calcite + quartz + plagioclase ± biotite ± microcline. The growth of diopside results from breakdown of amphibole:

$$\text{Ca-amphibole} + \text{calcite} + \text{quartz} \rightarrow \text{diopside} + H_2O + CO_2 \qquad [5.16]$$

Ferry points out that, again, this sequence of mineral assemblages can only develop if the calcareous schists are continuously being infiltrated by H_2O. The reason for this is that zoisite + quartz is not stable in the presence of a CO_2-rich fluid, and at higher grades even small amounts of CO_2 will cause reaction [5.15] to proceed to the left. This is illustrated in Fig. 5.6. In other words, the release of CO_2 that accompanies the prograde metamorphism of these rocks should, if the fluid composition were internally buffered, lead to a relatively CO_2-rich fluid phase and preclude the possibility of zoisite being present at the higher grade zones. The growth of zoisite with increasing temperature can only be accounted for by infiltration of a water-rich fluid causing a shift in fluid composition similar to that indicated by the dotted arrows on Fig. 5.5b.

Chemical changes caused by fluid infiltration:
The calcareous layers of the Vassalboro Formation not only change mineralogically with increasing metamorphic grade, it is also clear that the metamorphism is not isochemical and the higher grade rocks are depleted in K and Na in particular (compare the assemblages of the ankerite zone, with albite and muscovite both abundant, with those of the higher grade zones which lack Na-minerals and from which potassic phases may also be absent). This depletion is also ascribed by Ferry (1983a, b) to the effects of infiltrating water. It is a phenomenon that may be quite widespread in calc-silicates.

SUMMARY AND DISCUSSION

The study of calcareous metasediments emphasises the importance of yet another possible variable in metamorphism: the composition of the fluid phase. Despite this additional variable, marbles in particular appear to have mineral assemblages that vary regularly with temperature (and to a lesser extent with pressure) in much the same way in many different metamorphic terranes. This can be accounted for if the fluid composition is in fact controlled by the mineralogy of the rock, and cannot therefore vary independently of pressure and temperature. On the other hand some calcareous rocks, notably calc-silicates, show evidence of fluid compositions varying under external influences, so that particular mineral assemblages can appear at different temperatures according to the local fluid composition.

The precise extent to which moving fluids externally control the fluid composition during metamorphism is currently controversial. The fluid flow might be taking place equally through all lithologies but only producing observable effects in the calcareous rocks; alternatively, it may be focused quite specifically into carbonate horizons. Decarbonation reactions lead to a much larger decrease in the volume of the solid phases than other metamorphic reactions, and this could cause an increase in the porosity and permeability of calcareous layers and hence help them act as channelways of fluid flow (Rumble *et al.*, 1982). A further possibility, however, is that some of the chemical changes in calc-silicates and their pore fluids may not in fact be caused by fluid flow at all. Simple diffusional exchange between calc-silicate layers and other rock types nearby could also account for the entry of water into these layers, and removal of alkalis. Diffusion along the network of grain boundaries or through a static fluid phase in pores is a much less efficient method of transferring matter than bulk flow of fluid through the rock, but since many calc-silicate layers are only a few centimetres in thickness, diffusion could nevertheless play an important role.

POSTSCRIPT

A final factor, which has not been considered in most studies of metamorphic fluids in calcareous rocks, is the extent to which CO_2 and H_2O are actually miscible. In the foregoing treatment of calcareous rocks it has been assumed that H_2O and CO_2 can mix to form a single fluid phase. In fact pure CO_2 and H_2O are miscible only at temperatures above about 275 °C, and so this assumption may not be valid at very low metamorphic grades. Furthermore, the miscibility gap between H_2O and CO_2 fluids is greatly extended if dissolved salts are present in the fluid (as is almost invariably the case). The possibility of immiscibility and its implications for metamorphic reactions in carbonates, has been considered recently by Sisson *et al.* (1981), Bowers & Helgeson (1983) and Skippen & Trommsdorf (1986). Clearly, if an additional fluid phase is present in carbonate rocks, then from the phase rule the number of degrees of freedom will be reduced by one, and, for example, divariant equilibria will become univariant. To date, there is little evidence for the widespread occurrence of immiscible fluids in greenschist facies rocks and at higher grades, but some examples have been reported. Trommsdorf, Skippen & Ulmer (1985) found immiscible H_2O–$NaCl$ and CO_2 fluids in veins in marble at moderately high grades, within the amphibolite facies.

6 METAMORPHIC TEXTURES AND PROCESSES

In Chapters 2 to 5 the emphasis has been on the attainment of chemical equilibrium in metamorphism, because it is only by identifying assemblages of minerals that have coexisted together in equilibrium that the $P-T$ conditions of metamorphism can be determined. This is one of the major objectives of metamorphic petrology, and has been especially emphasised in recent years, in response to new developments in experimental and theoretical petrology and the tremendous opportunities for studying natural rocks brought about by the development of the electron microprobe. However, equilibrium studies tell us only about the $P-T$ conditions prevailing when a particular assemblage formed, they cannot tell us anything about the rock's history before or after the determined $P-T$ conditions were attained.

The study of the textures of metamorphic rocks provides a complementary line of evidence about the events to which the rock was subjected. Textures are very important to the study of metamorphism because they often indicate deviations from equilibrium which allow us to see the way in which a rock was recrystallising towards an equilibrium assemblage. In this way something of the metamorphic history of the rock may be inferred.

Metamorphic textures may be broadly divided into two types: those that preserve information about the metamorphic reactions that have taken place; and those that are related to deformation that occurred during metamorphism. The study of the first type reveals something of the sequence of assemblages, and hence history of metamorphic conditions, while the study of the second type tells us about the history of deformation and the relative chronology of deformation and metamorphic mineral growth in a particular region. These are ideal aims of course, and in practice some rock types or individual specimens preserve far more historical information than others.

METAMORPHIC TEXTURES – THE UNDERLYING PRINCIPLES

Whereas the study of equilibrium mineral assemblages is based on the properties of minerals as perfect crystals, textural studies require an appreciation of the properties of ordinary, imperfect materials. To understand the growth and breakdown of crystals during metamorphism it is necessary to take into account such factors as the effects of mineral surfaces on the way in which new grains grow, or to consider precisely how atoms may move through the bulk of a rock, or through its constituent mineral grains. Such studies are of great importance in materials science, but until recently their application to geology has been largely restricted to the field of structural geology.

Energies of crystal surfaces

Atoms at or near the surfaces of a crystal are not bonded in such a stable way as those in the interior; they have unsatisfied bonds and as a result there is an excess energy associated with them, known as **surface energy**. Surface energy can be defined as the energy needed to increase a surface by unit area, and is therefore measured in units of J/m^2. The presence of this extra energy results in a tendency for foreign atoms to be attracted to mineral surfaces and loosely bound to them or **adsorbed**.

The magnitude of the energy associated with a surface depends on the nature of the substance on each side. For example, a surface between a mineral and air may have a higher surface energy than one between the same mineral and water. The addition of a 'wetting agent' such as detergent to water results in a reduction in the surface energy of the interface between solids and water, so that the water is able to spread more thinly, i.e. with a larger area of interface. Surface energies usually make only a very small contribution to the total Gibbs free energy of a mineral in a rock, and so they are unlikely to affect the $P-T$ conditions for equilibrium between mineral assemblages. However, when grains are extremely small a high proportion of their constituent atoms occur near surfaces, and the surface free energy becomes significant. A very fine grained mineral assemblage will have a larger free energy than a coarse grained assemblage, and this could affect the temperature and pressure at which a reaction takes place. For a sphere of quartz of 1 cm radius the surface free energy $\sigma = \sim 6.3 \times 10^{-4}$ J; if the same mass of quartz is broken up into spheres of radius 10^{-3} cm, then $\sigma = \sim 0.63$ J, while if it is reduced to spheres of 10^{-7} cm radius, $\sigma = \sim 6280$ J and surface energy will make a significant contribution to the total free energy of the quartz.

Defects in crystals

The mineral lattice may contain imperfections or be disrupted within individual crystals as well as at their surfaces. Indeed if defects were not present, crystals would be very much stronger than is actually the case.

Crystal defects can be grouped into three classes: point defects, line defects and surface imperfections.

Point defects are usually centred around a single lattice site or a pair of sites and may be due to: (a) an empty site or **vacancy**; (b) **substitution** of an impurity atom in place of the species that would normally occupy the site; or (c) **insertion** of an extra atom into a hole in the structure that would not normally be occupied.

Line defects or **dislocations** are one of the most important types of defect and play a major role in facilitating the deformation of grains. Dislocations can be visualised as resulting from the slip of one block of the lattice past another along a plane. Imagine a prism made up of a single crystal of a mineral with a simple cubic lattice as shown in Fig. 6.1(*a*). A cut is made along its length from the outside to the centre and represents the **slip plane**. If the lattice on one side of the slip plane is pushed up by exactly one unit cell dimension there will be no disruption across the slip plane; however, the ends of the prism are now spiral ramps instead of planar surfaces. Near the centre of the prism, where the slip plane comes to an end, the lattice must be somewhat distorted and this zone of distortion extends along the entire length of the prism, hence the classification as a line defect.

The type of dislocation shown in Fig. 6.1(*a*) is known as a **screw dislocation**. An alternative type of dislocation is illustrated in Fig. 6.1(*b*), whereby an extra plane of atoms is inserted in the right-hand half of the block which is absent from the left. The lattice is distorted along the edge of this extra 'half plane' and this constitutes an **edge dislocation**. Real dislocations can be intermediate between these two extreme types and can change character and orientation along their length. For example it can be seen that the insertion of an extra horizontal plane of atoms in the left half of the block shown in Fig. 6.1(*a*) would

a)

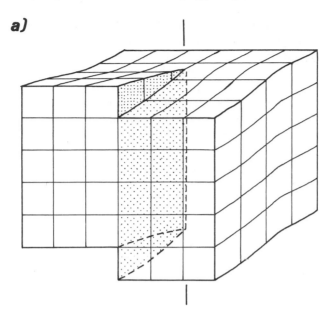

b)

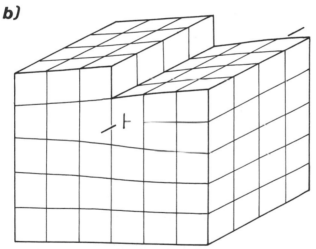

Fig. 6.1 Schematic representations of dislocation types in a material with an idealised cubic lattice. *a)* Screw dislocation passing throughout the length of the block. The slip plane is stippled, but note that distortion is restricted to the immediate vicinity of the line of the dislocation, which has produced spiral ramps in the top and bottom surfaces. *b)* Edge dislocation. An extra horizontal plane of atoms is present in the right half of the block.

terminate the slip plane, allowing lower planes of atoms to continue across the block uninterrupted. This would give rise to a slip plane terminated by a screw dislocation along one edge and an edge dislocation along the other.

The movement of dislocations through a crystal causes it to become distorted, or undergo **permanent strain**. Progressive movement of dislocations through a crystal allows it to deform while only breaking a small number of atomic bonds at a time, and this is why real crystals that contain dislocations are very much weaker than would be the case for perfect crystals.

Surface imperfections include any change in the orientation, spacing or composition of atom planes across a planar boundary. The most extreme type is the **grain boundary**, which separates crystals with unrelated lattice orientation, and often different chemistry. The ordinary grain boundaries in rocks are sometimes known as high angle grain boundaries. The precise nature of grain boundaries in metamorphic rocks has been the subject of speculation for many years, but recent studies using high resolution transmission electron microscopy are beginning to provide a picture of grain boundary structure. According to White and White (1981), the mineral lattices adjacent to a grain boundary are somewhat distorted and contain impurities. They found that the plane of the boundary appears to contain many small voids, while tubular holes lie along the junction of two or more grain boundaries and probably give rise to an interconnecting porosity throughout the rock.

Some boundaries between grains of the same mineral species involve only a slight atomic mismatch and are known as **low-angle** or **tilt boundaries**. They are most often seen in quartz and olivine in thin section, and appear when a large and apparently uniform grain is viewed in cross-polars and rotated to extinction. The large grain may then prove to be made up of a mosaic of smaller grains, or **sub-grains**, each of uniform but slightly different extinction position and separated by low-angle boundaries.

Other types of surface imperfections include **twin boundaries**, which may form as a crystal grows or be produced by deformation of an originally uniform lattice, and **stacking faults**, which result from individual planes of atoms being omitted or repeated out of turn in the mineral structure. In silicate minerals, which typically have large unit cells, crystal defects may be made up of a combination of different elements. For example Smith (1985) reported that many defects in garnet are a combination of a 'partial dislocation', i.e. one on which the amount of displacement across the slip plane is not exactly one unit cell, so that the lattice does not match across it, and stacking faults, whereby the continuity of the lattice is restored by omission or duplication of specific lattice planes on one side of the slip plane.

DIFFUSION IN SOLIDS

Diffusion is the process by which atoms, ions or molecules are transported through matter. Even in a crystalline solid, where atoms are strongly bonded, the continuous thermal vibrations mean that individual atoms are in motion, exchanging positions within the crystal. These random motions within a chemically homogeneous crystal may be detected in experiments by the use of a distinctive isotope of one of the constituent elements as a tracer, and are known as **self-diffusion**. Where compositional gradients exist within a crystal, for example in a zoned grain of a solid solution mineral, there will be a tendency for diffusion to occur to make the grain homogeneous. In the case of a zoned olivine crystal for example, Fe will tend to diffuse away from the fayalite-rich portions towards the parts enriched in forsterite, while Mg will tend to diffuse in the opposite direction. Indeed the opposing movements must be exactly balanced to prevent the local development of charge imbalances. This type of diffusion is known as **interdiffusion**, and both self-diffusion and interdiffusion are examples of **volume diffusion** because they involve movement of atoms through the bulk of a crystal, rather than around its margins. Since most of the atomic bonds in crystal lattices are strong, volume diffusion is often sluggish compared with the diffusion of material through a rock along grain boundaries, known as **grain boundary diffusion**, but becomes dominant as crystals approach their melting temperatures.

The simplest way of looking at interdiffusion is to imagine it as a response to concentration gradients within the crystal. The rate at which diffusion takes place is measured in terms of the **flux** of matter, $\mathcal{J}$. Flux means the mass of material diffusing across a unit area of an imaginary surface within the crystal in unit time. The flux is proportional to the concentration gradient that is driving the diffusion, so if we represent concentration by C and the direction in which C changes is designated x we have:

$$\mathcal{J} \propto dC / dx$$

If we designate a constant of proportionality, D, then:

$$\mathcal{J} = -D. (dC / dx)$$

The constant D is known as the **diffusion coefficient** and the second equation represents Fick's first law of diffusion. In many instances where diffusion occurs, the concentration gradient dC / dx changes as diffusion proceeds, and many equations have been developed to express this and measure changes in composition with time in bodies of different shapes. A more thorough treatment of the subject is beyond the scope of this book, but further references are given at the end of this chapter, and a similar problem involving diffusion of heat through rocks is outlined below on page 178.

The amount of material that can be moved by diffusion depends on the time available and also on the size of the diffusion coefficient D. In minerals D is very strongly temperature dependent and typically varies exponentially with $1/T$. As a result it is possible for diffusion within a crystal to be negligible at one temperature, but very rapid, in geological terms, if the temperature is raised by only 50 or 100 degrees. For example garnet crystals in pelites from the garnet and staurolite zones typically display strong chemical zonation whereas those from migmatites are more or less unzoned. One reason for this is that diffusion in garnet apparently becomes effective at the temperatures of the higher grade metamorphism (Yardley, 1977a).

Diffusion in fluids is very much more rapid than through minerals, and the presence of water in grain boundaries is believed to greatly enhance diffusion rates through rocks. Water present in defects in crystal lattices may also speed up volume diffusion in some minerals, although in others, diffusion rates are apparently unaffected by the presence and pressure of water in experiments.

The occurrence of diffusion must be seen as clear evidence for deviations from chemical equilibrium. Although it is most readily envisaged as a process that takes place in response to concentration gradients, it is more strictly correct to consider it as a response to gradients in chemical potential. The two concepts are very similar as long as we only consider diffusion within a single grain, but where diffusion is occurring in more complex situations, for example between several different minerals reacting together, it is no longer a simple response to concentration gradients (e.g. see Joesten, 1974; Fisher, 1977).

NUCLEATION AND GROWTH OF MINERAL GRAINS

Nucleation Having explored some of the properties of real crystals, it is possible to analyse what actually happens when a mineral grows. It is simplest in the first instance to consider a reaction that involves only breakdown of one reactant phase and growth of one product, but the same arguments apply to reactions involving assemblages of minerals. If the reactant R is heated to a temperature T_r so that the product P is more stable, then we can define the drop in the Gibbs free energy of the system as ΔG_v per unit volume of P that is produced. Clearly the value of ΔG_v depends on the amount by which the equilibrium temperature has been overstepped (Fig. 6.2).

With the exception of certain types of polymorphic transitions, any reaction requires chemical bonds in the original phase to be broken and the constituent atoms to be

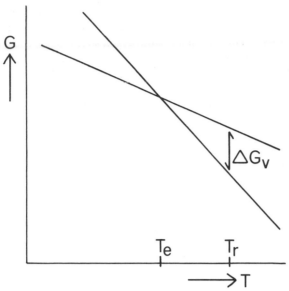

Fig. 6.2   Schematic illustration of the variation in the free energy change of a reaction (here defined as ΔG_v the change in Gibbs free energy per unit volume of product generated) with the extent to which the equilibrium temperature (T_e) has been overstepped (cf. Fig. 2.2).

rearranged to produce a new structure. Suppose that at temperature T_r, random vibrations in the lattice of phase R result in the formation of a small volume of material or **nucleus** with the structure of phase P. The energy of the system is reduced by an amount $V_n.\Delta G_v$, where V_n is the volume of the nucleus. However, because of its small size, most of the atoms in the nucleus are close to its surfaces and so it has a high surface energy which is equal to $A_n.\sigma$, where A_n is the surface area of the nucleus and σ the surface energy. Provided that:

$$-V_n.\Delta G_v > A_n.\sigma$$

the total free energy of the system has been reduced by the formation of the nucleus of P and so it will be stable. If, however, the extra free energy associated with the newly created surfaces of the nucleus is greater than the reduction in energy due to the conversion of phase R into phase P, the nucleus will be unstable and will spontaneously break down again. In this case its formation led to an increase in the total free energy of the system. As the equilibrium conditions for coexistence of R and P are progressively overstepped, so ΔG_v becomes larger, and the nucleus is more likely to be stable. This is why reactions that involve the formation of a new crystalline phase (and hence involve nucleation) do not take place at the equilibrium temperature but only when it has been overstepped, whereas reactions that produce only phases with no definite crystal structure, such as congruent melting, can take place at almost exactly the equilibrium temperature.

The formation of a stable nucleus of a product mineral will be favoured by the following factors:

1. Increased overstepping of the equilibrium conditions for the reaction concerned, resulting in an increase in the value of ΔG_v.
2. Reduction in surface energy σ, which may be achieved by growth of the nucleus on particular mineral substrates.
3. Increased size of the nucleus, so that its surface energy contribution becomes less significant.

The ideal case of nucleation, for which the nucleus is spherical, has a uniform surface energy everywhere and forms independently of other crystalline material, is known as **homogeneous nucleation**. It may sometimes be a reasonably good model for the crystallisation of certain phases from melts, but in metamorphism there will always be a wide variety of crystalline substrates on which a nucleus may develop, and so nucleation is always **heterogeneous**. In other words the nucleus forms on a mineral substrate, and therefore has a different surface energy on different faces. The effect of this is to make nucleation easier, especially on mineral substrates whose lattices match that of the nucleating phase.

Growth   Once a stable nucleus has formed, there is a chemical potential gradient between the reactant grains and the product nucleus, and this will drive diffusion of material towards the newly formed grains of the product minerals, causing them to grow at the expense of the reactants. In regions of the rock that are too far away from the initial nuclei for material to diffuse from them, additional nuclei of the product grains are likely to form (Fig. 6.3). It is clear that the formation of a grain of a new mineral species involves two distinct steps: an initial nucleation episode, followed by a period of growth. In order for nucleation to occur, the equilibrium conditions for the reaction must be overstepped by a significant, but generally unknown, amount. However, once nuclei are present, growth may take place even if the amount of overstepping is relatively small, i.e. less than was required for the initial nucleation. The rates at which the nucleation and growth steps proceed depend on different factors. In some cases many nuclei form but none of them grows into large grains; this might be the case where nucleation is facilitated by the presence of an existing mineral in the rock whose lattice provides a particularly favourable substrate for the new mineral to nucleate on, or where diffusion is especially sluggish, e.g. because the rock contains no

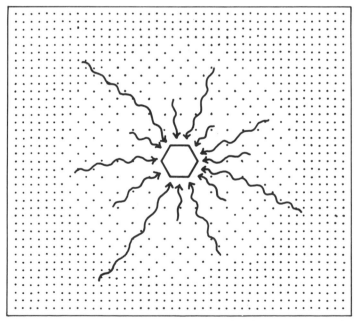

Fig. 6.3   Schematic representation of the diffusion of material towards a newly formed nucleus. Close to the nucleus the rock has begun to equilibrate with it, and so further nucleations are unlikely, but further away, little material has begun to move towards the nucleus and new nuclei may still form here. This stage in the reaction is said to be transport controlled (Fisher, 1978); its importance is discussed further below (page 184).

fluid. Less commonly in metamorphism, abundant nuclei may form where a reaction has been overstepped by a large amount so that ΔG is large. The opposite conditions: easy diffusion, difficult to nucleate product minerals, will tend to result in a small number of relatively large grains (i.e. porphyroblasts.)

THE TEXTURES OF METAMORPHIC ROCKS

FACTORS CONTROLLING THE WAY IN WHICH GRAINS GROW

Some mineral grains form during metamorphism because a metamorphic reaction is proceeding and is producing a new mineral or increasing the amount of an existing mineral in the rock. In this case they may be said to be produced by **crystallisation** of the mineral. In other cases, however, grains form entirely at the expense of pre-existing grains of the same mineral, and this process is called **recrystallisation**. Recrystallisation can proceed independently of any conventional metamorphic reactions, it is analogous to the process of annealing in synthetic materials.

TEXTURES OF RECRYSTALLISATION

Recrystallisation can be driven either by the difference in energy between large grains and small grains, which favours the growth of large grains at the expense of smaller ones, or by strain energy in older, deformed grains, which makes them less stable than newly formed grains that have not been deformed. Similarly, grains that have an irregular shape or are riddled with inclusions have an anomalously high surface area for their volume and are also susceptible to recrystallisation.

For a rock in which one mineral predominates, such as quartzite or marble, large grains can grow very easily at the expense of small ones because atoms need only move across the grain boundary, without appreciable transport (Fig. 6.4). This permits

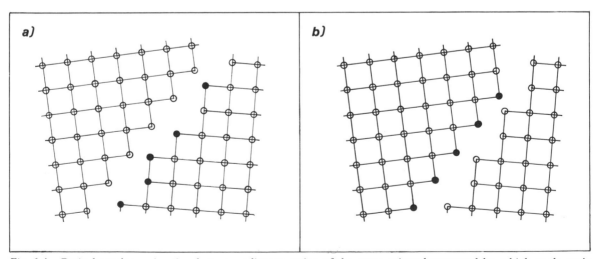

Fig. 6.4   Grain boundary migration between adjacent grains of the same mineral separated by a high angle grain boundary. *a*) The atoms represented by solid circles are part of the right hand grain. *b*) The marked atoms have now become part of the left hand grain with only minimal movement while the grain boundary has moved through the material as a result.

recrystallisation by the process of **grain growth**, driven by the free energy differences between large and small grains even though these are very small and would not be sufficiently large to drive effective diffusion. The process can be envisaged as one in which the grain boundaries move through the rock while the atoms remain more or less stationary but break their bonds to one crystal surface and are reattached to an adjacent one. Robinson (1971) was able to demonstrate that this process had taken place during contact metamorphism of a sequence of limestones, because the pure layers had recrystallised to a coarse equigranular marble, while graphitic limestone layers remained quite fine grained because the presence of graphite impurities along the grain boundaries had apparently inhibited movement of atoms across them.

The minimum surface energy configuration for a given grain size is ideally attained when all grains are of the same size and have planar boundaries intersecting at approximately 120°. This is known as a **granoblastic polygonal** texture and is illustrated in Fig. 6.5(*a*). Because surface energies are usually small, this texture is seldom attained in silicate rocks except at very high grades of metamorphism in the granulite facies, but it is common in marbles; once formed it is very stable.

An essential feature of granoblastic polygonal texture is that none of the individual grains can have a well-developed crystal form. However, some minerals such as garnet or amphiboles show a strong tendency to form **idioblastic grains** (i.e. with well-developed crystal faces). This may be because the rational crystal faces have a much lower surface energy than surfaces in any other orientation, or because molecules add on to the growing grain on certain surfaces selectively, resulting in the formation of rational faces. Rocks very rich in amphibole or mica seldom develop a true granoblastic polygonal texture because the grains are usually elongate, but in the absence of deformation they may recrystallise to an interlocking network of elongate or platy grains aligned in all directions and bounded by rational crystal faces. This is known as **decussate texture** (Fig. 6.5(*b*)).

Recrystallisation in response to strain does not produce such regular textures and the driving forces may be very much larger. When a crystal is deformed or strained many defects are created in the lattice, notably dislocations, and the distortion around these defects increases the free energy of the crystal. Strained grains of some minerals, such as quartz, can be recognised in thin section by their **undulose extinction** (Fig. 6.5(*d*)); as the stage is rotated different parts of the grain go into extinction in slightly different positions because the lattice has been bent. Sometimes a strained crystal will release much of its strain energy by recrystallising into sub-grains, whereby the dislocations move and align into low-angle grain boundaries that separate unstrained sub-grains, each with a slightly different lattice orientation. Release of strain energy by movement of dislocations through the crystal is known as **recovery**, and can only take place if the temperature is sufficiently high for a limited amount of volume diffusion to occur. This is because some diffusion must take place for dislocations to move through the lattice. At high temperatures, diffusion in minerals such as quartz may be so fast that they can recover as fast as they are deformed and older grains are continuously replaced by a fine mosaic of new, unstrained grains as the deformation proceeds, a process known as **syntectonic recrystallisation** (Fig. 6.5(*c*)). In other cases, new grains are able to nucleate at the margins of the deformed grains and grow at their expense (Fig. 6.5(*d*)).

In monomineralic rocks or veins, strained grains are sometimes observed to have grown into one another, developing a sutured grain boundary (Fig. 6.5(*d*)). Since the surface area of this texture is very large it is clear that strain energy contributions to the free energy of grains can be much larger than surface energy contributions. Thus the energy saved in recovery was greater than the additional surface energy created. Another

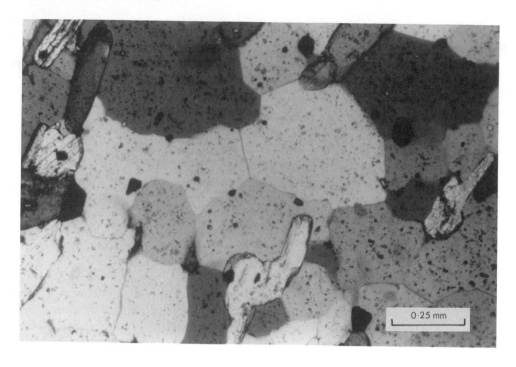

0·25 mm

Fig. 6.5   Photomicrographs of metamorphic textures described in the text. *a*) Granoblastic polygonal texture of quartz in kyanite quartzite. Kofi Mountain, north-east Tanzania, courtesy of R.A.
Cliff. *b*) Decussate texture of interlocking micas (primarily muscovite). The high relief grain
(arrowed) is a relic inclusion of staurolite in muscovite. Pelitic schist from the sillimanite-muscovite
zone, Connemara, Ireland.

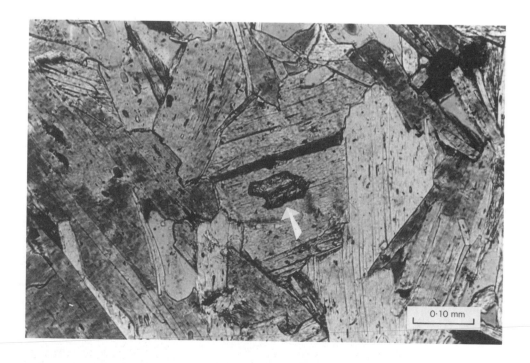

0·10 mm

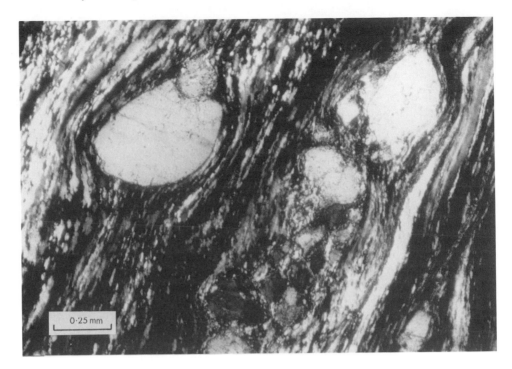

Fig. 6.5 (continued)

c) Porphyroclasts of feldspar in a matrix of fine grained, syntectonically recrystallized quartz displaying ribbon texture. Note the rounded, abraded margins of the porphyroclasts and the incipient recrystallisation of some feldspar in the lower central part of the photograph. Skagit Gneiss, Washington, USA. Photo R.J. Knipe. *d*) Deformed quartzite with large deformed grains displaying strained extinction and sutured grain boundaries (arrowed) due to strain-induced grain boundary migration. These older grains are set in a matrix of finer grained recrystallized quartz approaching a granoblastic polygonal texture. Photo R.J. Knipe.

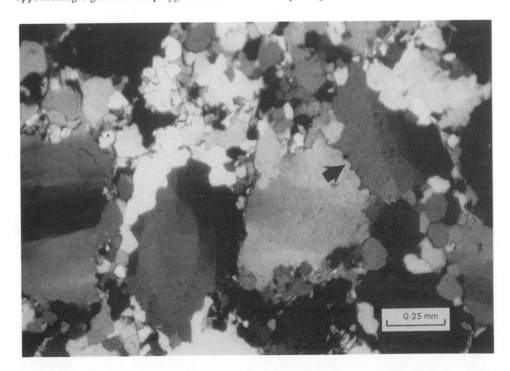

consequence of the relatively large strain energy contributions is that recovery of deformed grains may result in unstrained grains of different sizes, that may then undergo an episode of grain growth to minimise surface energy. Bouchez and Pecher (1981) have published a series of photomicrographs that illustrate the progressive deformation of quartz-rich rocks with the production in the first instance of textures of syntectonic recrystallisation, to eliminate strain energy, and their subsequent replacement by textures produced by grain growth.

Mylonites   Recrystallisation in response to strain is particularly well displayed in mylonites. These are relatively fine grained rocks produced as a result of grain size reduction in zones of intense deformation such as shear zones. This may be achieved by brittle cracking of grains (cataclasis), but under most metamorphic conditions grain size is often reduced by plastic deformation accompanied by syntectonic recrystallisation (p. 155). Different minerals show very different responses to deformation even at the same $P-T$ conditions. Except at very low grades, quartz deforms readily and undergoes syntectonic recrystallisation, ultimately producing 'ribbon texture' (Fig. 6.5(c)). Carbonates, and, at very high grades, olivines also commonly undergo syntectonic recrystallisation. Feldspar and garnet are relatively rigid and are likely to undergo brittle failure, although ductile deformation and syntectonic recrystallisation can occur in feldspar. Sheet silicates commonly deform by kinking.

Mylonite textures therefore depend on both mineralogy and the amount of strain, as well as on $P-T$ conditions. Often, mylonites display a **mortar texture** with a matrix, or mortar, of fine, syntectonically recrystallised material enclosing larger, fractured and strained relics of pre-existing grains of resistant minerals such as feldspar or garnet. These relics are known as **porphyroclasts** (Fig. 6.5(c)). In **protomylonites** the porphyroclasts are still the dominant constituents of the rock, while in **ultramylonites** they make up less than 10 per cent. If the porphyroclasts have themselves undergone syntectonic recrystallisation the rock is a **blastomylonite**. Phyllosilicate-rich rocks produce more markedly platy mylonite known as **phyllonite**.

TEXTURES OF CRYSTALLISATION

In general, the products of metamorphic reactions can only begin to grow once a stable nucleus has formed, although cation exchange reactions and some continuous reactions provide exceptions to this rule because they do not lead to the appearance of a new solid phase. The textures produced will therefore reflect both the nucleation and growth characteristics of the minerals involved.

Influence of nucleation characteristics   In the simplest case, reaction products need not form a new nucleus on which to grow because the same mineral is already present in the rock, having been produced by previous reactions. In this instance the new material may form a distinct overgrowth or rim to the pre-existing grains of the same phase, or new and old material may recrystallise together to form new grains. In some instances overgrowths can be readily identified (Fig. 6.6(a)), but in others their presence is much more equivocal.

It is apparent from the discussions above of surface energy and nucleation, that nuclei will form most readily on the particular substrate and in the particular orientation that minimises the surface energy of the interface. In some instances there may be sufficient similarity between the structures of new and pre-existing phases for the new mineral to grow by replacing certain atoms only and leaving much of the structure of the substrate intact, thereby minimising the surface energy at the interface. This type of replacement is known as **topotaxy**, and common examples include the replacement of biotite by chlorite or intermediate plagioclase by albite. Even where the new phase has a structure that does

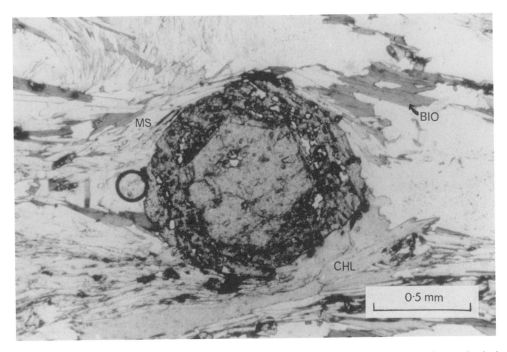

Fig. 6.6 a) Photomicrograph of garnet zone pelitic schist in which an inclusion-free euhedral garnet core has been overgrown by a distinct rim of inclusion-rich garnet. Morar, Invernesshire, Scotland. (The dark circle at the left of the porphyroblast is an air bubble!) Courtesy of R.A. Cliff. *b*) Complex intergrowth of biotite and fibrolite replacing an original garnet and displaying epitaxial growth of fibrolitic sillimanite on a biotite substrate. Conncmara, Ireland. From Yardley (1977b).

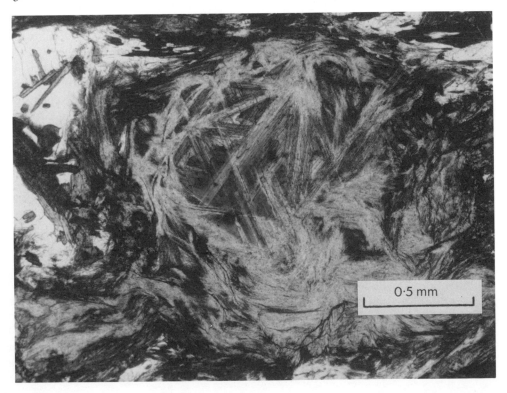

not closely match that of any of the existing phases, it may be that there are certain orientations where the surface energy is minimised. For example Chinner (1961) described how sillimanite preferentially nucleates on biotite in certain orientations (Fig. 6.6(*b*)). This type of crystallographically controlled preferential nucleation gives rise to **epitaxial growth**, i.e. in a particular orientation on a particular substrate. Kyanite and staurolite can also form epitaxial overgrowths on one another.

In many cases, however, it is impossible to determine reliably on what specific site a particular grain originally nucleated. Many minerals appear to nucleate more or less at random throughout the rock. This has been demonstrated by Kretz (1966, 1973) who has carried out careful studies of the three-dimensional distribution of metamorphic minerals within rock samples.

TEXTURES REFLECTING THE INTERACTION OF NUCLEATION, GROWTH AND DIFFUSION CHARACTERISTICS

Many of the different ways in which metamorphic minerals grow can be explained qualitatively in terms of differences in their nucleation and growth characteristics, although we do not understand these sufficiently well to attempt quantitative description of metamorphic textures at the present time.

Relative rates　　Whether a metamorphic mineral grows as a porphyroblast or forms matrix grains is dependent in large part on the relative rates of nucleation and growth, as we have seen above. Some minerals tend to form much the same types of textures in a wide range of rock types and metamorphic environments, which suggests that a particular characteristic is so pronounced as always to dominate. For example andalusite occurs almost invariably as porphyroblasts, whereas sillimanite usually forms a large number of very small grains, except in some high temperature granulites. One possible explanation for these observations is that nucleation of andalusite is always a difficult step, so that further growth will occur on the first-formed nuclei, whereas nucleation of sillimanite is presumably a relatively rapid step, and it is often easier for sillimanite to grow as new nuclei rather than add on to existing grains. It has been recognised for many years that certain minerals, such as andalusite, garnet, staurolite and kyanite typically occur as porphyroblasts, whereas others, such as muscovite, quartz and feldspar, usually do not; this is presumably related to the relative ease with which nuclei of the different phases can form in common rock types. Nevertheless, it is important to be aware that there are many exceptions to such simple rules; biotite, chlorite and feldspar are all typically matrix minerals but can form porphyroblasts in some instances, likewise hornblende, sillimanite and lawsonite are common examples of minerals that may occur in either manner. In many high grade rocks that are uniformly quite coarse grained the 'porphyroblast' minerals such as garnet often prove to be much the same size as matrix minerals.

Crystal growth and diffusion in the rock matrix　　The addition of material to a stable nucleus involves three steps: dissolution of the reactant grains; diffusion to the surface of the growing grain; and transfer of atoms on to the crystal surface. These steps also appear to play an important role in the development of metamorphic textures, and especially in determining the density of inclusions trapped in a mineral, i.e. whether it grows as a porphyroblast or a poikiloblast. Porphyroblasts without inclusions are more stable than poikiloblasts because they have a smaller surface area and hence lower free energy. However, poikiloblasts may be able to grow more rapidly because the greater area of surface means that atoms can add on to the growing grain at a greater rate.

In addition, the amount of mass transfer that is involved in poikiloblast growth is likely to be less than for growth of a porphyroblast with few inclusions. To envisage this, imagine two grains of the same mineral in two rocks of the same composition, one grain occurs as a poikiloblast, the other as an inclusion-free porphyroblast. If we draw an imaginary line around the porphyroblast then the composition of the region that we enclose has precisely the chemical composition of the porphyroblast mineral itself, whereas if we draw a comparable line around the boundary of the poikiloblast the composition of the enclosed region is intermediate between that of the poikiloblast mineral and that of the inclusions within it. Since the bulk composition of the rock is the sum of the composition of its constituent minerals, it is apparent that the composition of the region within the boundaries of the poikiloblast will be closer to the average rock composition, and therefore less mass transfer is involved in the growth of the poikiloblast texture than for the porphyroblast. Mass transfer by diffusion is of course a time-dependent process and so this observation suggests that relatively rapid growth of a mineral is likely to cause it to form a poikiloblast, rather than a porphyroblast. This treatment is simplistic because it assumes that many other factors are equal, for example the grain size of the rock matrix. This is significant because diffusion takes place much more rapidly along grain boundaries than through mineral lattices, and a fincr grained rock therefore has a larger number of possible pathways for diffusion. Perhaps this is why large porphyroblasts are not uncommon in hornfelses, despite the fact that they presumably grew relatively quickly, because such rocks often have a much finer matrix grain size than regionally metamorphosed rocks.

Anisotropic growth Some anisotropic minerals appear to grow more rapidly on some faces than on others, and this influences the final shape of the grain. For example if the rate of growth of prism faces is slower than that of the faces that terminate the crystal form, then the mineral will grow into elongate prismatic or **acicular** grains. (It is possible to demonstrate this phenomenon readily in the laboratory by melting a small quantity of para-dichlorobenzene (moth balls) on a microscope slide and watching it crystallise under a petrological microscope set up with crossed polarisers.) Anisotropic growth is especially important in determining the textures of metamorphic rocks at low grades, where minerals such as amphiboles or pumpellyite are frequently acicular and may occur as bundles of radiating crystals. At higher grades most minerals form more equidimensional grains.

DISEQUILIBRIUM TEXTURES

The textures of many metamorphic rocks preserve evidence of deviations from equilibrium that are of great importance in interpreting the metamorphic history of the rock. These may take the form of the preservation (e.g. in the cores of porphyroblasts) of minerals that were formerly present throughout the rock but were no longer stable during the main period of metamorphic crystallisation. Alternatively, textures may preserve evidence of incomplete reaction between minerals. Note, however, that incomplete reaction does not always indicate disequilibrium because many reactions are continuous, and even if equilibrium is closely maintained they will not go to completion unless the $P-T$ conditions continue to change until one reactant is entirely consumed.

CHEMICAL ZONATION IN MINERALS

Solid solution minerals often form grains that vary in composition between the core and rim. For some minerals, such as tourmaline or amphibole, chemical zonation is accompanied

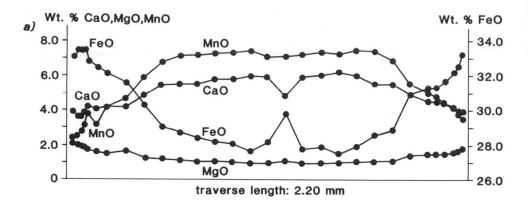

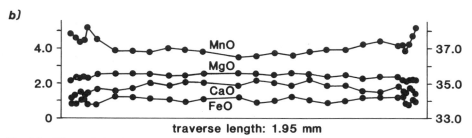

Fig. 6.7   The variation in composition across typical regional metamorphic pyralspite garnets from pelitic rocks, as determined by electron microprobe. Traverses are shown across two crystals and in each case the data for all four major divalent cations is shown. Spots represent individual analyses. *a)* Typical lower amphibolite facies garnet from a pelitic schist in the upper staurolite zone, Connemara, Ireland. Note the 'bell-shaped' profile with core enriched in Mn and Ca, and rim enriched in Mg and Fe. The anomalous points in the core occur at the edge of an inclusion. *b)* Garnet from the sillimanite–K-feldspar zone, in the same region and from the same stratigraphic horizon as *a*). This profile is typical of higher grade garnets. Any primary zoning has been eliminated by diffusion, and the fluctuations at the edge of the crystal are 'retrograde effects', produced by limited diffusion at the crystal edge as the rocks slowly cooled. After Yardley (1977a).

by colour differences that are readily observed under the microscope, but in other minerals, especially garnet, zoning can only be detected by careful analysis of a series of points across the crystal using an electron microprobe analyser (see Fig. 6.7).

    Zoning may form in a variety of ways, but the fact that it is present means that at best the zoned grain has only partially equilibrated with the other minerals in the rock.

Growth zoning   This is zoning that develops as a crystal grows, most typically during progressive metamorphism, and hence if it is preserved today we can conclude that, even at the peak of metamorphism, volume diffusion in the particular mineral was too sluggish to eliminate the original compositional gradients. When zoning in garnet was first discovered with the advent of electron microprobe studies in the mid-1960s, one of the features found most frequently was a high concentration of Mn in the cores of garnet grains from amphibolite facies pelitic schists (Fig. 6.7(*a*)). This was explained by Atherton and Edmonds (1966) and Hollister (1966) in terms of the fractionation of Mn into garnet when it first appeared. Hollister showed that the process was analogous to that of Rayleigh fractionation, a model originally developed to describe the condensation of liquid droplets from a multicomponent vapour, but since widely applied to problems involving the fractionation of trace constituents into a particular phase growing in a system. The basis for Hollister's treatment was two-fold. Firstly, that the distribution coefficient (K_D) for distribution of Mn between garnet and each of the other Fe–Mg–Mn minerals that occur in pelitic schists is invariably large, which means that most of the Mn in the rock occurs in garnet. Secondly, diffusion in garnet at medium grades appears to be negligible, so that the central parts of the grain are isolated from subsequent processes. Clearly for a rock containing a given amount of Mn, the less garnet there is in the rock, the higher must be the Mn-content of those garnets that are present. In other words, the small amounts of garnet that first appear, scavenge Mn from the rest of the rock and are Mn-rich. As garnet growth continues, the Mn-content of the reactant phases has become much lower, and therefore even if K_D remains constant, the Mn-content of the later-formed parts of the garnets will also be lower as they cannot re-equilibrate with the Mn-rich cores.

The Rayleigh fractionation model can explain the formation of zoned garnets by reaction at constant temperature, however garnet is often produced by continuous reactions taking place over a range of P–T conditions, or by several successive and distinct reactions. In these cases zoning will also be produced, because the K_D values for Fe–Mg and Fe–Mn exchange between garnet and other minerals present are of course temperature sensitive (page 56).

Retrograde zoning   Zoning patterns may also be influenced by volume diffusion in minerals. For example, it was noted above that garnets from very high grades (upper sillimanite zone and above) are typically more or less homogeneous because they form at temperatures at which volume diffusion in garnet appears to be effective. However, the rims of high grade garnets often exhibit a certain amount of zoning which is believed to form as a result of diffusive exchange with matrix minerals during cooling, and is therefore known as **retrograde zoning** (Grant and Weiblen, 1971) (Fig. 6.7(*b*)). Loomis (1983) provides a valuable review of the many and complex factors that control the development of zoned crystals.

The nature of the chemical zoning in minerals such as garnet or amphibole provides a guide to the way in which P–T conditions changed as the zoned porphyroblast grew. For example many authors have described amphibole grains in which a core of sodic amphibole is rimmed by a margin of calcic amphibole. This implies that metamorphism progressed from blueschist (or transitional blueschist/greenschist) facies conditions to greenschist facies conditions. Interestingly, the reverse zoning pattern has been described from other areas.

RELIC MINERALS

Chemical zoning preserves remnants of mineral compositions that were formerly present in the rock by surrounding them with a shell of essentially the same mineral through which volume diffusion is ineffective. Core and rim are structurally continuous despite the difference in composition. Armoured relics (or simply relics) are preserved rather similarly in that they are remnants from an earlier metamorphic episode which are preserved enclosed within later porphyroblasts after they have completely broken down in the rest of the rock. Figure 6.5 (*b*) illustrates a relic staurolite inclusion in muscovite from a sillimanite zone schist. Other common examples would include inclusions of chloritoid in garnet from staurolite schists, or inclusions of staurolite in garnet from the upper sillimanite zone. Although garnet is a common host to relic inclusions they are also found in a variety of other minerals. Relic minerals also owe their preservation to the sluggishness of volume diffusion through the host porphyroblast, which serves to isolate them from other phases in the rock matrix with which they would otherwise react.

An example of a study that made use of relic inclusions in garnet to deduce something of the metamorphic history of the host rock is the work of Thompson and others (1977) on large garnet porphyroblasts from the Gassetts Schist of Vermont, USA. These workers carried out microprobe analyses of a close spaced grid of points on a garnet crystal from a staurolite–kyanite schist, and also studied the distribution of inclusions in the garnet.

Chloritoid and staurolite inclusions occurred throughout the garnet except in the innermost core and outer rim, but chloritoid did not occur elsewhere in the rock and so was a relic phase. They concluded that the bulk of the garnet had been produced by the reaction:

$$23 \text{ chloritoid } + 8 \text{ quartz} \rightarrow 4 \text{ staurolite } + 5 \text{ garnet } + 21 \text{ H}_2\text{O} \qquad [6.1]$$

This is a continuous reaction, and so the progressive growth of garnet over a temperature interval would account in part for its chemical zonation. Kyanite inclusions are present in the outer parts of the garnet, suggesting that they were produced by the reaction:

$$6 \text{ staurolite } + 11 \text{ quartz} \rightarrow 14 \text{ garnet } + 23 \text{ kyanite } + 3 \text{ H}_2\text{O} \qquad [6.2]$$

REACTION RIMS AND SYMPLECTITES

Reaction rims are textural evidence for incomplete reaction and consist of zones of product minerals separating grains of the reactant phases, which never occur in contact with one another. They develop most commonly in coarse grained rocks which have undergone a later metamorphic reaction under conditions in which diffusion was not effective over sufficiently long distances for the cores of the coarse reactant grains to be able to participate. Corona textures, described in Chapter 4, are a particularly well-developed type of reaction rim texture. Because the reaction has been inhibited, it is possible to see the reactant phases and often to determine their mutual textural relationships. For example in the case of the corona textures from Norway, described on page 111, it is possible both to determine the conditions of the high grade metamorphism and also to deduce that the rocks were gabbros unaffected by other metamorphic events up to the time when the corona textures began to form.

Symplectite is another distinctive texture often associated with the growth of coronas. It is used to describe the intimate intergrowth of two or more minerals that have nucleated and grown together, comprising a single shell of a reaction rim or corona. The texture therefore has a high surface energy but can form with less mass transfer than if the phases had grown separately.

METAMORPHIC TEXTURES AS A GUIDE TO THE MECHANISMS OF METAMORPHIC REACTIONS

For many years, one of the paradoxes of metamorphic petrology was the observation that, although we can write metamorphic reactions indicating the growth of some specific mineral from a particular reactant or group of reactants, there is often no textural evidence to confirm that the reaction products have indeed grown from the supposed reactants. Because of such observations, some petrologists have questioned whether metamorphism is truly progressive.

This paradox was resolved by Carmichael (1969), who pointed out that local replacements could be identified in one part of a thin section which were balanced by other replacements in adjacent parts of the sample so that the overall change corresponded to a simple metamorphic reaction. As an example he considered the polymorphic transition from kyanite to sillimanite, which was notorious for the fact that sillimanite rarely grows directly from kyanite.

Typically, kyanite is rimmed by muscovite as it breaks down, and this reaction, supposing it takes place at a contact between kyanite and quartz, can be described by the reaction:

$$3 \text{ kyanite} + 6 \text{ quartz} + 2 \text{ K}^+ + 9 \text{ H}_2\text{O} \rightarrow 2 \text{ muscovite} + 2 \text{ H}^+ + 3 \text{ Si(OH)}_4 \quad [6.3]$$

In this reaction K^+, H^+, and Si(OH)_4 are general ways to denote the occurrence of these components in solution in the pore fluid phase. Since the reaction is not balanced in terms of conventional solid or fluid components and requires the participation of ionic species, it is known as an **ionic reaction**. There is usually no one correct way to balance ionic reactions; reaction [6.3] is written in such a way that the muscovite produced occupies approximately the same volume as the quartz and kyanite that it replaces (which is indicated by thin section studies), and so that Al does not have to be added or removed in solution (because it is commonly believed to have a low solubility in natural fluids). Alternative ways of writing the reaction are almost limitless, e.g.:

$$3 \text{ kyanite} + 3 \text{ quartz} + 2 \text{ K}^+ + 3 \text{ H}_2\text{O} \rightarrow 2 \text{ muscovite} + 2 \text{ H}^+ \quad [6.3a]$$

$$2 \text{ kyanite} + 3 \text{ quartz} + 2 \text{ Al(OH)}_3 + 2 \text{ K}^+ \rightarrow 2 \text{ muscovite} + 2 \text{ H}^+ \quad [6.3b]$$

Once kyanite has been completely rimmed by muscovite so that it is no longer in contact with quartz, it can continue to break down by the reaction:

$$3 \text{ kyanite} + 3 \text{ Si(OH)}_4 + 2 \text{ K}^+ \rightarrow 2 \text{ muscovite} + 2 \text{ H}^+ + 3 \text{ H}_2\text{O} \quad [6.4]$$

which is essentially the same as reaction [6.3a], except that Si must now gain access to the reacting surface of the kyanite through the fluid phase. Suppose, however, that the rim of muscovite around the kyanite becomes sufficiently thick that the diffusion of K and Si through it is inhibited. The reaction at the kyanite surface might then be:

$$4 \text{ kyanite} + 3 \text{ Si(OH)}_4 + 2 \text{ K}^+ \rightarrow 2 \text{ muscovite} + 1 \text{ sillimanite} + 2 \text{ H}^+$$
$$+ 3 \text{ H}_2\text{O} \quad [6.5]$$

and this would produce rims of muscovite containing crystals of sillimanite. Precisely this texture has been described by Chinner (1961) from the sillimanite zone in Scotland (Fig. 6.8). If the metamorphism is isochemical overall, then there must be complementary ionic reactions taking place in adjacent parts of the rock; for example to liberate K and act as a sink for H:

$$2 \text{ muscovite} + 2 \text{ H}^+ \rightarrow 3 \text{ sillimanite} + 3 \text{ quartz} + 2 \text{ K}^+ + 3 \text{ H}_2\text{O} \quad [6.6]$$

This reaction would lead to the replacement of muscovite in the rock matrix by sillimanite + quartz, and again this is a texture that may be observed in thin section. The occurrence of reactions [6.3] and [6.6] in nearby parts of a rock, coupled with diffusive exchange between these two domains and the solution and reprecipitation of quartz according to:

$$\text{Si(OH)}_4 \rightleftharpoons 1 \text{ quartz} + 2 \text{ H}_2\text{O} \quad [6.7]$$

can account for the growth of sillimanite in different parts of the rock from those where kyanite is breaking down, despite the fact that the overall reaction, obtained by adding reactions [6.3], [6.6] and $3 \times$ [6.7], is simply:

$$3 \text{ kyanite} \rightarrow 3 \text{ sillimanite} \quad [6.8]$$

All the other participants then cancel out. (To perform this addition of the reactions, simply write out one large reaction containing everything that appears on the left of the three participating ionic reactions on one side, everything that appears on the right on the other. Cancel all participants that appear on both sides according to their stoichiometric coefficients.)

Since Carmichael's study, many authors have described comparable reaction cycles, usually from pelitic rocks, whereby the reaction taking place at a particular mineral surface involves a local change in composition, but this is balanced by changes taking place simultaneously elsewhere in the rock. The precise reason why a reaction should proceed in such a complex fashion is not always clear, but it is probably related to the selective nucleation of the reaction products on particular substrates; for example sillimanite does not appear to nucleate on kyanite as readily as it does on micas, and this may account for the cycle shown on Fig. 6.8. Yardley (1977b) has described a comparable cycle of ionic reactions leading to the replacement of garnet by sillimanite fibres, and suggests that this was initiated by the selective nucleation pattern of sillimanite.

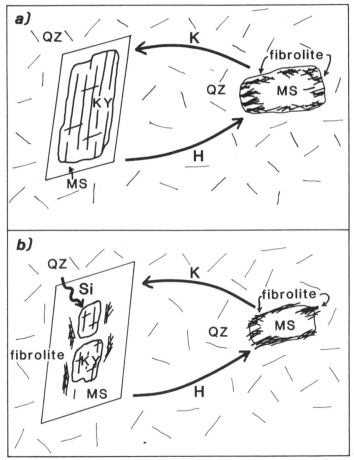

Fig. 6.8 Schematic representation of an ionic reaction cycle linking breakdown of kyanite to growth of sillimanite. *a*) A corroded kyanite on the left, in a quartz-bearing matrix, has been mantled by newly-grown muscovite according to reaction [6.3], while on the right muscovite is breaking down and being replaced by fibrolite sillimanite according to reaction [6.6]. *b*) The kyanite is now partially screened from matrix quartz by the mantle muscovite, and the reaction at the kyanite surface is now [6.5], leading to growth of fibrolite within the muscovite mantle. Based on Carmichael (1969).

THE INFLUENCE OF ROCK DEFORMATION ON METAMORPHIC TEXTURES AND PROCESSES

DEFORMATION OF MINERALS AND ROCKS

Rock and mineral deformation is principally the field of the structural geologist and geophysicist, but it is also a very important aspect of regional metamorphism. The treatment here is necessarily brief and aimed very largely at the influence of deformation on metamorphic textures and reactions.

During most types of metamorphism, ductile or plastic deformation is dominant. In **ductile deformation** the rock body as a whole changes shape, and often the individual grains are also deformed, whereas **brittle deformation** is localised along discrete fracture planes. Brittle deformation can also occur in metamorphism, even at high temperatures. Ductile deformation occurs in two principal ways: **intercrystalline deformation** involves the movement of individual grains past one another by **grain boundary sliding**, whereas **intracrystalline deformation** changes the shape of individual grains.

There are several distinct mechanisms of intracrystalline deformation, and they may be divided into three types: those in which units of the crystal move relative to one another on discrete planes, thereby distorting the grain as a whole; those in which some constituent atoms move independently by diffusion while other parts of the crystal are entirely unchanged; and syntectonic recrystallisation in which new grains continuously replace older ones, but there need be little relative movement between nearby atoms (Fig. 6.9).

Examples of mechanisms of intracrystalline distortion include the formation of **twins** (commonly seen in calcite), the development of **kink bands** (especially in micas) and the movement of dislocations causing 'flow' or 'creep'. **Dislocation flow** is the most important of these mechanisms, but its effectiveness varies between different minerals and according to the metamorphic conditions. It is aided by increased temperature and in some minerals by the presence of even small amounts of water, because these make

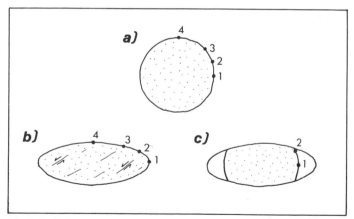

Fig. 6.9 Schematic representation of the distinction between grain deformation by dislocation flow and by pressure solution. An original grain in *a*) has four points labelled on its surface. After dislocation flow in *b*) these markers remain on the surface in the same relative positions, although distances between them have changed. In contrast, pressure solution *c*) totally eliminates part of the original grain, including points 3 and 4, while points 1 and 2 now lie within the grain, having been mantled by newly precipitated (unstippled) material. After Elliott (1973).

volume diffusion easier and thereby make it possible for the dislocations to travel through the crystal lattice and to bypass obstructions (such as other dislocations) that they may encounter. At lower temperatures or high strain rates, dislocations may become concentrated together into narrow **deformation lamellae** which are sometimes visible in thin section as lines of anomalous refractive index. Even at high temperatures the movement of dislocations is restricted to certain crystallographic planes in any mineral species. This means that when a rock undergoes ductile deformation, the distortion of one grain may not readily be accommodated by the grain next to it if the second grain is lying in a different orientation or is of a different mineral. Hence dislocation flow is usually accompanied by grain boundary sliding or other deformation mechanisms, to accommodate the different distortions of adjacent grains.

Deformation by diffusive mass transfer occurs most commonly as **pressure solution**. This is a process by which material diffuses around grain margins from highly stressed points to less stressed portions of the same or adjacent grains. It is probable that in most geological settings this mechanism is only important where there is a pore fluid phase present to facilitate the diffusion. Although pressure solution phenomena were recognised by Sorby in the nineteenth century and studied experimentally by Correns (1949), it is only recently that it has been recognised as a major natural deformation mechanism, especially at low metamorphic grades where dislocation flow is ineffective. The work of Voll (1960), Durney (1972) and Elliott (1973) has been particularly influential, and Elliott's paper figures superb photomicrographs of pressure solution phenomena in rocks. Elliott showed that the mechanism of intracrystalline deformation is dependent on grain size as well as on temperature, with fine grain size favouring pressure solution because the distance around the outside of the grain from its highly stressed to its less stressed portion, is of course smaller. In general, however, pressure solution is most important in quartz-bearing rocks at low grades, up to the greenschist facies, above which other mechanisms dominate.

The third mechanism of intracrystalline deformation, **syntectonic recrystallisation**, is closely linked to dislocation flow because it occurs as a result of the development of large numbers of dislocations within a grain. It has been treated above on page 155.

TEXTURES PRODUCED BY DEFORMATION DURING METAMORPHISM

The effects of deformation may vary greatly according to the mineralogy and grain size of the rocks concerned, temperature, availability of fluid, strain rate and the pre-existing texture. They include the production of tectonic fabrics (foliations, lineations) and recrystallisation of minerals in response to strain. In addition, metamorphic reactions and deformation processes may interact with one another. This interaction may be purely passive, as for example when a poikiloblast mineral grows while deformation is proceeding, so that the inclusion trails within it record the evolving schistosity of the rock. Alternatively, there may sometimes be an active interaction between deformation and metamorphism, whereby deformation actually promotes metamorphism or vice versa.

Recrystallisation and deformation

We have already seen that recrystallisation is often a response to deformation, being triggered either by grain size reduction or by the increased strain energy of deformed grains (page 155).

The development of mineral preferred orientations

One of the most characteristic features of deformed metamorphic rocks is the presence of fabrics such as **cleavage**, **schistosity** or **mineral lineations**. These fabrics reflect the alignment of all or some of the constituent mineral grains in particular **preferred**

orientations. Platy minerals such as micas, or elongate minerals such as amphibole are especially likely to become aligned during deformation, but intracrystalline deformation can give rise to preferred orientations of other minerals such as quartz or calcite also. These may be crystallographic orientations only, or may also involve the formation and alignment of elongate grains.

Preferred orientations may develop as a result of:

1. Physical rotation of originally asymmetric grains (e.g. mica flakes) into new orientations, usually accompanied by slip along grain boundaries or pressure solution to accommodate their changing orientations.
2. Recrystallisation or crystallisation of minerals to form new grains growing directly in the preferred orientation.
3. Change of shape of grains to a new, aligned, asymmetric shape.

Physical rotation is the easiest of these mechanisms to envisage, and provides the classical explanation of the origin of cleavage. The process appears to operate during the formation of **crenulation cleavage** by buckling of a pre-existing foliation (Fig. 6.10). The limbs of the crenulation become tightened up until minerals lying within them form a set of spaced parallel cleavage planes. Physically to rotate all the platy grains in the rock into parallelism is however more difficult and requires very large strains indeed, and so this mechanism alone is unlikely to be able to produce a truly penetrative foliation. One factor which will facilitate rotation of platy grains is the removal of some of the intervening material. This

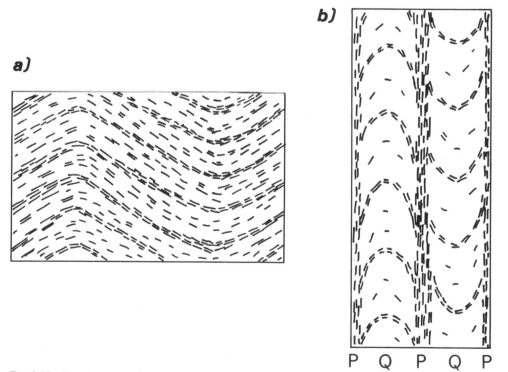

P Q P Q P

Fig. 6.10 Development of a spaced, crenulation cleavage by buckling of an earlier planar fabric. As the buckles tighten, platy grains in the limbs of the buckles form continuous new spaced cleavage planes parallel to the fold axial surfaces. This process is often accompanied by segregation of phyllosilicates into the fold limbs (known as P-domains because they are phyllosilicate-rich) and quartz into the fold hinges (Q-domains).

seems intuitively improbable, but Wright and Platt (1982) were able to use graptolites as markers of the deformation in a series of very low grade slaty rocks, and found evidence suggesting a 50 per cent volume loss during formation of the first cleavage from the original shale. In such circumstances, grain rotation could be a much more potent mechanism for cleavage formation, but such large volume losses probably occur only during burial and diagenesis, when pore water is expelled.

Detailed microchemical studies of slates by Knipe (1981) suggest that growth of new grains is often the dominant factor in cleavage formation. Initially a new fabric develops by crenulation of a pre-existing one, but as low grade metamorphic reactions take place, the reaction products crystallise preferentially in the orientation of the newly developing cleavage. As the new phyllosilicate grains continue to grow, and the old ones break down, so the new cleavage becomes pervasive and completely replaces the earlier one. Knipe was able to demonstrate that the phyllosilicates aligned in the new cleavage direction resulted from new growth because they were chemically distinct from the older grains. In higher grade rocks, any such distinctions are likely to have been obliterated subsequently by diffusive exchange, but it is certainly true that comparable textures are observed in schists.

Changes in the shape of individual grains are especially important for the development of preferred orientation in minerals such as quartz, olivine and calcite which readily deform by dislocation flow, although they may also be produced by pressure solution (Fig. 6.9, see also Fig. 1.10(*a*)).

Metamorphic segregation layering

Some regionally metamorphosed rocks exhibit a layering, usually parallel to the schistosity, that is caused by segregation of minerals. For example, layers rich in quartz and feldspar may alternate with layers rich in micas. In some cases such layering may reflect original sedimentary banding, but in others it is of metamorphic origin. Metamorphic segregation layering is best known from high grade rocks; it is characteristic of gneisses and a frequent feature of migmatites. However, the same phenomenon is also known at lower grades, especially if deformation is intense. In extreme cases, metamorphic segregation layers may become indistinguishable from veins concordant with the foliation.

The thermodynamic rationale for segregation layering is probably that grain boundaries between crystals of the same, or a structurally similar phase, are likely to have a lower surface energy than those between wholly unlike phases. Hence by segregating the rock into layers, each dominated by one structural type, the free energy of the system is reduced. However, this provides no information about the way in which the layering originates. It is apparent from Fig. 6.10 that the development of a crenulation cleavage can lead to the segregation of mica-rich and mica-poor layers, known as P and Q domains respectively. In other cases, segregation of a melt phase or movement of pore fluid may be important. It appears that this is a common phenomenon that may originate in several different ways, not all of which are well understood.

RELATIONSHIPS BETWEEN METAMORPHISM AND DEFORMATION

METAMORPHIC TEXTURES AND THE RELATIVE TIMING OF METAMORPHISM AND DEFORMATION

Poikiloblasts, and also most porphyroblasts, contain inclusions, and these often reflect the fabric of the rock at the time when the porphyroblast grew, either by occurring in distinct planes corresponding to the foliation of the rock matrix prior to its being grown over by the expanding porphyroblast, or by the alignment of individual elongate inclusions. Further deformation taking place after the growth of the porphyroblast may completely destroy the original foliation in the rock and replace it with a new one, but the inclusions inside the porphyroblast preserve information about the earlier fabric. With the recognition that many metamorphic rocks have been affected by more than one fold episode, the study of inclusion fabrics in porphyroblasts acquired considerable importance both for the information it gives about the occurrence of earlier deformation events, and because it allows us to determine the relative ages of porphyroblast growth and the various deformations.

The major textural criteria for determining the relative ages of deformation and metamorphic mineral growth were developed in the late 1950s in a series of studies, e.g. by Rast (1958), Zwart (1962), Voll (1960), although details of their interpretations have since been questioned. There are three main types of relationship between the schistosity preserved by the pattern of inclusions in a porphyroblast (known as the **internal schistosity**, S_i), and the dominant foliation in the rest of the rock, (known as the **external schistosity**, S_e). Variations on these relationships are illustrated in Fig. 6.11.

The post-tectonic growth of porphyroblasts leads to simple patterns in which the inclusion trails define a fabric that is parallel to, and continuous with, the external schistosity. This type of pattern is particularly common in thermal aureoles, if the aureole rocks underwent regional deformation prior to the thermal metamorphism, but it is also not unusual in some types of regional metamorphism. Where the external schistosity that has been overgrown was not perfectly planar but had been previously crenulated so that there are small folds in the inclusion trails within the porphyroblast, it is said to exhibit **helicitic texture** (Fig. 6.11(b)).

In contrast, if porphyroblast growth was pre-tectonic, the internal fabric, S_i, is likely to be discordant with S_e. In addition, the foliation in the matrix often wraps around the porphyroblast because it has behaved as a rigid body as the matrix deformed. An angular discordance between S_i and S_e is the best indication of pre-tectonic porphyroblast growth, but in the absence of a suitable internal fabric in the porphyroblast other indications may be present, including cracking or straining of the large crystal, or the development of a **pressure shadow** beside it (Fig. 6.11(c)). Pressure shadows are areas of coarsely recrystallised material, usually quartz, that develop in the region around the porphyroblast that was shielded from the maximum compressive stress during deformation by its proximity to the rigid porphyroblast. Intense wrapping of the foliation around a porphyroblast is also commonly considered to indicate pre-tectonic growth, but some bending of the schistosity is sometimes seen around porphyroblasts that are post-tectonic by other criteria. In these cases the bending of the schistosity has been ascribed by some authors to the 'force of crystallisation' of the growing porphyroblast, but is more commonly put down to strain resulting from relatively minor, later deformation. It is important to note that the converse of this argument is not true. It appears to be possible for pre-tectonic porphyroblasts to survive through a subsequent deformation without the new schistosity being bent around them, depending on the mineralogy of the matrix and the nature of the deformation itself (Fig. 6.11(d)). For this reason it is not possible to use the absence of distortion of the external schistosity about a porphyroblast as evidence that the porphyroblast was post-tectonic (Ferguson and Harte, 1975).

Examples of porphyroblasts that grew in rocks while they were deforming are not uncommon. This process is known as syn-tectonic growth and the classic type of texture produced is 'snowball structure' (Spry, 1963b). This consists of a spiral pattern of inclusions within the porphyroblast, classically interpreted as the result of the porphyroblast

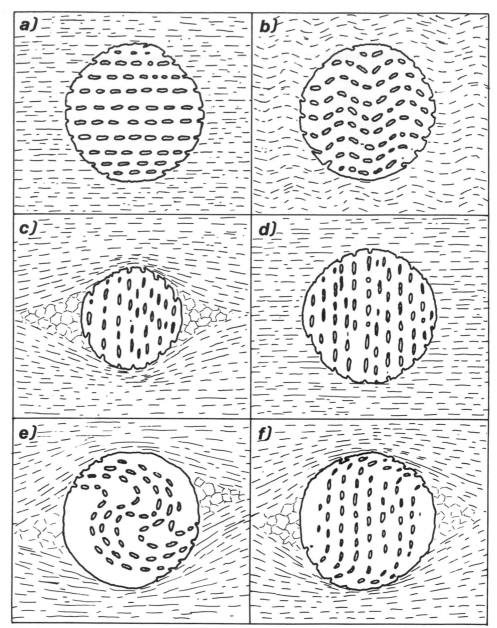

Fig. 6.11 Relationships between internal schistosity in poikiloblasts or porphyroblasts, and external schistosity. *a*), *b*) Examples of post-tectonic porphyroblast growth in which Si is continuous with,
and parallel to Se. *c*), *d*) Porphyroblasts formed prior to the external schistosity and preserving an
internal schistosity that is oblique to it. In *c*) there is marked flattening of the later schistosity around
the porphyroblast, which has a pressure-shadow around it. Rarely, however, as in *d*) there is little or
no flattening and no pressure shadow. This type of pattern has been figured by Ferguson & Harte
(1975) and appears to develop where the matrix is almost exclusively of phyllosilicate. *e*), *f*)
Syntectonic porphyroblasts: *e*) is a classic 'snowball garnet' with about 180° rotation during growth,
while *f*) is a more common rotational garnet, probably produced by recrystallisation of the schistosity in a new orientation, rather than 'rolling' of the porphyroblast.

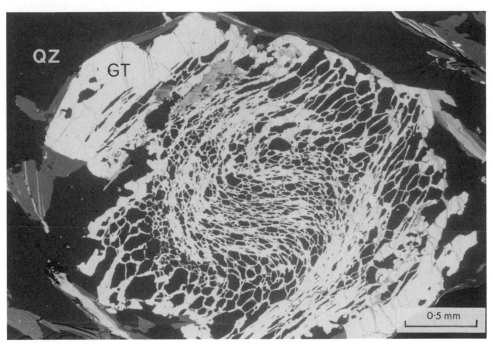

Fig. 6.12   a) Backscattered electron micrograph of a 'snowball' garnet from metachert, Saas-Fee, Switzerland. Brightness in this picture depends on the mean atomic number of the mineral concerned. Hence garnet (high at. no.) appears white, quartz is dark grey and micas mid-grey. Note that the garnet stopped growing before the external schistosity was fully developed, so that S_e is discordant to S_i. Photo courtesy D.J. Prior. *b*) Refolded fold in metasediments. The axial plane of the earlier fold is indicated by the short-dashed line, that of one of the later folds by the longer dashes. Much of the rock has a pronounced planar fabric, parallel to the axial planes of the later folds, which is defined by aligned mats of knotted fibrolite, demonstrating that the metamorphic peak in the area coincided approximately with the second phase of folding. Connemara, Ireland.

being rolled by shear along the schistosity plane as it grew (Figs 6.11(*e*) and 6.12(*a*)). The spiral pattern will only be seen in thin sections cut nearly at right angles to the axis of rotation, and Powell and Treagus (1967) have described some of the alternative patterns that would be seen in a 'snowball crystal' cut in other orientations. Snowball crystals, most commonly garnets, have been described with spirals of inclusions indicating rotations considerably in excess of 180°, but these are rare (Spry, 1963b; Rosenfeld, 1970). Usually the amount of rotation is less than 90°, and the curved S_i trails could equally result from recrystallisation of the schistosity into a new orientation during continuing deformation, without the crystals being rolled at all (Ramsay, 1962; see Fig. 6.11(*f*)). Bell (1985) has pointed out that inclusion trails in pre- or syn-tectonic porphyroblasts are often almost parallel through large volumes of rocks, and even across folds (Fig. 6.13), demonstrating that porphyroblasts often do not roll during deformation. Other syntectonic growth patterns are also known and include examples in which S_i is undisturbed in the centre of the porphyroblast but becomes increasingly tightly folded towards the margins as a result of porphyroblast growth during the formation of a crenulation cleavage. All syntectonic growth patterns are characterised by S_i being curved in some way, usually more or less symmetrically about the centre of the porphyroblast. Where porphyroblast growth continued to the end of deformation or outlasted it, S_i passes out of the crystal into S_e without a break, as for post-tectonic growth, but if deformation outlasted porphyroblast growth there will be a discordance between S_i and S_e (e.g. Fig. 6.12(*a*)).

There are a number of accounts in the literature of porphyroblasts whose growth was of long duration relative to deformation; for example Spry (1969, p. 255) has figured crystals that range from pre-tectonic through syn-tectonic to post-tectonic growth. Most examples of such prolonged growth appear to be garnets, and this may reflect the fact that garnet is frequently produced by continuous reactions over a large temperature

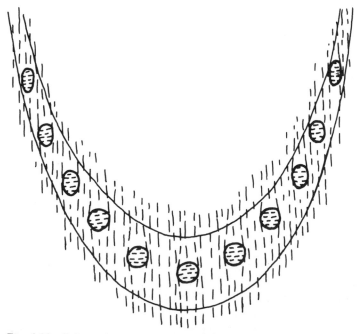

Fig. 6.13   Schematic representation of the evidence that porphyroblasts do not normally 'roll' during deformation. Pre-tectonic porphyroblasts occur around a similar-style fold, and have parallel internal schistosites throughout. Based on Ramsay (1962) and Bell (1985).

interval, and may in fact be produced by a number of separate reactions in sequence (Ch. 3).

Fabric minerals and timing of metamorphism
Many minerals that define metamorphic fabrics recrystallise too readily, and are stable over too wide a range of conditions, to provide clear constraints on *P–T* conditions during deformation. However, the amphiboles provide a clear exception to this, as does sillimanite. An example of a rock with a specific fold episode related to synchronous growth of sillimanite is shown in Fig. 6.12(*b*).

INTERACTIVE RELATIONSHIPS BETWEEN METAMORPHISM AND DEFORMATION

It has been appreciated for many years that deformation may facilitate metamorphic reactions, and more recently it has been recognised that in certain circumstances metamorphic reactions may promote deformation. At one time it was believed that deviatoric stress was actually necessary to stabilise certain minerals (Harker, 1932), but this is now known not to be the case because all the so-called 'stress minerals' have been synthesised in the laboratory under controlled conditions. Nevertheless, deformation can accelerate reactions in the following ways:

1. Juxtaposing reactant mineral grains as they move past one another.
2. Increasing the free energy of reactant grains by straining them or breaking them into smaller pieces.
3. Providing possible sites for preferential nucleation on newly created fracture surfaces.
4. Enhancing diffusion through movement on grain boundaries, increasing the area of grain boundary available and causing influx of fluid.

Examples of all these effects have been discussed by Brodie and Rutter (1985).

In addition, movement of fluid into zones of deformation can result in reactions involving hydration or metasomatism, that would not otherwise have taken place. Beach (1973) has described the mineralogical changes associated with the formation of shear zones cutting basic and acid gneisses in north-west Scotland. The major transformations in the shear zones are:

– hornblende → biotite
– garnet → biotite + Al-silicate
– orthopyroxene → biotite
– Ca-plagioclase → Na-plagioclase
– feldspars → muscovite (granitic rocks).

These transformations imply both hydration and metasomatic changes, notably addition of K and removal of Ca. It is very commonly the case that retrograde hydration reactions occur in zones that have also been affected by late-stage deformation, and so it seems likely that the two processes aid one another.

The possibility that metamorphic reactions could themselves cause deformation to occur, by weakening the rock so that it yields in response to smaller deviatoric stresses, has been discussed by White and Knipe (1978) and a possible field example has been described by Rubie (1983). Even under high grade metamorphic conditions, rocks have a finite strength which the applied stresses must exceed before plastic deformation can occur, but a reaction can reduce the strength of the rock in several ways:

1. Producing new minerals that are weaker than those originally present.

2. Producing large numbers of small grains, and so making it easier for the rock to deform by grain boundary sliding.
3. Setting up local stresses as a result of the volume change of the reaction, which can then act in combination with the tectonic stresses to permit deformation in response to a smaller applied stress.

The phenomenon of extensive deformation taking place in response to an anomalously small applied stress is known as **superplastic flow**.

Rubie described examples of rocks metamorphosed under eclogite facies conditions from the Sesia–Lanzo zone of northern Italy. Jadeite + quartz have grown from sodic feldspar in parts of the rocks that were intensely deformed and have mylonitic fabrics, and Rubie concluded that it was the deformation that had become concentrated in those parts of the rock in which feldspar had first begun to break down, rather than the other way round. This reaction is an ideal one for causing weakening because the products form fine grained intergrowths, quartz is more readily deformed than feldspar under most conditions, and the reaction involves a large volume change. According to Rubie, once the reaction had gone to completion there was a period of recrystallisation by grain growth which coarsened the rock and inhibited further deformation.

Fluid release and rock deformation: veins in metamorphic rocks

One of the most widespread features of regionally metamorphosed rocks at all grades is the occurrence of veins of minerals, most commonly dominated by quartz, which also occur in the adjacent wall rocks. It has been suggested by many authors that these syn-metamorphic veins are the result of extensional failure of the rock caused by the build-up of pore fluid pressure as devolatilisation reactions proceed. Most of the mineral material is quite locally derived, although not always from the immediate margins of the vein (Fig. 6.14).

We have seen in Chapter 2 (page 41) that it is commonly assumed that in prograde metamorphism fluid pressure is approximately equal to lithostatic pressure. Hence the fluid tends to push grains apart and maintain some porosity and permeability. If further fluid is released by reaction faster than it can escape, fluid pressure increases above lithostatic pressure until the rock is burst apart by the process of hydraulic fracturing. The amounts of overpressuring of the fluid required to cause this are probably small, and the process is the most widespread type of brittle failure in high temperature rocks. The fractures produced become pathways by which fluid can migrate out of metamorphic rocks, driven by fluid overpressures (relative to what the fluid pressure would be it if were caused by an overlying column of fluid rather than rock); these overpressures appear to be the norm in prograde metamorphism. Ultimately, fluid pressure drops, the cracks reseal, and there will be a period of quiescence before fluid pressure builds up again to repeat the process, but in the meantime material from solution has been precipitated in the crack to form a vein (Etheridge, 1983; Yardley, 1986).

Veining is most common in siliceous metasediments; it appears that calc-silicates often develop sufficiently large porosites during their devolatilisation that fluid escapes by pervasive flow through the layer without fracturing (e.g. Rumble *et al.*, 1982). The existence of deviatoric stresses is also a factor since they lower the fracturing threshold; veins are rare in many thermal aureoles (Yardley, 1986).

Fig. 6.14 Outcrop of psammitic schist cut by quartz veins. Note how part of the vein crosscuts the schistosity of the host rock while smaller veinlets are concordant. Connemara, Ireland, from Yardley (1986).

RATES OF METAMORPHIC PROCESSES

Much of the material of this chapter has skirted around the fascinating problem of how quickly metamorphic rocks can form. In the early days of geology it was assumed that all metamorphic rocks must be older than unmetamorphosed sediments, however with the discovery of occasional fossils in metamorphic rocks, and more recently with the development of age dating techniques, it has become apparent that many metamorphic rocks are relatively young. Regional metamorphism is therefore a process that occupies a finite time span which need not be particularly long in geological time.

There are two ways of approaching the question of how quickly metamorphism occurs. Either the duration of a metamorphic cycle of heating and cooling can be measured or calculated, or the kinetics of a specific mineral transformation can be investigated. The first approach has been the most successful to date, and involves either direct age dating studies of metamorphic assemblages, sediments or magmatic rocks whose formation brackets a particular metamorphic episode, or indirect estimation of the duration of a metamorphic cycle from knowledge of the thermal conductivity of the rocks and the likely rate at which they were heated.

THE DURATION OF A METAMORPHIC CYCLE

THERMAL MODELLING STUDIES

Contact metamorphism

Jaeger, in a series of papers summarised in a 1968 review article, has provided a series of calculations demonstrating the time taken for magmatic intrusions of different sizes to cool, and the temperatures that would be attained in the surrounding country rocks. Despite the many approximations that have to be made, these calculations probably yield estimates of the duration of contact metamorphism that are correct to within a factor of perhaps 2 at worst, and so they provide a valuable indication of the sorts of time spans over which the metamorphic cycle takes place. However, they are based on the assumption that heat passes through rocks only by the process of conduction (analogous to the process of diffusion for matter), whereas the bulk movement of hot or cold material, e.g. a circulating metamorphic fluid, can produce much faster changes in temperature through convective heating or cooling. For this reason conductive heat flow models are not always appropriate for heating and cooling around high level intrusions where convective circulation is often important (Norton and Knight, 1977).

An important feature of the equations for heat flow is that there is a simple relationship between the size of the intrusion and the size and period of heating of the thermal aureole. As a result it is possible to write all the equations in terms of ratios and apply them equally to any size of intrusion. The pattern of heating and cooling in a thermal aureole does depend, however, on the shape of the intrusion, e.g. whether it is a sheet-like body, a sphere or a cylinder, and equations are available for all these possibilities. Outlined in the box are some of the equations describing the contact metamorphism in the vicinity of a sheet-like intrusion, and these provide a simple example to illustrate the types of calculation that are possible.

HEATING IN THE VICINITY OF A SHEET-LIKE INTRUSION

The quantities that we need to define in order to calculate the development of this ideal type of thermal aureole are as follows:

Symbol	Unit	Quantity
y	m	half-thickness of intrusion
x	m	distance from centre of sheet to the point of interest
T	°C	temperature at point and time of interest
T_o	°C	initial temperature of magma
k	cm^2/s	thermal diffusivity of the rock (a function of thermal conductivity, density and specific heat)
t	a	time in years

The quantity k has been determined for a wide range of rocks and only varies by a factor of about 3 between extremes; it is therefore convenient to take a value of 0.01 cm^2/s, equivalent to 31.5 m^2/a, which is a good approximation for many rocks. The calculations are performed using two ratios:

$$d = x/y$$
$$r = kt/y^2$$

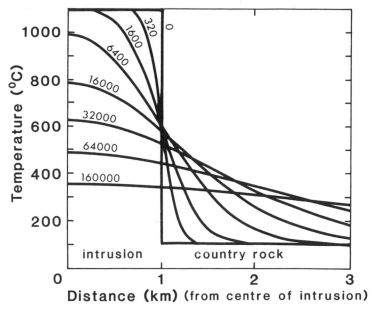

Fig. 6.15 Temperature profiles showing the thermal evolution of a 2 km-thick sheet-like intrusion and its aureole. The initial intrusion at time 0 was at 1100 °C, emplaced into country rocks at 100 °C. Successive thermal profiles are labelled with the time (in years) after the date of the intrusion at which they are attained, assuming conductive cooling only. From data of Jaeger (1968).

Table 6.1 Values of φ as a function of d and r. (After Jaeger, 1968)

d	r					
	0.01	0.04	0.20	1.0	4.0	10.0
0	1.000	1.000	0.886	0.520	0.276	0.177
0.4	1.000	0.983	0.815	0.503	0.274	0.176
0.8	0.921	0.760	0.622	0.455	0.266	0.174
1.2	0.079	0.240	0.376	0.384	0.253	0.171
1.6	—	0.017	0.171	0.303	0.237	0.166
2.0	—	—	0.057	0.222	0.217	0.160
4.0	—	—	—	0.017	0.106	0.119
6.0	—	—	—	—	0.032	0.073

Table 6.2 Values of T_{max} and r_{max} for country rocks to an intrusive sheet. (After Jaeger, 1968)

d	1.0	1.2	1.5	2.0	3.0	4.0	5.0
r_{max}	0	0.500	0.932	1.820	4.33	7.83	12.33
T_{max}	0.500	0.407	0.324	0.242	0.161	0.121	0.097

In both cases the units cancel out and the ratios are dimensionless. The temperature at any point in the aureole specified by the distance ratio d, after a period of time t from intrusion of the magma (which is assumed to be instantaneous), is given by a mathematical function of d and r:

$$T/T_o = \phi\,(d,\,r)$$

Values of the function ϕ are given in Table 6.1 for a limited range of values of d and r; for a more complete treatment and more extensive tables, the references should be consulted. It is interesting to note that at the immediate contact of the intrusion ($d = 1$), the maximum temperature that can be attained is only half that of the magma. This assumes that the initial temperature of the country rock is 0 °C. If significantly greater than this, half the actual initial temperature must be added to the calculated value of T. Even so, it is clear that aureole temperatures will only exceptionally approach those of the magma, even for short periods of time. The fact that very high temperature rocks do occur in the innermost parts of many aureoles suggests the model is oversimplified, e.g. the cooling magma may convect.

Another parameter of great interest is the time that it takes for an aureole rock to attain its maximum temperature; again, this is a function of d. Values of T_{max}, the maximum temperature attained, and r_{max}, the value of r when $T = T_{max}$ and from which t at T_{max} can be calculated, are tabulated in Table 6.2. Fig. 6.15 illustrates a series of successive temperature profiles across the contact of a vertical, sheet-like intrusion. Note how country rocks progressively further from the contact attain their maximum temperatures at progressively later times after the intrusion occurred.

It is worth reiterating that the treatment given here and illustrated in Fig. 6.15 involves many simplifications, for example the heat of crystallisation given off by the magma is ignored or assumed to be balanced by the heat of reaction for the metamorphic reactions in the aureole. Nevertheless, even simple calculations of this type yield results that are correct to better than an order of magnitude and may therefore provide useful constraints on the geological development of a region. They show that the timescale for contact metamorphism should often be measured in thousands rather than millions of years.

Regional metamorphism

In recent years thermal modelling calculations have been applied to regional metamorphic settings (e.g. England & Richardson, 1977) and have provided some invaluable information about the way in which metamorphic conditions must evolve with time. They also indicate the sorts of times (of the order of tens of millions of years) that are required for orogenic belts to heat or cool. The subject of thermal modelling of regional metamorphic terranes is returned to in Chapter 7.

AGE DATING

Age dating provides important guidelines to the overall duration of metamorphic episodes, at least over the past 1000 Ma. In New Zealand, the Haast Schists (see Fig. 4.1) and their precursor rocks include sediments ranging up to Jurassic in age (about 150 Ma). K–Ar ages for their regional metamorphism are mostly in the range 100–140 Ma, while there is evidence for extensive Upper Cretaceous erosion, exposing most of the grades now seen, by 60–80 Ma (Landis & Coombs, 1967). Thus the duration of the cycle of heating, metamorphism, and return to the surface is no more than 90 Ma, and the duration of the metamorphic episode itself must be less than this.

A shorter time span is quite precisely defined for the Acadian orogeny of the north-east USA. Naylor (1971) has shown that the syn-metamorphic folding was completed within

30 Ma of the deposition of Lower Devonian sediments, and Sleep (1979) demonstrated that the observed folding of the isograds while the rocks were still hot meant that the major folds had in fact formed in times of only about 0.5 Ma within this period. This apparently rapid regional deformation implies movement rates of a few centimetres per year, similar to rates of plate motion.

For many examples the duration of regional metamorphism is broadly comparable, and can be measured in tens of millions of years. While the total time that elapses between sedimentation and the exposure of metamorphic rocks at the surface is often as long as 100 Ma, individual metamorphic episodes may be as short as 10 Ma, and individual metamorphic reactions must often proceed in very much shorter times than this.

RATES OF METAMORPHIC REACTIONS

The specific problem of, for example, 'how long did this particular garnet crystal take to grow?' is not solved by knowing that the metamorphic cycle took 60 Ma overall. However, the study of the kinetics of metamorphic reactions is still very much in its infancy, despite the pioneering approach of Fyfe, Turner & Verhoogen (1958) and much interest in recent years.

The most fundamental problem in understanding reaction rates is that of knowing which of the various steps involved in a reaction will be the slowest in a natural geological setting, and will therefore control the rate of the overall reaction. The slowest step is known as the **rate limiting step**.

When a reaction is taking place under conditions that are not close to those for equilibrium between the particular reactants and products, possible rate limiting steps could be:(a) the rate of **formation of nuclei** of the product grains (a very difficult step to give a satisfactory treatment to in metamorphism because we are only able to study the reaction because nuclei did in fact form!); (b) the rates of **surface steps**, i.e. the migration of atoms from the surface of the reactant crystals into a less ordered intermediate phase, such as an intergranular pore fluid, or conversely the precipitation of atoms on to the growing crystals; (c) the rate of the **transport step** between reactants and products, e.g. diffusion of material through a pore fluid phase or through a zoned crystal; (d) for reactions that are not isochemical, the rate of **supply or removal of reactants or products** by metasomatic processes.

In addition, in the case of reactions taking place under conditions that are close to equilibrium, the rate of **supply or removal of heat energy** may control the overall reaction rate. This is because many metamorphic reactions are strongly endothermic, especially those that release volatiles, while a few prograde reactions that take place at high pressures are exothermic. The amount of heat required for the reaction is often quite large compared to the heat capacity of the rock, for example the heat given off when feldspar breaks down to produce eclogite minerals will raise the temperature of the rock by about 1 °C for every 1 per cent feldspar that breaks down. If conditions are close to equilibrium this heat of reaction must be supplied or removed externally in order for the reaction to proceed, but if the equilibrium conditions are strongly overstepped the rock itself can act as a source or sink for heat, changing temperature as the reaction takes place. The heat of reaction effect will not control reaction rates for polymorphic transitions, and other reactions for which $\triangle H$ is small, nor will it be important if the conditions of the reaction have been grossly overstepped. In many other cases, however, the rate of metamorphic reactions must be analogous to the rate at which water boils: it depends on how fast you heat the system!

The reaction whose kinetics have been studied most closely is the polymorphic reaction: aragonite $\rightleftharpoons$ calcite. The equilibrium conditions for these phases to coexist were shown in Fig. 4.7, and this demonstrates that calcite is the thermodynamically stable form at the earth's surface under all conditions. However, it was also noted in Chapter 4 that metamorphic aragonite is sometimes found in areas of low temperature, high pressure metamorphism, though it is often partially converted to calcite. An understanding of the kinetics of this reaction would therefore enable us to understand something of the way in which high pressure metamorphic rocks are cooled and brought up to the surface, because they must do this in a way that permits aragonite to be preserved.

Brown, Fyfe & Turner (1962) studied this reaction in the presence of water and found that even at temperatures as low as 200 °C there was 10 per cent replacement of aragonite by calcite within six days. This result is perhaps not surprising in view of the widespread diagenetic replacement of biogenic aragonite by calcite in sedimentary rocks, but makes it difficult to understand how metamorphic aragonite can ever be preserved. Furthermore, in the hydrothermal experiments calcite replaced aragonite by a mechanism of solution of aragonite and redeposition of calcite as new grains, whereas in natural examples calcite forms a topotactic replacement of aragonite, involving very little atomic movement (Carlson and Rosenfeld, 1981, Figs 1–7 therein). A likely conclusion is that rocks that contain relict aragonite cannot have had a pore fluid phase present when they were uplifted and subjected to conditions of the calcite stability field, because Brown *et al.* also found that the reaction was very much slower in the absence of water.

More recently, Carlson and Rosenfeld (1981) devised an elegant series of experiments to study further this reaction under dry conditions, and reproduced the natural replacement textures. They concluded that the nucleation step did not control the overall rate of the reaction, because it is common to find partially replaced aragonite crystals surrounded by large numbers of small calcite grains, and so they set about measuring the growth rate of the calcite grains once they had nucleated. They did this by direct observation of the size of the calcite grains while they were maintained at known temperatures on a heating stage attached to a petrological microscope. By this technique they were also able to compare growth rates in different crystallographic directions. The results that they obtained on sections of aragonite parallel to 001 are reproduced in Fig. 6.16(*a*). This type of plot is known as an Arrhenius plot, and is appropriate for describing the rates of many sorts of geological processes, such as diffusion in minerals. The logarithm of the rate constant (in this case the rate of growth of calcite crystals) varies linearly with $1/T$; as a result it is possible to extrapolate experimental results to conditions that may be more appropriate for slower, natural processes, with some reliability. Figure 6.16(*b*) has been compiled from the experimental data and the extrapolation in Fig. 6.16(*a*). It shows the growth rates that will result if aragonite-bearing rocks at different temperatures are uplifted and subjected to pressures at which calcite is stable. If the uplift takes place at a temperature of about 250 °C (path A–A^1), calcite will replace aragonite at a rate of 1 m/Ma, so that even the largest aragonite grains will be totally replaced by calcite unless there is also very rapid cooling. On the other hand at 100 °C (path B–B^1) the rate of replacement would be only 10^{-3} mm/Ma and aragonite would remain unchanged over geological time periods.

The experiments seem to accord well with observations of natural rocks in that they demonstrate that metamorphic aragonite can be preserved during subsequent uplift to the surface, but that it is only likely to be found in very low grade blueschists, despite the fact that many high temperature blueschists may also have been metamorphosed at sufficiently

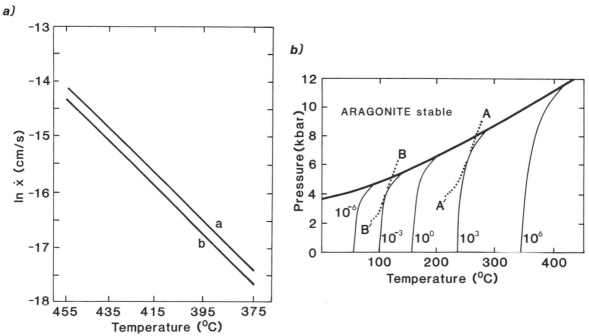

Fig. 6.16 (a) Experimentally determined Arrhenius plot showing the rate of growth of calcite crystals growing by topotactic replacement of aragonite, as a function of crustallographic orientation and temperature. Logarithm of growth rate in cm/s ($\dot{x}$) is shown for growth parallel to the a and b axes of aragonite (lines a and b respectively) in a section cut parallel to 001 of aragonite. (b) Calculated growth rates of calcite from aragonite within the calcite field. Light curves are lines of constant growth rate, labelled with the rate in mm/Ma. Bold curve is the aragonite–calcite boundary. Dotted lines A–A' and B–B' are possible uplift paths discussed in the text. Both parts of this figure are after Carlson & Rosenfeld (1981).

high pressures to have contained aragonite, rather than calcite, at the peak of metamorphism.

The example of the aragonite–calcite reaction serves to emphasise the importance of temperature in controlling reaction rates. This works in two ways. Firstly, as T deviates from the equilibrium temperature T_e, the free energy change of the reaction, $\triangle G_v$ becomes larger (Fig. 6.2) and this increases the probability of nucleation of the product minerals. It also increases the driving force for diffusion of material from reactants to products where this is necessary. Secondly, temperature controls the rate of steps involving movement of atoms, either across an interface as in topotactic replacement or by diffusion through an intervening phase. In this case the effect of increased temperature is always to increase the rate of movement. The two temperature effects work together in the case of prograde reactions to increase the reaction rate with temperature, but for reactions taking place on cooling or as a result of pressure changes the reaction may be effectively inhibited as a result of sluggish atomic movement at low temperature, even when $\triangle G_v$ is very large. Putnis and McConnell (1980) provide a proper introduction to this subject.

The reaction of calcite + quartz to wollastonite   The reaction: $CaCO_3 + SiO_2 \rightarrow CaSiO_3 + CO_2$ is probably a much better analogue for metamorphic reactions in general than the calcite–aragonite reaction, and was studied experimentally by Kridelbaugh (1973). He found that the wollastonite always grew as rims around the quartz grains in his experimental charges, analogous to natural reaction rims, and once formed they could only continue to grow if Ca was able to diffuse to the surface of the quartz grain, and Si could diffuse to the outer surface of the wollastonite rim. Thus diffusion through the reaction rim of wollastonite became the rate limiting step for the reaction, and was strongly temperature dependent. Kridelbaugh calculated from his experimental results that a 1 mm thick rim of wollastonite around a quartz grain in a limestone would grow in 2.5 years at 950 °C, 50 years at 800 °C, but would take 400 000 years to form at 600 °C, close to T_e.

These experiments reproduce a process that is found in nature; reaction rims of wollastonite around chert nodules in limestone have been described from a number of thermal aureoles (Joesten, 1974), but it is also clear from the experiments that in order to calculate how long they took to form it is necessary to know the temperature of metamorphism with a high degree of accuracy. Furthermore, the reaction rims are made up of a number of grains, and these may recrystallise as metamorphism proceeds; if diffusion is dominantly along grain boundaries, the diffusion rates may change as the reaction rim develops. To date, no independent estimate has been made of the duration of a metamorphic episode from the development of reaction rim structures, even though considerable progress has been made in understanding the theory of the process (Fisher, 1977), and diffusion rates can in principle be measured experimentally.

KINETICS OF NATURAL REACTIONS

The experiments on the kinetics of simple reactions emphasise the importance of understanding which step in the reaction controls its rate in nature. Hydrothermal experiments on conversion of aragonite to calcite give meaningless results when applied to the natural topotactic replacement reaction, apparently taking place in the absence of fluid.

Fisher (1978) considered the kinetics of reactions involving the formation of more or less spherical segregations (e.g. a porphyroblast or a cluster of grains), mantled by a halo depleted in the constituents of the segregation, and recognised three stages in the reaction which would proceed at different rates. The first stage is **reaction controlled**, i.e. controlled by the rate of addition of atoms to the surface of a small initial grain or cluster from a relatively large volume of matrix in which the pore fluid is supersaturated with respect to the growing phase. This step proceeds until the immediate vicinity of the segregation has become depleted in the constituents of the growing phase, while the surface area of the segregation has increased, facilitating addition of atoms. There is then a transition to a second stage of **transport-controlled** growth limited by the rate of diffusion of material to the growing segregation (as illustrated in Fig. 6.3). Many reactions are markedly endothermic (e.g. devolatilisation reactions), and for these a third stage of reaction follows. At the time the reaction began (i.e. when nucleation occurred) the equilibrium temperature must have been significantly overstepped. If the first two stages of an endothermic reaction proceed to consume heat faster than it is supplied to the rock, the temperature of the rock must drop during stages 1 and 2 until the equilibrium temperature is restored. At that point the reaction can proceed no faster than heat is supplied to drive it, and this gives rise to the third stage of **heat-flow-controlled** growth.

What remains controversial is the relative importance of these different stages during different sorts of reactions in nature. When the overall prograde metamorphic history of sediments is considered, the large amounts of volatiles to be driven off by endothermic devolatilisation reactions require that the rate of metamorphism is linked to the rate of heat supply (Walther & Orville, 1982; Yardley, 1986). Furthermore, normal prograde reactions in pelites are more likely to produce the sort of complex ionic reaction cycles discussed above (page 165) than segregations surrounded by depleted haloes, and Yardley

(1977b) argued that these reaction cycles are a consequence of the selective nucleation of reaction products such as sillimanite on the most favourable sites only. The implication of this is that the equilibrium temperature was overstepped by only a minimal amount, and hence it is likely that reaction proceeded rapidly to being heat-flow controlled.

Where larger amounts of overstepping are possible, reaction-controlled or transport-controlled growth steps may predominate. Ridley (1985) has pointed out that if the equilibrium temperature for a reaction has to be overstepped by an amount of the order of $20\,°C$ before a porphyroblast phase can nucleate, then extensive reaction can proceed once the nucleus is present, independent of heat supply, making use of heat already in the rock and lowering its temperature towards T_e.

Wood and Walther (1984) have compiled experimental data on rates of solution and precipitation of a wide variety of minerals and conclude that this step is always rapid. Their results indicate that reactions will proceed rapidly at temperatures only a few degrees in excess of T_e, provided nuclei have formed, and so reaction-controlled growth is unimportant. However, they assume that a fluid phase is present, and we have seen above that in the case of the retrograde metamorphism of aragonite-retaining, low grade blueschists this assumption cannot be correct. Even in the presence of a fluid phase, reactions with a very small $\triangle G$, such as the transitions between the Al_2SiO_5 polymorphs, may need more substantial overstepping before significant replacement can occur. In contrast to the view of Wood and Walther, Loomis (1983) has concluded that the surface dissolution or precipitation step may often remain the dominant control on reaction rates. He points out that it is not uncommon for two Al_2SiO_5 phases to persist together in rocks that show evidence for extensive reaction and mass transfer involving other minerals, and argues that the unstable polymorph can only be preserved in such conditions if the rate at which atoms could leave its surface was very slow.

The development of reaction rims or coronas is often associated with metamorphism under conditions far from equilibrium, where the reaction is inhibited by the physical separation of different reactant phases. It is commonly considered that the transport step between different reactant phases, e.g. diffusion along grain boundaries, is likely to be the rate limiting step for the formation of such temperatures, and this is the basis of the treatment developed by Fisher (1977). Indeed for some geometries of reaction zone, Fisher (1978) argues that transport-controlled growth may take over as the dominant step after a period of heat-flow-controlled reaction. On the other hand Loomis (1983) has argued that reaction zone textures could also be developed if the surface reaction step was controlling the overall reaction rate.

Corona textures are often developed in very high grade rocks, and it is evident from the results of Wood and Walther that the reactions would have gone rapidly to completion if an aqueous fluid phase had been present throughout the metamorphism. Hence coronas in high grade rocks may indicate unusually dry conditions in which reactions were inhibited.

CONCLUSIONS

The study of metamorphic kinetics is a developing and exciting field, and anything written here will soon become outdated. Nevertheless, it seems clear that reaction rates in the presence of an aqueous fluid phase are probably quite fast in geological terms except at very low temperatures. The preservation of rocks with their peak-metamorphic assemblages little altered by retrogression is therefore good evidence that a fluid phase is not present in such rocks while they are cooling down. The heat of reaction of many of the reactions encountered during prograde metamorphism of sediments is sufficiently large that the overall rate of metamorphism is effectively linked to the heating rate. In principle this can be calculated for many tectonic situations from heat flow estimates, and such calculations are therefore very important for metamorphic studies.

Nevertheless, even in the presence of an adequate heat supply, reaction rates may be limited by other factors, and in particular it seems possible that the importance of surface reaction rates, emphasised in the work of T.P. Loomis, has been underestimated in many studies. Lasaga (1986) has modelled the breakdown of muscovite + quartz in a thermal aureole on the basis of experimental data for the rate of the surface reaction steps involved, and predicts that the notionally univariant assemblage muscovite + quartz + K-feldspar + sillimanite + fluid will persist over a finite temperature range, at least for relatively rapid heating rates, because of the sluggishness of the surface reaction step. This will result in the four-phase mineral assemblage outcropping across a distinct zone in the aureole instead of being restricted to a discrete isograd. This is in accordance with what has been found in many field studies, including the classic description of this isograd in Maine by Evans & Guidotti (1966). The phenomenon has usually been explained qualitatively as being due to the presence of additional components such as Na substituting for K, or CO_2 in the fluid, which would make the reaction divariant. Lasaga's calculations provide a refreshing alternative explanation for the observations which also fits them quantitatively. They are a benchmark showing that kinetic effects which have hitherto been considered as not susceptible to quantification, and have therefore been ignored, can become part of the everyday parlance of metamorphism.

FURTHER READING

Borradaile, G.J., Bayly, M.B. & Powell, C.McA. (eds) 1982. *Atlas of Deformational and Metamorphic Rock Fabrics*. Springer-Verlag, Berlin.

Crank, J., 1975. *The Mathematics of Diffusion*. Oxford University Press, Oxford.

Hofmann, A.W., Giletti, B.J., Yoder, H.S. & Yund, R.A. (eds) 1974. *Geochemical Transport & Kinetics*. Carnegie Institution of Washington, Publication 634.

Spry, A., 1969. *Metamorphic Textures*. Pergamon, Oxford.

Urai, J.L., Means, W.D. & Lister, G.S., 1987. Dynamic recrystallisation of minerals. *In* B.E. Hobbs & H.C. Heard (eds) *Mineral and rock deformation: laboratory studies*. Geophysical Monograph American Geophysical Union **36**, 161–200.

Vernon, R.H., 1975. *Metamorphic Processes*. Halsted, New York.

Yardley, B.W.D. & Harte, B. (eds) 1986. Mechanisms of metamorphic reactions. *Mineralogical Magazine*, **50**, 357–480.

Yardley, B.W.D., MacKenzie, W. S. & Guilford, C. 1990. *Atlas of Metamorphic Rocks and their Textures*. Longman, Harlow.

7 THE RELATIONSHIPS BETWEEN REGIONAL METAMORPHISM AND TECTONIC PROCESSES

One of the most important results that emerges from the study of metamorphic rocks is the insight that is provided into the past thermal structure and tectonic behaviour of the earth. It has been recognised for a long time that regionally metamorphosed rocks have usually also undergone complex deformation and occupy elongate **metamorphic belts** that may be thousands of kilometres in length but only tens to hundreds of kilometres across. Even within deeply eroded Precambrian shield areas, distinct **mobile belts** can be recognised. These are metamorphic belts composed of remobilised pre-existing metamorphic rocks, and/or metasediments, flanked by older metamorphic basement that was not disturbed appreciably when the mobile belt was formed. Sooner or later, the geologist working with metamorphic rocks is likely to question why and how a particular belt has originated, and how it is that rocks that have recrystallised at depth have come to be exposed today at the surface.

This chapter concerns the broad relationships between metamorphism on the large scale and orogeny. Metamorphic petrology provides factual information about depths of burial and temperatures that prevailed when particular rocks formed, to help evaluate tectonic models, while tectonics provides a unifying framework for metamorphism and suggests possible modern analogues for the setting of ancient metamorphism.

The extensive folding and crustal thickening that often accompanies regional metamorphism implies that it is usually associated with compressive tectonic regimes, and for this reason most attempts to link regional metamorphism to plate tectonic setting have concentrated on regions where plates are converging and on continental collision zones. The first part of this chapter describes the characteristics of metamorphic belts believed to have developed at convergent margins, and discusses how the metamorphism in these belts can be interpreted in the light of modern geophysical results about the nature of convergent processes today.

Studies of contemporary tectonic processes yield only information about instantaneous events, whereas rocks can retain characteristics acquired over a long history of changing conditions. Hence the extra dimension of variation in metamorphic conditions with time is considered also. In the next part of the chapter examples of metamorphism asssociated with continental collision are described, and separate sections discuss the metamorphic processes on the sea floor and during ophiolite emplacement, which may dictate, or indeed dominate, the final metamorphic character of ophiolite slices within mountain chains. Despite the undoubted importance of convergent margins for regional metamorphism, it is possible that some types of metamorphism take place in strike slip or extensional zones, and some of the evidence for this is also discussed.

METAMORPHISM, GEOTHERMAL GRADIENT AND PAIRED METAMORPHIC BELTS

It was inherent in Eskola's original definition of metamorphic facies that some rocks attained high temperatures at relatively shallow depths of burial, whereas others were subjected to greater pressures without ever attaining comparable temperatures. The metamorphic mineral assemblage therefore acts as an indicator of the thermal gradient in the crust at the time when the metamorphism took place. With the extension of detailed metamorphic studies to a wider range of metamorphic belts worldwide, it gradually became apparent that there could be systematic regional associations of metamorphic facies, identifying metamorphic belts with distinct thermal gradients during metamorphism. For example, Zwart (1967) distinguished Alpinotype metamorphic belts, which included extensive areas of high pressure metamorphism and had extensive flat-lying nappe structures, from Hercynotype belts with little or no high pressure metamorphism and predominantly upright structures.

In 1961, A. Miyashiro published a seminal paper on metamorphic belts in which he pointed out the different successions of facies in different Japanese metamorphic belts. He coined the term **facies series** for a characteristic association of facies and described two different facies series from Japan:

prehnite–pumpellyite $\rightarrow$ blueschist $\rightarrow$ greenschist *or* amphibolite
greenschist $\rightarrow$ amphibolite $\rightarrow$ granulite

Each facies series characterises a **baric type** of metamorphism, that is, there is a characteristic P/T ratio for all metamorphic grades. The two facies series described originally from Japan correspond to high P and low T (high pressure type), and to low P with high T (low pressure type). In addition, Miyashiro recognised a third intermediate pressure facies series, essentially equivalent to 'Barrovian metamorphism'. The thermal gradient deduced from the mineralogy of the intermediate pressure type corresponds to the average thermal gradient of continental crust today (Richardson & Powell, 1976). The characteristics of the different baric types are summarised in Table 7.1.

Miyashiro also recognised that in Japan contrasting high and low pressure belts of the same age occur next to one another, separated by major faults. These associations of contemporaneous belts of contrasting baric type he termed **paired metamorphic belts**. For example in south-west Japan (Fig. 7.1) the high pressure Sanbagawa belt is adjacent to the low pressure Ryoke belt and both are of Jurassic–Cretaceous age. Similarly the Permo-Triassic Sangun (high pressure) and Hida (low pressure) belts are also adjacent. Since Miyashiro's original classification of the Japanese belts, considerable new work has been carried out, resulting in revision of some of his conclusions. Banno (1986) provides an English language update based on more recent work.

In particular it has become apparent that many of the examples of paired metamorphic belts identified by Miyashiro around the Pacific margin do not fulfil the strict criteria. In many cases it has transpired that either the metamorphism in the two belts of a supposed pair was not in fact contemporaneous, or the metamorphism of each belt does not correspond simply to a single baric type as was originally supposed. In other instances it appears that the belts were widely separated at the time of metamorphism and have only subsequently been juxtaposed. Hence, although the paired belt concept has been profoundly important in encouraging geologists to attempt to correlate metamorphic and tectonic processes, it is clearly not a full explanation for different types of metamorphism and their distribution. (Similarly Zwart's distinction between Alpinotype and Hercynotype

Table 7.1 Minerals and mineral assemblages characteristic of different baric types of metamorphism

Baric type	Rock type	
	Metapelites	*Metabasites*
Low pressure	Andalusite, cordierite, staurolite, sillimanite, ilmenite	Hornblende + plagioclase + epidote Hornblende + cummingtonite Plagioclase + olivine Orthopyroxene + clinopyroxene
Medium pressure	Kyanite, sillimanite, staurolite, ilmenite + rutile, cordierite + garnet, garnet + biotite	Garnet + hornblende + plagioclase Epidote + hornblende + plagioclase Hornblende + staurolite *or* kyanite Orthopyroxene + clinopyroxene + garnet Ilmenite + rutile Hornblende + cummingtonite
High pressure	Lawsonite, kyanite, garnet, talc, jadeite/omphacite, Mg–Fe carpholite, crossite/glaucophane, rutile, *no* biotite	Lawsonite, zoisite, crossite/glaucophane barroisite, jadeite/omphacite, garnet, rutile

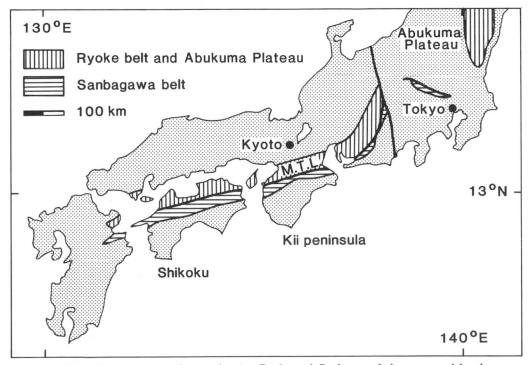

Fig. 7.1 Map of south-western Japan, showing Ryoke and Sanbagawa belts separated by the Median Tectonic Line (M.T.L.), and the Abukuma Plateau.

belts has become blurred by later work.) Nevertheless, examples of truly paired metamorphic belts probably do exist, including the classic example of the Sanbagawa and Ryoke belts.

THE SANBAGAWA BELT

This belt exhibits many typical characteristics of high pressure metamorphic belts around the Pacific, even though it will be apparent from the discussion of the mineral assemblages in Chapters 3 & 4 that the blueschists of the Sanbagawa belt did not form at such high pressures as some blueschists elsewhere.

The dominant metasedimentary rock types are pelitic to psammitic schists and phyllites, with rarer metamorphosed cherts and sporadic calcareous horizons (Ernst, Onuki & Gilbert, 1970). Conglomerates occur in the north with a decrease in sedimentary grain sizes to the south. Abundant volcanogenic rocks are predominantly of mafic composition and include both metamorphosed lava flows and volcaniclastic rocks. There are also tectonically emplaced slices of ophiolite. Post-tectonic acid intrusions, common in many metamorphic belts, are entirely absent.

The metamorphic grade is highest in the north in general, i.e. near the Median Tectonic Line (Fig. 7.1); to the south the phyllites pass into only very weakly metamorphosed upper Palaeozoic sediments. There are also metamorphic variations along strike. The most conspicuous isograd in the field is due to the appearance of porphyroblastic albite, but detailed studies have identified a series of zones in both pelitic and basic lithologies (c.f. Fig. 3.10). One of the most complete zonal sequences is that described by Banno (1964) from the Bessi–Ino district in central Shikoku, based on the mineralogy of the metabasites:

Zone A   The lowest grade metabasites are characterised by pumpellyite + chlorite. Other common minerals include glaucophane, epidote, actinolite, muscovite, albite and quartz. Lawsonite has also been reported.
Zone B   Coexisting actinolite + chlorite replaces the pumpellyite + chlorite association. Glaucophane and epidote are common.
Zone C   Here a barroisitic green amphibole (Na-rich hornblende) is present, but glaucophane persists.
Zone D   The assemblage albite + epidote + hornblende + chlorite is typical and sodic amphibole is absent. At this grade both garnet and biotite occur in pelitic schists.
Zone E   At the highest grades albite is replaced by a more calcic plagioclase (typically oligoclase). The amphibole may be blue–green hornblende or a sub-calcic hornblende and garnet may be present.

There are variations in the mineral assemblages between different sections across the Sanbagawa belt. For example amphibolite facies rocks are of very restricted occurrence, while in some regions sodic amphiboles are restricted to the lowest grade zones only. Truly high pressure assemblages are rare, and jadeite + quartz does not occur although omphacite has been reported. A distinctive feature of the pelitic assemblages in this belt is that almandine garnet appears at lower grades than biotite. This implies significantly higher pressures than for Barrovian metamorphism because the stability field of biotite becomes restricted to higher temperatures with increased pressure, whereas that of garnet probably expands at high pressure (Fig. 3.11).

THE RYOKE BELT

The rocks of the Ryoke belt differ from those of the Sanbagawa belt in both their type of metamorphism and in the nature of the rock types present. Basic volcanic rocks are rare and the dominant rocks are aluminous pelites and psammites, with minor quartzite and limestone. In addition, there are abundant granitic intrusions while ultrabasic rocks are absent.

The metamorphic grade in the Ryoke belt shows a general decrease to the north, although the thermal maximum is not always adjacent to the Median Tectonic Line. Metamorphic zones have been defined on the basis of pelite assemblages, and Miyashiro (1973) quotes the following sequence from the work of Ono:

Zone 1: *Chlorite–biotite zone*   Fine grained slaty rocks containing chlorite, phengitic muscovite, quartz, sphene, ilmenite, graphite and minor biotite together with sodic feldspar. Albite and oligoclase may both be present.
Zone 2: *Biotite–andalusite zone*   In this zone biotite replaces chlorite and oligoclase replaces albite as the rocks recrystallise to schists. Both andalusite and K-feldspar are present in the highest grade parts of this zone, which implies the onset of muscovite breakdown.
Zone 3: *Sillimanite zone*   The appearance of sillimanite characterises this zone, although andalusite may persist. Muscovite persists in the lower grade part of the zone but is absent at higher grades, where cordierite appears. The highest grade assemblages, which may include coexisting garnet and cordierite, are typical of high grade migmatitic gneisses in many parts of the world, and some melting has almost certainly occurred.

The assemblages in the Ryoke belt provide firm constraints on the pressures of metamorphism because muscovite + quartz begin to break down within the stability field of andalusite. This is characteristic of bathozone 1 in the scheme of D.M. Carmichael (page 88) and indicates depths of burial less than 10 km.

PLATE TECTONIC INTERPRETATION OF PAIRED METAMORPHIC BELTS

With the advent of modern theories of plate tectonics in the 1960s, some of the reasons for the variations in thermal gradient in the crust became apparent. Present-day variation in heat flow across different types of plate boundary was illustrated in Fig. 1.5. In summary, high heat flow is a consequence of steep thermal gradients and indicates areas where low pressure metamorphism could be occurring today. Such areas are found in regions of igneous activity where rising magmas convectively transfer large quantities of heat into the middle and upper crust, and in regions of crustal stretching, even where magmatism is absent. High heat flow also occurs in regions undergoing rapid uplift and erosion. This is because rocks are good insulators and lose heat only slowly compared to the rate at which they are being unroofed, hence hot rocks from depth can rise close to the surface before they cool appreciably, so that the near-surface thermal gradient is very steep. Regions of crustal stretching are areas of high heat flow because tectonic thinning has a similar effect to erosion, bringing hot rocks nearer the surface.

Low surface heat flow is typical of areas above subduction zones and accretionary prisms, where the thrusting of cold lithosphere down into the mantle lowers the thermal gradient and transports cool rocks to high pressures before they can warm up. Surface heat flow is also low in other areas such as above sedimentary basins. High pressure, low temperature conditions may be generated by overthrusting wherever relatively cool rocks have another rock mass thrust over them causing a rapid rise in pressure. If the overthrust mass is relatively hot however, it may give rise to the formation of inverted metamorphic zones in the underlying rocks.

In the early 1970s, the descriptions of paired metamorphic belts from around the Pacific were reinterpreted in terms of the new global tectonics. Oxburgh & Turcotte (1971) produced thermal models for processes at convergent plate margins, and were able to link low temperature, high pressure conditions to the trench region, and the development of high temperatures near the surface to the arc, above the region where the descending lithosphere begins to cause melting. They also emphasised the importance of accompanying back-arc spreading. Miyashiro (1972) also reinterpreted high pressure belts as representing metamorphism in a subduction zone setting, while the low pressure baric type he considered to develop in the associated volcanic arc where rising magmas caused additional heating. Since much of the magma is erupted, the thermal effects in the arc are not restricted to immediate contact aureoles around individual intrusions, but are of more regional extent.

Recent results have allowed Miyashiro's original ideas about the tectonic setting of Japanese belts to be further refined. For example Banno, Sakai & Higashino (1986) have shown from a study of garnet composition that the highest grade zones in the Sanbagawa belt actually formed at lower pressures than some of the lower grade ones. They suggest that this pattern reflects heating of the rocks of the downgoing slab by the hot base of the overriding plate, causing higher temperature rocks to form above cooler ones.

MODERN CONVERGENT MARGINS: IMPLICATIONS FOR METAMORPHISM

SUBDUCTION ZONES

The most important feature of subduction zones from the point of view of the mineral assemblages developed is the low heat flow, and indeed many geologists would assign all blueschist facies terranes* to an origin associated with subduction. In this chapter the term **subduction** is used in a broad sense to denote deep underthrusting of rock and sediment at a convergent plate margin, including materials that are subsequently returned to the surface with a high pressure metamorphic imprint.

In more detail there are a number of important questions about subduction zone metamorphism remaining; for example in precisely what parts of the subduction zone does the high pressure metamorphism occur, and how do such rocks return to the surface? The main types of model for subduction zone metamorphism of sediment have been reviewed by Miyashiro, Aki & Celas Sengor (1979), who divide them into three classes. In class 1, high pressure metamorphism takes place in rocks which are part of the downgoing slab, whereas in class 2 the metamorphism occurs in material that has been accreted to the base of the overlying plate. Class 3 models assume that the metamorphism takes place in the

---

* The term **metamorphic terrane** has recently become popular to denote a group of metamorphic rocks that has behaved tectonically as a distinct, single entity during an orogenic episode; it may include a wide range of rock types and metamorphic grades. Where metamorphic belts are precisely defined and have a broadly homogeneous character, the terms belt and terrane may be synonymous, however many metamorphic belts are more complex than this and are made up of a number of terranes.

downgoing slab, but the high pressure rocks are then incorporated into the upper plate by a shift in the location of the Benioff zone.

Recent geophysical studies, especially by seismic reflection profiling, have begun to provide a picture of processes currently going on at active convergent margins which should help in the evaluation of alternative models for old metamorphic belts. Karig (1983) has reviewed the evidence for the way in which sediments are accreted on to the upper plate in a variety of modern convergent margins and points out several different patterns of behaviour occurring in different parts of the world today. These are illustrated in Fig. 7.2. At the Cascade Arc, thick turbidite sequences deposited on the downgoing plate are caught up in thrusting and build up an accretionary prism at the front of the upper plate. Hence, the bulk of the sediment is effectively decoupled from the downgoing plate at a shallow level, and is not carried down deep enough to be subjected to high metamorphic pressures. In contrast, the thick (<6 km) sequence on the floor of the Arabian Sea is largely carried along with the downgoing plate when it enters the Makran Trench, and only about the upper third is being scraped off to form an accretionary prism. In both cases pelagic sediments remain attached to the downgoing plate and it is the behaviour of the overlying clastics that varies. The oceanic crust being subducted at the northern Lesser Antilles Arc carries predominantly pelagic sediments and here, although the bulk of the sediments remain attached to the downgoing plate, some pelagic material is accreted at the trench.

The accretion of high pressure metamorphic rocks, formed as part of the downgoing plate, on to the base of the overlying plate requires a process known as **subcretion** or **underplating**. It is only very recently that direct geophysical evidence has been obtained to show that this process is actually taking place today. Seismic reflection profiling across

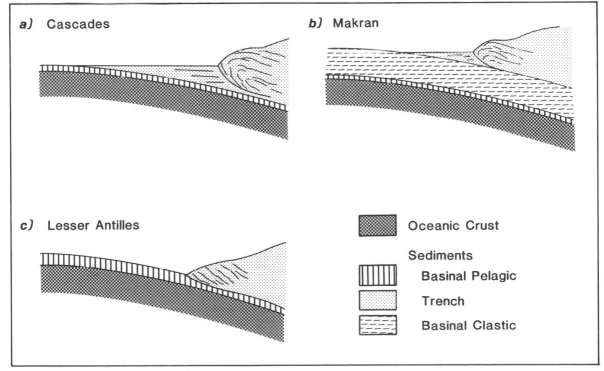

Fig. 7.2   Cross-sections of contemporary convergent plate margins displaying contrasting sediment behaviour. From Karig (1983).

the southern part of Vancouver Island as part of the Canadian LITHOPROBE project has identified mafic rocks derived from the downgoing Juan de Fuca plate *above* the presently active Benioff zone (Clowes *et al.*, 1987). These mafic rocks are believed to be blueschists, and they are now probably rather nearer the surface (20–30 km) than at the time they were metamorphosed. The same traverse shows sediment currently being underthrust to depths in excess of 30 km. The work of Walcott (1987) on active subduction beneath New Zealand (described below) provides a different kind of evidence for underplating.

Although most emphasis has been on the subduction of oceanic crust and sediment, there is also some geological evidence to suggest that continental crust can be deeply underthrust. This is seen most clearly in the breakdown of plagioclase in granite to produce jadeite + quartz, reported from the Sesia zone of northern Italy (below). Although normal continental crust is too buoyant to be subducted, stretching at continental margins can greatly thin the continental crust. In the Bay of Biscay, the European continental crust thins westwards to as little as 6 km; if such thin crust were subsequently to converge with another continent as part of the downgoing plate, it might well be underthrust to considerable depths, where metamorphic reactions will in any case increase its density. Even where evidence of subduction of continental crust is lacking, blueschists do not necessarily result from subduction of purely oceanic or trench-fill materials, as was implied in the early models of subduction zones; they may still have continental affinities. Forbes, Evans & Thurston (1984) showed that this was the case for blueschists of the Seward peninsula, Alaska, and argued that the high pressure metamorphism resulted from the overthrusting of unknown material over sediments of continental origin that were already 200 Ma in age. Although associated with plate convergence, this process need not correspond to straightforward subduction.

One of the most interesting results of modern geophysical studies is that they clearly show that broadly continental material can be subducted to depths of the order of 50 km, and exceptionally to still greater depths. These depths are in exellent agreement with what is inferred from the mineral assemblages of high pressure metamorphic rocks, assuming that pressure is due to overlying load only. This effectively precludes the need to invoke additional factors such as 'tectonic overpressures' to account for the mineral assemblages of blueschists and related rocks.

Metamorphism in the accretionary wedge   Sedimentologists and structural geologists have identified sequences in some metamorphic belts that appear to have formed in ancient accretionary prisms (cf. Fig. 7.2(*a*)). An example is the lower Palaeozoic sequence of the Southern Uplands of Scotland, which has been demonstrated by Leggett *et al.* (1979) to have a number of features found in recent active margins. These rocks are now slates with metabasites that sometimes display prehnite–pumpellyite facies metamorphism (Oliver, 1978). This low grade of metamorphism, with extremes of neither pressure nor temperature, is consistent with what is known of the conditions prevailing in modern accretionary prisms, where the depth does not exceed 20 km and temperatures are low. In convergent zones such as the Makran Trench (Fig. 7.2(*b*)) where most sediment is subducted, the distinction between the accretionary prism and sediments that have been truly subducted may be a gradational one however.

METAMORPHISM AT A CONVERGENT MARGIN TODAY: THE EXAMPLE OF NEW ZEALAND

If we are to apply the principle of Uniformitarianism to the interpretation of metamorphic belts, then it follows that paired metamorphic belts are currently forming in a number of parts of the world. One area whose recent geological development is especially well studied is New Zealand, where motions over the past 20 Ma are constrained by studies of sea-floor magnetic stripes, palaeomagnetic declination and contemporary motions determined from repeated triangulations. Walcott (1987) has combined the information obtained from such sources to give a picture of the types of metamorphism that are likely to be going on now.

Subduction is taking place today along the Hikurangi Trough and Kermadec Trench, to the east of the North Island (Fig. 7.3), and much of the North Island is underlain by subducted lithosphere. Further south, the motion on the plate boundary gradually changes to predominantly strike-slip. Andesite volcanoes are developed in the central North Island.

The structural zones that can be recognised are, in order westwards from the trench:

1. *Frontal Wedge*   An accumulation of Miocene and younger sediments that are continuing to accrete and are actively deforming into a fold and thrust belt.
2. *Forearc Basin*   Rotation of the Pacific Plate relative to the Australian Plate around a pivotal point near the south end of the North Island has caused a forearc basin to develop in the Hawke Bay region, with extension perpendicular to the trend of the trench. Part of the forearc basin has recently begun to undergo uplift, even though stretching is continuing.
3. *Axial Zone*   This is a region of Quaternary uplift, at rates in excess of 1 mm per year, and corresponds to the arc. Mesozoic basement rocks are now exposed but the crustal thickness is in fact greatest beneath this zone.
4. *Central Volcanic Zone*   Another region of extension linked to the relative rotation of the plates is the site of much of the contemporary volcanic activity on the island. This zone has been particularly active over the past 1 Ma, during which time 2 km of subsidence of the Mesozoic basement has been more than compensated by the accumulation of volcanic products. Active stretching has greatly thinned the continental crust, and this region has been interpreted as the on-land extension of the back-arc stretching taking place beneath the ocean to the north.

Walcott has shown that the volume of sediments accreted is less than that which has been deposited in the area since the Miocene, and proposes that the missing sediments have been carried beneath the accretionary wedge and underplated on to the base of the crust as illustrated in Fig. 7.3(*b*). The thickness of underplated sediment would then account for the rise of the axial zone. The underplated sediment has reached its present position at depths of 20–30 km in only 2–3 Ma, and so must still be relatively cool ($\ll 400\,°C$). This material is almost certainly undergoing high pressure metamorphism today. In contrast the Central Volcanic Region is an area of high heat flow in which low pressure, high temperature metamorphism is taking place, though whether the metamorphic rocks that are forming there will be preserved will depend on what happens next to this very thin piece of crust.

CONCLUSIONS

At this stage it is possible to draw three major conclusions from the study of present-day tectonics, which are probably very important when it comes to interpreting old metamorphic belts.

There is no such thing as a typical convergent plate margin and therefore we should not expect all metamorphic rocks that formed in a broadly convergent setting to be the same. Oxburgh and Turcotte (1971) showed that the rate of convergence can dictate whether or

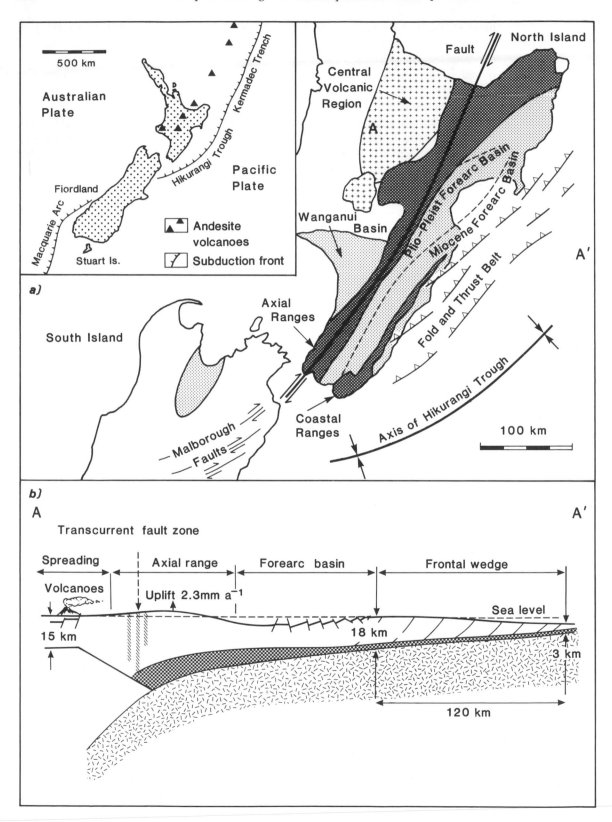

not paired metamorphic belts develop. Back-arc spreading and even the arc itself may be absent according to rate and angle of subduction.

Important changes in plate configuration can take place very rapidly. Walcott (1987) points out that major changes in New Zealand tectonics have taken place in the past 1 Ma, while there was a previous major change at 20 Ma.

When plate motions continue in an approximate steady state for periods of a few million years, progressive metamorphism of the downgoing slab takes place. It follows that when we obtain a date for the 'age of high pressure metamorphism' of old metamorphic rocks formed in a subduction zone setting, what we may be dating is the time at which such a period of steady state metamorphism ceased, due to changes in plate motions bringing a part of the 'conveyor belt' to a stop.

TIME AS A VARIABLE IN METAMORPHISM

One of the most important aspects of the work by Oxburgh and Turcotte (1971) was that they considered how the metamorphic conditions of a particular group of rocks would change over a period of time as subduction and accretion progressed. In the simplistic view of paired metamorphic belts, the metamorphism found simply reflects a prevailing thermal gradient that has been somehow 'frozen-in' at a particular moment in time. In practice, however, rocks may retain some mineralogical features inherited from their early history, while they may also undergo further changes before being exhumed at the surface. In addition, it may not be correct to assume that the metamorphism across a range of zones has necessarily been 'frozen-in' at the same instant.

The main impetus for investigations along these lines has come from the work of the Heat Flow Group set up by Oxburgh with S.W. Richardson at Oxford in the early 1970s. Their models, based on heat flow calculations, predicted a number of features of metamorphic rocks which have since been found by petrologists carrying out field-based studies. The principles of the procedures can be found in the equations of Jaeger, which were shown in Chapter 6 to allow the modelling of heating rates around cooling plutons. High speed computers permit more complex problems to be addressed.

Oxburgh and Turcotte (1971) not only characterised the thermal regimes in the vicinity of trench and arc, they also considered what would happen to the first-formed metamorphic rocks near the trench if steady state accretion were to continue for a long period of time. They showed that any individual rock would move progressively away from the low heat flow zone with time, and indeed might ultimately end up in the high heat flow arc region (Fig. 7.4). Hence, rocks that initially had high pressure mineral assemblages would tend to be progressively overprinted by intermediate pressure and even low pressure assemblages. This accords well with, for example, the common observation of calcic amphiboles typical of intermediate pressure metamorphism rimming glaucophane in blueschists. More extremely, some metamorphosed successions can be shown to have been subjected to high, intermediate and low pressure metamorphism at different times (e.g. Yardley, Barber & Gray, 1987).

Fig. 7.3 opposite Contemporary tectonics of northern New Zealand (see inset for general tectonic setting). *a*) Map showing the location of areas of uplift (dark stipple), subsidence (light stipple), and contemporary volcanism (+ symbol). *b*) Speculative cross-section along the line A–A', heavy stipple denotes underplated sediment. After Walcott (1987).

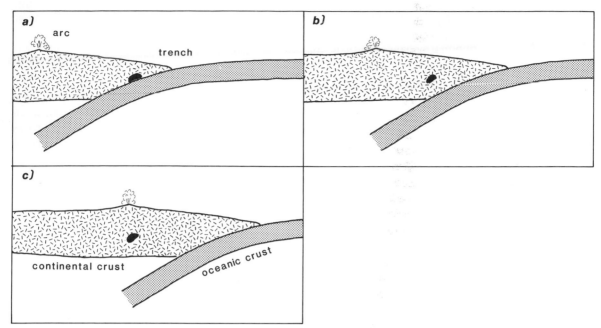

Fig. 7.4   Cartoons to illustrate the model of Oxburgh & Turcotte (1971) to account for changes in the baric types of metamorphism affecting an individual rock with time. *a*) The body of rock shown in black has been underplated on to the upper plate in a low heat flow setting, preserving high pressure assemblages. *b*) With further migration of the trench and underplating, the same rock mass now experiences intermediate pressure metamorphism. *c*) After further underplating the labelled rock body is now beneath the arc, experiencing high temperature metamorphism.

The steady state model is clearly not a unique interpretation of such occurrences; they could equally result from a change in plate configuration imposing a new thermal regime on older metamorphic rocks. For example Mesozoic basement rocks which have been subjected to low pressure metamorphism in the Central Volcanic Zone of New Zealand over the past million years are likely to have previously undergone a separate low grade intermediate pressure metamorphism during the Cretaceous Rangitata Orogeny. Nevertheless, it is quite clear from the Oxburgh and Turcotte model that high pressure metamorphic rocks will only be returned to the surface unchanged if something happens to interrupt the original pattern of plate motions, or if the steady state motions include a 'return mechanism' to bring them up to the surface.

The general thermal consequences of crustal thickening on metamorphism were considered by England and Richardson (1977). One particular model that they investigated was a simple but very general one in which the continental crust is thickened by overthrusting of another 30 km thick crustal block. The rocks beneath the thrust undergo a rapid increase in pressure but although they also warm up, they do so more slowly because of their insulating qualities and the slow supply of heat. Hence the initial thermal gradient is very low and with time not only do the deeper rocks begin to warm up, there will also be uplift and erosion leading to a drop in pressure. Eventually, as rocks are exhumed towards the surface, they will finally begin to cool.

The way in which the metamorphic conditions of a specific body of rock change with time as a result of being overthrust in this way is shown in Fig. 7.5 and is known as a **pressure–temperature–time path** or *P–T–t* path. Note that the maximum temperature that the rock attains is reached much later than the maximum pressure; it is not really

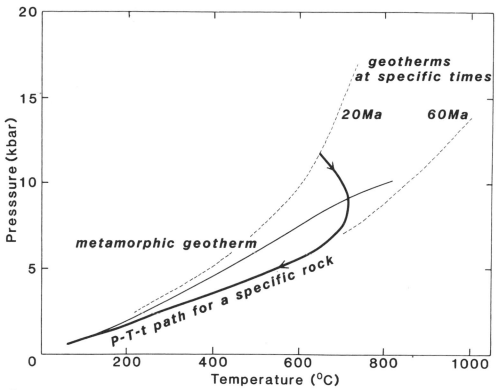

Fig. 7.5 P–T diagram to illustrate the relationship between geotherms in the crust at different times (broken lines), the changing P–T conditions to which an individual rock is subjected (P–T–t path), and the metamorphic geotherm, or array of peak-temperature P–T conditions obtained from rocks of successive metamorphic zones. After England & Richardson (1977).

meaningful to talk about 'peak metamorphic conditions' in this case without specifying whether it is peak pressure or peak temperature that is meant.

Another very important feature of the England and Richardson analysis emerges when the P–T–t paths of two different rock bodies, representing different zones within the overthickened mass, are compared. They showed that the point at which the upper body reaches its maximum temperature and begins to cool is reached while the lower body is still heating up. In other words maximum temperature conditions are not reached at the same time throughout the overthickened mass. Now if we consider what we actually find in the field, it is a series of metamorphic zones whose mineral assemblages normally reflect the maximum temperature conditions to which a rock has been subjected. By joining the points representing the P–T conditions at T_{max} calculated for each individual metamorphic zone we can define a line on the P–T diagram which has the appearance of being the geothermal gradient during metamorphism. In fact this line, termed a **metamorphic geotherm** by England and Richardson, has no real physical significance because the points that it connects do not represent P–T conditions that obtained simultaneously in the crust. The term **metamorphic field gradient** is often used in place of metamorphic geotherm where the geotherm is based on field data, not theory.

Similarly, it is clear from Fig. 7.5 that the sequence of metamorphic conditions deduced from a study of metamorphic zoning (i.e. the metamorphic geotherm) does not, in this case, correspond to the sequence of P–T conditions that the highest grade rocks passed through during progressive metamorphism. This does not invalidate the concept of progressive metamorphism, in the sense that rocks recrystallise progressively during heating rather than passing directly from a pre-metamorphic assemblage to a peak metamorphic one, but it does show that the sequence of zones through which a rock has passed need not correspond exactly with the zonal sequence that can be mapped today.

The simple overthrust model used by England and Richardson is not, it could be argued, a very good model for the formation of metamorphic belts. This would be to ignore the great value of their modelling, which is to demonstrate how the $P-T$ conditions and baric type of metamorphism can change with time within a metamorphic belt. In fact, further modelling, notably the extensive studies of England & Thompson (1984) and Thompson & England (1984) have confirmed that the types of $P-T-t$ paths illustrated in Fig. 7.5 are likely to be followed whenever continental crust is overthickened and subsequently cooled by being uplifted and eroded *en masse*.

FIELD AND TEXTURAL EVIDENCE FOR CHANGING BARIC TYPE WITH TIME

One of the conclusions of thermal modelling studies is that the metamorphism of classic paired metamorphic belts such as the Sanbagawa belt and the Ryoke belt may not correspond to a single metamorphism in a specific geothermal gradient in the way that it appears to from earlier work. In fact Ernst, Onuki & Gilbert (1970) record that in many parts of the Sanbagawa belt sodic amphibole is partially replaced by green calcic amphibole, probably indicative of a rise in temperature or drop in pressure leading to a shift from blueschist to greenschist facies. Low pressure metapelites from the Abukuma plateau, an extension to the north-east of the Ryoke belt (Fig. 7.1), yield very rare kyanite (Kano & Kuroda, 1968), also suggesting that in detail the metamorphism does not correspond to a single baric type. It is always possible to interpret these occurrences where minerals typical of more than one baric type occur together as indicating that, by coincidence, metamorphic conditions were intermediate between the baric types; however, textural evidence frequently indicates that in fact metamorphic conditions have changed with time.

Changing geothermal gradient in the Haast Schists, New Zealand

The bulk of the Haast Schists of the South Island of New Zealand have been metamorphosed under greenschist facies conditions, although locally higher and lower grades are preserved. The metamorphic zones are defined on what are probably peak temperature assemblages and their distribution is shown in Fig. 4.1. The sequence of zones (from Turner, 1981) is:

$$\text{prehnite–pumpellyite} \rightarrow \text{pumpellyite–actinolite} \rightarrow \text{chlorite} \rightarrow \text{biotite} \rightarrow \text{garnet}$$

It has been known for many years that sodic amphiboles are occasionally present in the metabasites, and Landis and Coombs (1967) used these as indicators of relatively high metamorphic pressures. When the textures are examined in detail however, it is apparent that the sodic amphiboles are almost invariably relics, either mantled by calcic amphiboles or preserved as inclusions in albite porphyroblasts. Yardley (1982) reinterpreted these occurrences in terms of changing metamorphic conditions and inferred that, although there are no metamorphic zones in the sequence in which sodic amphibole occurs as an equilibrium member of the peak metamorphic assemblage, the $P-T-t$ path of the middle grade metabasites had nevertheless passed through conditions in which sodic amphiboles grew. In other words this example illustrates the distinction that was made by England and Richardson between the metamorphic geotherm, indicated by the peak metamorphic $P-T$ conditions of formation of successive metamorphic zones exposed today, and the $P-T-t$ paths followed by specific rocks during their metamorphic history (illustrated in Fig. 7.6). The metamorphic geotherm lies at moderate pressures in the prehnite–pumpellyite and

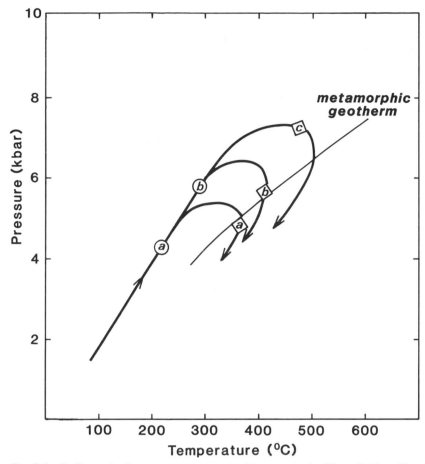

Fig. 7.6   P–T–t paths for successive metamorphic zones in the Haast Schists, New Zealand (Fig. 4.1). From Yardley (1982). Also shown is the metamorphic geotherm indicated by the peak metamorphic assemblages.

greenschist facies, whereas the *P–T–t* paths of individual rock units passed through a higher pressure region of the *P–T* diagram, transitional to the blueschist facies. (See however Jamieson & Craw, 1987, for a critique of some of the detailed estimates of *P–T* conditions along the *P–T* paths shown.)

CONCLUSION

When account is taken of the way in which *P–T* conditions of metamorphism can change with time within a given body of rock, it is clear why the zones in many metamorphic belts do not correspond to a simple high pressure or low pressure facies series, even if the rocks were metamorphosed in an arc or trench setting at some time in their history. Many metamorphic rocks have experienced more than one tectonic setting of metamorphism in their history, while all that are now exposed at the surface may show effects formed during uplift superimposed on those formed during burial.

PRESERVATION OF HIGH PRESSURE ROCKS AFTER METAMORPHISM

High pressure metamorphism requires an anomalously low geothermal gradient, and therefore mineral assemblages formed in such an environment are liable to react to form assemblages typical of intermediate pressure metamorphism when a normal thermal gradient is restored. We are therefore faced with a considerable problem in explaining how it is that well-preserved examples of high pressure metamorphic rocks come to be found at the surface today. It could be argued that the occurrence of granulites at the surface presents a similar problem of the preservation of assemblages formed under extreme conditions once they return to the upper crust. However, the retrogression of granulites is much more strongly dependent on the infiltration of the water necessary to form low grade assemblages than is the case for high pressure rocks, and so high grade granulite assemblages are preserved with relatively few changes during uplift provided they remain dry. While examples of partial to complete overprinting of blueschist by greenschist are not uncommon, e.g. the Haast Schists described above, it is nevertheless true that many blueschists display little or no overprinting. Furthermore, quite apart from the metamorphic problems of preserving blueschists under conditions in which they are no longer stable, there is also a tectonic problem in driving the uplift of rocks from the base of the crust to a position outcropping at the surface and typically overlying a normal thickness of crustal rocks.

There is general agreement by many workers that blueschists must often be returned to the surface rapidly, and this may itself place kinetic constraints on further reaction. Quite apart from the fact that rapid return to the surface may be necessary for preservation of the mineral assemblages, there is direct evidence for rapid uplift from the very young ages of some blueschists (<15 Ma). This uplift may be a response to changes in plate configuration or it could be a continuous process going on during steady state subduction.

The fact that in many metamorphic belts a distinct age of blueschist metamorphism can be recognised tends to favour the view that their uplift and preservation is a 'one-off' event consequent upon a change in plate motions. On the other hand, some high pressure belts do show multiple events of high pressure metamorphism, and there is also evidence in some areas that blueschists reached the surface while convergence was continuing.

The thermal problem of blueschist preservation amounts to this: if a large body of blueschists at, say, 450 °C and 10 kbar is uplifted rapidly so that little cooling can take place, to conditions of 400 °C and 5 kbar then it should react to form greenschist facies assemblages. Although some reactions may be kinetically unfavourable (e.g. those that also require addition of water), a steep uplift path like this may well cross dehydration reactions which are likely to proceed rapidly (Fig. 7.7). In order for blueschists to rise to the surface without extensive further reaction towards Barrovian assemblages, they need to cool as they rise (e.g. path A on Fig. 7.7) and this implies uplift in a region of low heat input, either slowly or with loss of heat into adjacent rock units.

The varying degrees of low pressure overprinting in different high pressure belts suggests that both types of process may operate. One additional factor that may be significant is the fact that not all blueschists rise up as the large coherent masses assumed by the simple sort of thermal model outlined above. Many descriptions of blueschist occurrences emphasise that they have been emplaced as imbricate thrust slices e.g. the Shuksan blueschists, north-west USA described by Brown (1986) or the Corsican blueschist described by Warburton (1986). Goffe and Velde (1984) have pointed out that if blueschists rise up in relatively thin slices they may lose heat to the cooler rocks into which

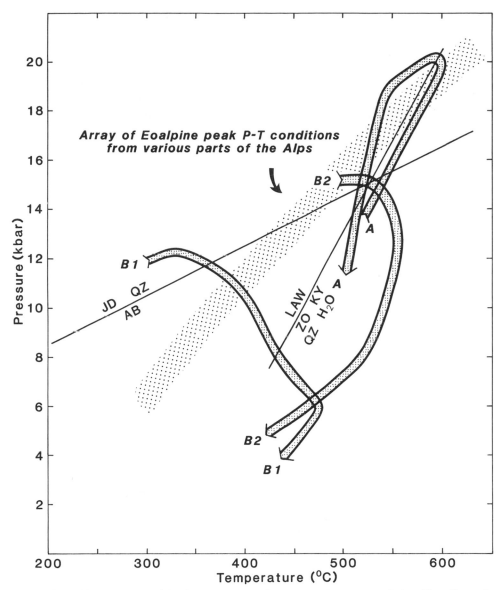

Fig. 7.7 P–T conditions of Eoalpine metamorphism in various parts of the Alps (from the compilation by Chopin, 1987), together with uplift paths proposed for specific high pressure rock units. Path A is the Zermatt–Saas zone (Fry & Barnicoat, 1987), path B1 is for the Vanoise (Platt, 1987) and path B2 for the Sesia–Lanzo zone (Rubie, 1984).

they are emplaced. To date, *P–T–t* paths described for the uplift history of blueschists include both those indicating warming during uplift, and others suggesting cooling. Contrasting examples from the literature on the western Alps (see below) are shown in Fig. 7.7, but it remains to be conclusively shown that the differences between these paths for nearby areas are real, and not the result of different approaches taken by different authors.

Various workers, following Cowan & Silling (1978) and Cloos (1982), have proposed models to account for the return of blueschists to the surface as part of continuing convergence at subduction complexes. A recent model by Platt (1986, 1987) is based on

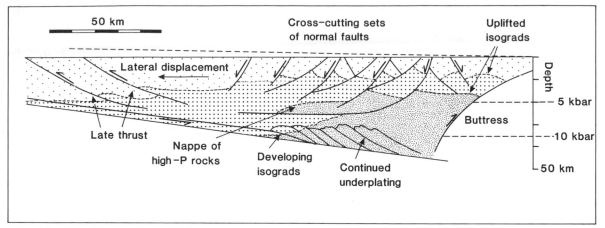

Fig. 7.8   Model for continuous high pressure metamorphism, underplating and return to the surface of high pressure rocks. From Platt (1987). A rigid buttress at the right constrains the behaviour of progressively accreting material. Sediment is underplated at high pressures in front of the buttress and causes uplift of earlier underplated high pressure rocks. Extension towards the trench causes some high pressure rocks to be thrust forward in nappes.

evidence from the geological history of both the European Alps and the Franciscan Formation of California, that blueschists formed early in a convergent episode can reach high crustal levels or even become exposed while the convergence continues. Platt's model is illustrated in Fig. 7.8, and has considerable similarities with Walcott's interpretation of present processes beneath New Zealand (Fig. 7.3). Continued underplating of high pressure rocks to the base of the crust causes it to become thickened and rise up, at the same time causing extension between the area of underplating and the trench. The high pressure rocks are forced upwards by the continued accretion of more metamorphic rocks beneath them, and may be thrust forward in nappes in the region of extension. An interesting feature of this model is that it predicts that older blueschists will occupy higher structural levels, and this is in accordance with field observations.

At the present time the subject of blueschist preservation remains controversial, but the necessary data to test alternative models are beginning to be accumulated from field-based studies.

TECTONIC SETTING OF LOW PRESSURE METAMORPHISM

Miyashiro (1972) showed low pressure, high temperature metamorphism as taking place in thickened continental crust of the volcanic arc, due to the introduction of large amounts of heat to the upper crust by ascending magmas. More recently emphasis has been placed on stretching and thinning of the crust as a cause for high heat flow, and hence, implicitly, this type of metamorphism. Since some low pressure metamorphic belts do not appear to conform to the paired metamorphic belt ideal, it is clear that considerable problems remain in their interpretation.

VOLCANIC ARCS AND LOW PRESSURE METAMORPHISM

There is no doubt that volcanic arcs provide a suitable heat flow setting for low pressure metamorphism, and there are many examples of low pressure metamorphism that are both temporally and spatially related to magmatism. Nevertheless, as indicated in Fig. 7.3, it is possible that the low pressure metamorphism may in fact take place in thinned crust adjacent to the arc as well as in the arc itself. An example of low pressure metamorphism developing on a regional scale adjacent to intrusions with volcanic arc affinities is provided by the Dalradian rocks of the Connemara region of Ireland (Fig. 7.9). These rocks are the along-strike correlatives of the Dalradian rocks of Scotland in which Barrow's zonal sequence was described (Ch. 1, page 7). Their metamorphic development during Ordovician times is described by Yardley, Barber & Gray (1987), who recognise an early phase of intermediate pressure metamorphism prior to the low pressure event, which followed without a break.

The early metamorphic episode was broadly comparable to Barrovian metamorphism, and resulted in the production of garnet, followed by staurolite which is rarely accompanied by kyanite. The reason for the scarcity of kyanite appears to be partly a lack of rocks of suitable composition, partly because, for the temperatures attained, pressures were a little less than in the classic Barrovian area. The metamorphic grade at this stage appears to have been rather uniform across the presently exposed area, but there followed a period of magma emplacement which resulted in further heating. At the same time, uplift occurred so that mineral assemblages develop which are indicative of lower pressures as well as higher temperatures. In one sample, small relics of kyanite occur within cordierite. The clearest evidence for declining pressures is textural; sillimanite and garnet are replaced by cordierite in migmatites, while in the sillimanite zone garnet has broken down extensively by reaction with muscovite to produce biotite and sillimanite. Both these reactions are strongly sensitive to pressure changes, but only weakly sensitive to temperature.

The metamorphic zones found today parallel the trend of a belt of calc-alkaline basic to acid intrusions in the southern half of the region and cross-cut complex structures (see also Fig. 6.12(*b*)). Hence, they amount to a thermal aureole developed on a regional scale since it extends for over 10 km from the intrusives. In part the large scale of the contact effects probably results from the rocks having been already at garnet grade temperatures prior to the onset of magmatism. The zonal distribution in Connemara is illustrated in Fig. 7.9, together with *P–T–t* paths for the rocks of the different zones. It is the fact that metamorphism was accompanied by uplift and erosion which accounts for the different paths followed by the different zones on the *P–T* diagram. This is in accord with the notion of thick crust during arc-related metamorphism in this instance.

The type of *P–T–t* path found in Connemara, indicative of high temperature, low pressure metamorphism overprinting earlier, higher pressure assemblages, is not uncommon, although the role of magmatism is not always as clearly defined. Vielzeuf (1980) has described spectacular low pressure granulite facies assemblages in pelitic migmatites from the French Pyrenees that are produced by depressurisation of earlier, kyanite-bearing parageneses.

CRUSTAL STRETCHING AND LOW PRESSURE METAMORPHISM

Although regional metamorphism is usually linked to convergent tectonic regimes, there are some instances where a case can be made for regional metamorphism caused by crustal extension. Probably the clearest example of this is the low grade (prehnite–pumpellyite facies) metamorphism in parts of the Andes (Aguirre & Offler, 1985). This occurs in volcanogenic sediments deposited in Mesozoic extensional marginal basins, and metamorphosed soon after deposition. Robinson (1987) has highlighted the distinctions between low grade metamorphism in this setting and in convergent zones. A steep

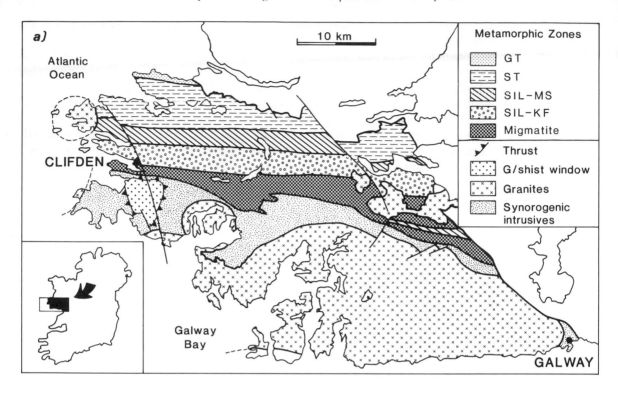

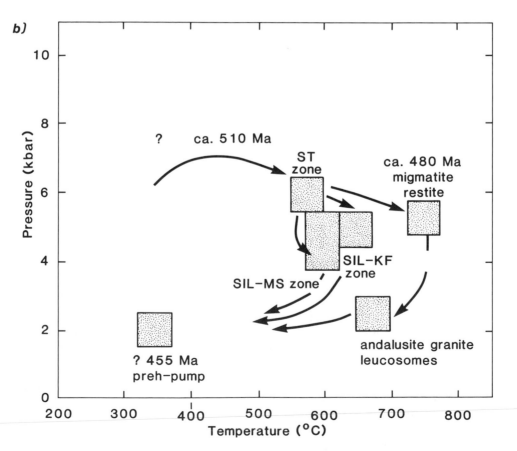

geothermal gradient, sometimes approaching that found in modern geothermal fields, is distinctive.

Additionally, some other low pressure metamorphic belts do not have a clear relation to contemporary arc magmatism, and nor can high pressure pairs be located for them. In the view of many geologists these belts have developed within continental crust without extensive oceanic involvement, although this is hotly contested by others who seek to interpret them in terms of more conventional plate tectonic models.

The Hercynian or Variscan orogenic belt of Europe and eastern North America is the best known belt of this kind. Although intermediate and even high pressure metamorphic rocks are present, most of the medium to high grade metamorphism is of low pressure type. Recently Weber and Behr (1983) have suggested that the granulite facies metamorphism that gave rise to the classic Saxony granulites during the lower Palaeozoic was the result of crustal stretching during rifting when the sedimentary basin, in which the Variscan sediments were subsequently deposited, first formed. The argument is that thinning of the continental crust by lithospheric stretching allows hot material of the aesthenosphere to approach closer to the surface, thereby causing an enhanced thermal gradient. It was initially put forward by McKenzie (1978) in the context of the origin of sedimentary basins. Wickham and Oxburgh (1985, 1987) have also argued that rifting was the cause of the high heat flow that gave rise to the upper Palaeozoic (Variscan) metamorphism in the Pyrenees. They have developed the model in considerable detail and show (1987) that even with the effects of rifting, some intrusion of magma at depth is also necessary to account for the observed metamorphism.

At the time of writing both the question of the tectonic origin of fold belts such as the Variscan or Hercynian, and the causes of their metamorphism, remain controversial; not least because new work is beginning to indicate that the Variscan may not be so different from other orogenic belts as has been thought. It is likely that considerable advances will be made in the next decade.

METAMORPHISM AND CONTINENTAL COLLISION

The idea that the continents must have moved together in some way to account for the structures observed in mountain ranges is an old one, dating back to the classic nineteenth-century studies by Suess and Argand in the European Alps. Of the early workers, Ampferer clearly recognised that there is insufficient basement present today in the Alps for the extensive cover sequences repeated by thrusting, and suggested that some had been lost by underthrusting. Early work in Scandinavia also demonstrated extensive crustal shortening, but it is only with the advent of plate tectonic theory that such processes have been placed in the context of movements elsewhere on the earth's surface.

When ocean basins close it is inevitable that the opposing areas of continental crust will be brought together. This may be a relatively passive process, with little overthrusting, or may lead to one mass of continental crust being thrust over the other to produce greatly overthickened crust as in the Himalayas today. If the continental crust on the downgoing plate has been appreciably thinned, there may be a period of deep underthrusting or

Fig. 7.9 opposite   a) Metamorphic zonation in the Connemara inlier, Ireland. *b*) *P–T–t* paths for successive zones in the Connemara Schists, showing uplift and erosion accompanied by heating. Also indicated are approximate dates for the different metamorphic stages. Both from Yardley, Barber & Gray (1987).

subduction of continental crust prior to collision. When we study collision belts today there is typically a baffling diversity of metamorphic phenomena because the rocks may include older metamorphic basement, metamorphic rocks formed in the same convergence event prior to collision, metamorphic rocks related to the collision itself and also the products of the contact effects of collision-related magmatism. Only with detailed field and geochronological studies is it possible to unravel this diversity.

METAMORPHISM OF THE ALPINE-HIMALAYAN CHAIN

The Alpine–Himalayan chain provides the most spectacular example of a collision orogeny from the relatively recent geological past; indeed collision is continuing today along parts of its length. The dramatic Alpine structures led Argand to recognise the major role of large-scale horizontal movements in the formation of the Alps long before the development of Plate Tectonic theory. Despite the fact that the European Alps have been the subject of many classic studies of metamorphism over the past 100 years, it has only been quite recently that a metamorphic map for the Alps has been prepared. Since the early 1970s, detailed structural and petrological studies have been carried out in the Himalayas, where many broad similarities in the structural and metamorphic evolution are becoming apparent.

The European Alps   The outer (northern and western) parts of the Alps comprise nappes of folded but only weakly metamorphosed sediments (up to prehnite–pumpellyite facies), together with slices of crystalline basement known as the external massifs. This comprises the Helvetic zone. Overlying these rocks are nappes of the Penninic zone, which include the highest grades of Alpine metamorphism. The zone includes both Mesozoic sediments and crystalline continental basement rocks (the internal massifs), together with ophiolite slices (for example, the Zermatt–Saas zone) which are remnants of the Mesozoic oceanic crust that floored the deep sedimentary basins. The Penninic zone is overlain by nappes of the Austroalpine series, again including both crystalline basement and Mesozoic sediments. These mostly occur in the eastern half of the Alps, but the lowermost of them, the Sesia–Lanzo zone, extends into the western Alps and displays a metamorphic history more comparable to that of the Penninic zone.

The major metamorphic divisions of the Alps are illustrated in Fig. 7.10. A distinctive feature of the Alpine chain is that high pressure metamorphism has affected extensive late Palaeozoic to Mesozoic platform sediments with abundant carbonates, together with the continental basement rocks. Apart from the scattered ophiolite fragments, the rocks involved are quite different from those of typical circum Pacific high pressure belts. In the Alps, high pressure metamorphism affects mature sediments, such as aluminous pelites and quartzites, and also granitic basement, whereas in the typical Pacific belts it affects immature greywackes.

Although conventional petrological and geological techniques provided a general overview of Alpine metamorphism, it was only with the advent of painstaking geochronological studies by several European groups that a coherent picture of the metamorphic development of the Alps began to emerge, and in particular that the timing of the high pressure metamorphism was established. Frey *et al.* (1974) have provided a major synthesis of Alpine metamorphism (see Fig. 7.10), and most of their conclusions remain unchanged by subsequent work. The specific problems involved in geochronological studies of metamorphic rocks have been outlined by Cliff (1985).

The earliest metamorphic events recorded in the Alps affect only the crystalline basement rocks. Although this metamorphism is usually referred to as Variscan

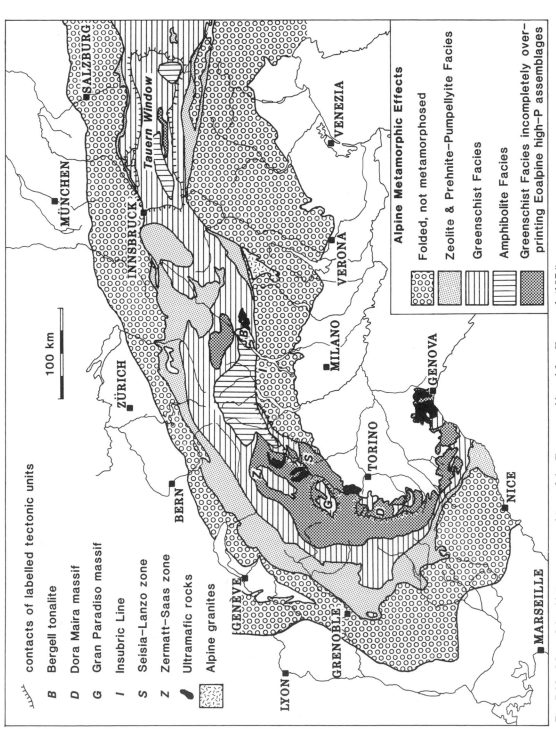

Fig. 7.10 Map of the Alpine metamorphism of the European Alps. After Frey *et al.* (1974).

(Hercynian), it probably includes earlier, Ordovician, effects. The Variscan metasediments were intruded by late Variscan granitoids in late Carboniferous to early Permian times providing a complex crystalline basement for subsequent sedimentation. The pre-Alpine metamorphism of this basement is specifically excluded from Fig. 7.10, so that some rocks shown as having been unaffected by Alpine metamorphism, are in fact granulites. Rifting and subsidence in the early Mesozoic were followed by the generation of new oceanic crust in the middle Jurassic through lower Cretaceous. This material is sparsely preserved in ophiolites which give a range of radiometric dates from mid-Jurassic to mid-Cretaceous. These may in part be dates of sea-floor metamorphism of newly formed oceanic crust.

The earliest Alpine regional metamorphism is late Cretaceous in age (90–65 Ma) and in the Penninic zone and Sesia–Lanzo zone comprises high pressure, relatively low temperature metamorphism. This event was responsible for the formation of eclogites and blueschists in both continental and oceanic basement and in Mesozoic sediments. It is well represented in both the western and eastern Alps (the Tauern Window of Austria), but has been completely obliterated by later events in the central Alps. The term **Eoalpine metamorphism** is used for this phase, and in addition to giving rise to high pressure rocks in the Penninic zone, metasediments of the Austroalpine nappes underwent moderate pressure metamorphism up to the amphibolite facies at this time. The relics of medium pressure Eoalpine metamorphism in the Austroalpine nappes are not distinguished on Fig. 7.10 but they include staurolite schists to both east and west of the Tauern Window that have been only partially overprinted by the Alpine greenschist event.

In the Sesia–Lanzo zone of northern Italy (described by Compagnoni, 1977), Eoalpine effects are well preserved in both polymetamorphic micaschists (the Eclogitic Micaschist Complex) and pre-Alpine granitoid intrusions. Although the high pressure metamorphism was usually accompanied by penetrative deformation, some areas remain undeformed and have recrystallised to high pressure assemblages while retaining their original textures. Perhaps the best known of these rocks is the metagranite of Monte Mucrone (Compagnoni & Maffeo, 1973), in which original K-feldspar is largely unchanged, while plagioclase has been completely replaced by pseudomorphs of fine grained jadeite with zoisite and quartz. Biotite grains are rimmed by fine garnets and quartz has recrystallised to polygonal aggregates, but the rock retains a typical granitic appearance in the less deformed hand specimens.

To the north-west of the Sesia–Lanzo zone, in the Penninic zone, both the Palaeozoic crystalline basement of the internal massifs (i.e. the Gran Paradiso, Dora Maira and Monte Rosa massifs) and the Mesozoic sediments (the Schistes Lustres) have been subjected to Eoalpine high pressure metamorphism (Dal Piaz & Lombardo, 1986). Chopin (1981) has shown that the unusual assemblage talc + phengitic muscovite is widespread, indicating the breakdown of chlorite at high pressure, e.g.:

$$\text{chlorite} + \text{quartz} \rightarrow \text{talc} + \text{garnet} + H_2O$$

Although originally most petrologists were conservative in their estimates of the pressures required for this and other high pressure reactions, suggesting figures of 'more than 7–8 kbar', it has become apparent recently that substantially higher pressures were actually involved. The most dramatic evidence of this is the discovery by Chopin (1984) of coesite, occurring as armoured inclusions in large pyrope garnets from metasediments of the Dora Maira massif. He has suggested that all the quartz in these rocks was converted to coesite at the high pressure peak of metamorphism, but inverted to quartz during uplift. However, those grains that were enclosed in garnet were unable to react to quartz because there was insufficient room inside the garnet for the necessary volume change: in other words the garnet acted as a rigid pressure vessel. In addition to coesite, these rocks contain a number of other unique phases because of their special combination of extreme pressures and sedimentary bulk compositions (Chopin, 1987). In view of the occurrence of coesite, metamorphic pressures are likely to have been of the order of 30 kbar, corresponding to depths of the order of 100 km. Not all the internal massifs were subjected to such extreme pressures, and part of the range of Eoalpine *P–T* conditions for different parts of the Penninic zone of the western Alps is indicated on Fig. 7.7.

A detailed study of the metamorphism of the Tauern Window by Miller (1974) showed that there were in fact three distinct metamorphic phases. The earliest phase was an eclogite facies event (restricted to one tectonic unit only), and this was followed by blueschist facies metamorphism. Lastly, there was greenschist facies overprinting of the high pressure assemblages. Further work by Holland (1979b) in this region demonstrated that the eclogite facies metamorphism took place at around 20 kbar and affected probable Mesozoic sediments as well as basement; hence it could not be a part of a separate, entirely pre-Alpine metamorphic cycle. Textures indicating the replacement of early formed eclogite by blueschist are also well known from the western Alps.

Eoalpine high pressure assemblages usually display varying degrees of overprinting during the main stage Alpine metamorphism or Lepontine event. This later event has been dated in the western Alps at around 40 Ma, i.e. late Eocene, and therefore postdates the onset of collision which was probably early Eocene in this area. Hence the Lepontine event is the first metamorphism that can be truly correlated with collision. The regional metamorphism in the central Alps is dominated by the Lepontine metamorphism and few Eoalpine relics remain. The metamorphic grade increases southwards until abruptly truncated at the Insubric line, and isograds have a relatively simple geometry, in contrast to the structural complexity. The zonal scheme for the metacarbonate rocks has already been outlined in Chapter 5 (page 131). Pelitic lithologies indicate an intermediate pressure facies series for Lepontine metamorphism, broadly comparable to that of the Barrovian zones of Scotland. However, the low grade zones are better developed and distinct stilpnomelane and chloritoid zones have been mapped. These pass southwards into a kyanite zone (with staurolite) and a sillimanite zone.

Although the peak of the main Lepontine phase of Alpine metamorphism is usually interpreted as having ended in the early Oligocene, many Miocene ages have been obtained. Low grade rocks in the Glarus Alps and fissure minerals elsewhere in the central Alps yield Miocene ages, and these may indicate a distinct later phase of metamorphism.

Causes of Alpine metamorphism

It is clear from what is known of the palaeogeography of the Tethyan area (Tethys being the ocean that broadly preceded the present Mediterranean) that at the time of the Eoalpine metamorphism there was subduction of oceanic crust, with or without continental fragments, beneath the southern margin of the European plate. Hence the high pressure Eoalpine event is subduction related, at least in a general sense. Oxburgh and Turcotte (1974) modelled the thermal development of Alpine metamorphic rocks with particular reference to the eastern Alps, but addressed the general problem of explaining the Eoalpine high pressure metamorphism followed by an intermediate pressure event about 25–30 Ma later. They concluded that major thrusting and development of nappes took place at about 65 Ma, and the high pressure rocks either formed as a result of this process or were incorporated into the nappe pile by it. With time, the thermal perturbations within the crust caused by the thrusting were eliminated and rocks at depth were warmed up to give rise to the Lepontine type of metamorphism. This is thus a precursor to the model developed by England and Richardson (1977) and discussed above.

Whether or not the nappe pile itself was thick enough to give rise to the Eoalpine high pressure metamorphism without subduction beneath an overlying plate has been controversial throughout the Alps, but the very high pressures determined recently from the western Alps make it seem unlikely. In any event it is clear that the major thrusts and nappes in the western Alps and Sesia–Lanzo zone postdate the Eoalpine metamorphism, since nappes of contrasting Eoalpine metamorphic grade are interleaved, whereas the Lepontine event is post-tectonic (Compagnoni *et al.*, 1977). There remains uncertainty about the precise causes of the Lepontine overprint. Many authors have ascribed it to the formation of a thermal dome, implying a locally enhanced heat flow perhaps linked to subsequent magmatism. On the other hand the metamorphism does not imply a steeper geothermal gradient than normal for continental crust so it might be thought that an enhanced heat flow was not required. This has been confirmed by heat-flow modelling studies by Oxburgh & England (1980).

Variations along the length of the chain   Although detailed field studies are still in their infancy in the Himalayas, it is now becoming apparent that there are general similarities in the metamorphic evolution to the Alps. Burg & Chen (1984) have summarised the development of the suture zone in southern Tibet, where again metamorphic effects produced in earlier orogenies and at continental margins have become juxtaposed and the effects of the Eocene collision of the Eurasian and Indian plates have been superimposed in addition.

In many respects a much greater contrast is provided by the rocks of the eastern Mediterranean. This is now a land-locked ocean basin, preserved while the continental plates to east and west collided due to irregularities in the plate boundaries. Le Pichon (1983) has described its current and likely future tectonic behaviour, and it is possible that paired metamorphic belts may be forming today in response to the active subduction, arc volcanism and back-arc spreading. Much younger high pressure metamorphic rocks are exposed today in and around the eastern Mediterranean than occur in the Alps, because of the continued subduction in the region. It provides a salutary example of the possibilities for major variations along the length of older metamorphic belts.

POSSIBLE ROOTS OF A CALEDONIAN COLLISION IN WEST NORWAY

In the Western Gneiss Region of Norway, introduced in Chapter 4 (page 107), high temperature, high pressure assemblages are present which appear to have formed under conditions corresponding to the lower part of continental crust overthickened by collision. The clearest demonstration of high $P-T$ conditions comes from eclogite bodies but the interpretation of these has been controversial, as was noted in Chapter 4 (page 107). On the other hand at least some of the eclogites retain relics of an original ophitic texture requiring the former presence of plagioclase, which shows that their precursors were high level crustal rocks, rather than mantle rocks; evidence which is supported by the dyke-like field relations of some eclogite bodies (Bryhni, Krogh & Griffin, 1977). Furthermore Krogh (1980) has found pyroxene + garnet assemblages in some of the acid gneisses that host the eclogites, indicating pressures of around 18 kbar, which approach those required for the eclogites.

If it is correct to assume that the eclogite bodies of the Western Gneiss Region have been metamorphosed in place and that the surrounding acid gneisses have also been subjected to the high $P-T$ metamorphism, then the bulk of this 300 km strip of western Norway has been metamorphosed at depths of around 60 km. On the basis of this interpretation of the metamorphism, Krogh (1977) has interpreted the region as having formed the lower plate of a continental collision zone, now believed to have formed during the Caledonian orogeny. Despite the uncertainties, the balance of the evidence does seem to support the interpretation of the eclogites having been metamorphosed in place, which is the basis for this tectonic model. Presumably comparable rocks are being formed below the Himalayas today.

METAMORPHISM RELATED TO OPHIOLITES

Ophiolite suites are a common feature of many orogenic belts, including the Alps, the Caledonian–Appalachian chain and many circum-Pacific belts. They are composed primarily of basic and ultrabasic rocks displaying many of the features of oceanic crust and mantle: pillow lavas and sheeted dykes overlying gabbro and ultramafic rocks. Hence, they are generally interpreted as slices of oceanic (or oceanic-like) crust that have been emplaced on to continental crust or **obducted**.

In many metamorphic belts ophiolites occur only as dismembered slices. They can display metamorphic effects resulting from a regional metamorphic overprint shared by the surrounding rocks, or may retain assemblages that formed on the sea floor. In some instances, however, metamorphic features are present that clearly relate to the emplacement of the ophiolite, although their interpretation remains controversial.

SEA-FLOOR METAMORPHISM OF OPHIOLITE ROCKS

The process of sea-floor metamorphism and the rocks that it produces have been described in Chapter 4 (page 120). Although believed to operate primarily at mid-ocean ridges, in principle it may also occur wherever igneous activity is taking place on the sea floor, although some differences might be expected where water depths are only shallow. Sea-floor metamorphism is the dominant metamorphic process occurring at constructive plate margins, but its primary significance for geologists working with rocks that outcrop today, is that they may retain mineral assemblages and textures produced beneath the sea bed, even after they have undergone subsequent deformation and regional metamorphism. The rocks of the Ligurian Alps outlined in Chapter 1 are an example of this, and many ophiolites retain evidence of some metamorphism on the sea floor; nevertheless, sea-floor metamorphism is by no means restricted to ophiolites, and can occur in other metavolcanic rocks within orogenic belts.

METAMORPHISM ACCOMPANYING OPHIOLITE EMPLACEMENT

Some of the most complete ophiolite sections occur in Newfoundland and are of Ordovician age; the best known is the Bay of Islands ophiolite. Williams and Smyth (1973) pointed out that the basal ultramafic portion of the ophiolite is usually underlain by 150–300 metres of metamorphosed basic volcanic rocks, known as the **metamorphic sole**. These were tectonically emplaced with the ophiolite and are unrelated to the underlying sediments. There is a rapid decrease in metamorphic grade away from the contact with the ultramafic base of the ophiolite, from pyroxene amphibolite through amphibolite and garnet amphibolite to greenschist, with the amphibolite facies zone only totalling about 90 metres in thickness. On the face of it, this sequence constitutes a metamorphic aureole caused by the overthrusting of the hot ophiolite. In detail, however, some puzzling anomalies emerge.

Jamieson (1981) studied the zonal sequence beneath the White Hills Peridotite, also in

Table 7.2   Progressive metamorphic zonation with increasing distance beneath the White Hills Peridotite, Newfoundland. (Compiled from Jamieson, 1981)

Rock unit	Typical metamorphic assemblage	P (kb)	T (°C)
White Hills Peridotite	—		
Green Ridge Amphibolite	Clinopyroxene + Mg–hornblende + plagioclase + sphene (contact rocks)	7–10	900–950
	Brown hornblende + andesine + clinopyroxene + quartz ± orthopyroxene ± rutile	3.5–5	800–900
	Green hornblende + andesine + quartz + sphene and/or ilmenite		650–800
Goose Cove Schists	Blue-green hornblende + epidote + albite ± oligoclase ± quartz ± sphene	2–4.5	500–650
	Actinolite + epidote + albite + chlorite ± quartz ± sphene		350–500

Newfoundland, and found a comparable zonal sequence which is tabulated in Table 7.2, together with the metamorphic temperatures and pressures that she calculated from the mineral chemistry and assemblages present. The remarkable variation in metamorphic pressures over a short distance, together with the fact that higher pressure rocks overlie those formed at lower pressures, demonstrates that this is no simple metamorphic aureole. Jamieson suggests that the rocks of the metamorphic sole became accreted to the base of the ophiolite progressively, i.e. the higher grade rocks became attached early in the emplacement history while the ophiolite was relatively deep; lower grade rocks were attached later and at higher levels.

Further complexities to the development of the metamorphic soles of ophiolites were suggested by the geochronological studies of the soles of some Greek ophiolites by Spray and Roddick (1980). They reported that the metamorphism of the sole dated from 170–180 Ma, whereas the age of emplacement was 140–150 Ma, and concluded that progressive accretion of the sole to the ophiolite took place due to movements in the oceanic lithosphere prior to obduction.

VARIATION IN METAMORPHISM THROUGH GEOLOGICAL TIME

Not all types of metamorphic rock are equally common throughout the geological record, and this may be an indication of variations in the thermal structure, thickness and/or the tectonic behaviour of the crust through time. Ernst (1972) pointed out that virtually all blueschists known then were Phanerozoic, and that truly high pressure blueschists with glaucophane + lawsonite rather than merely crossite + epidote were virtually all Mesozoic. This distinction has not strictly survived the intervening years, as more old blueschists have been discovered, however the general point is still a good one. Blueschists of Ordovician age are now widely documented from the Caledonides, and some of the oldest, on Spitzbergen, include jadeite + quartz assemblages (Ohta, Hirajima & Hiroi, 1986). What is less clear is the reason for the lack of old blueschists. It is a consequence of the England and Richardson (1977) thermal model that, once formed, blueschists have the option of either being rapidly uplifted to be destroyed by erosion, or of being left at depth to be overprinted by intermediate pressure assemblages. Hence, the preservation of ancient blueschists may not reflect their original abundance. On the other hand some high pressure minerals are not readily broken down in sediments and erosion should not always destroy them without trace. Glaucophane schist detritus has long been known from the Ordovician of southern Scotland (Kelling, 1961) even though outcrops of blueschist are minimal in the adjacent metamorphic terranes. Hence, if blueschists had been widespread in the Precambrian we might reasonably hope to have found more evidence of them from the sedimentary record.

In contrast there is little doubt that intermediate and low pressure metamorphic conditions have been attained throughout most of the Precambrian as well as more recently, and indeed attempts have been made to use metamorphic mineral assemblages to put limits on the thickness of early Precambrian crusts by determining their pressures of formation (Wells, 1977).

SUMMARY AND CONCLUSIONS

Much of the material of this chapter must be regarded as speculative, or at least of a preliminary nature, because the subjects with which it deals are being actively researched today. Nevertheless, a number of general trends are now well documented. It has been apparent for some time that different metamorphic belts may have different characteristics which are the result of metamorphism under different thermal gradients. These differences in thermal gradient can be related both qualitatively and quantitatively to the different heat-flow characteristics of different plate tectonic settings at the present day, and the pressures estimated for metamorphic rocks accord well with the depths at which we expect metamorphism to be taking place today in active tectonic zones.

Much more sophisticated studies of metamorphism are now being made in which the aim is not to assign a single metamorphic temperature and pressure to each rock, but to show how both these variables have changed during its progressive recrystallisation. Such studies involve careful examination of metamorphic textures and chemical zoning, and evaluation of the extent to which different minerals in the rock have been in equilibrium with one another at different times in its history. This approach has been strongly influenced by thermal modelling studies of the $P-T$ evolution of metamorphic rocks during burial, thrusting and uplift, which have predicted changes in $P-T$ conditions with time that have been documented subsequently from field studies. Despite the uncertainties implicit in geothermometry and geobarometry of metamorphic rocks, subtle details of their thermal evolution can be elucidated in favourable circumstances. This has been demonstrated in the work of Spear *et al.* (1984) who used garnet zoning as an indicator of changing metamorphic conditions. They were able to show that, for example, where two different rock slices were tectonically juxtaposed due to nappe formation, they both gave the same final conditions of metamorphism, but had arrived there by entirely different $P-T$ paths reflecting their very different $P-T$ conditions prior to thrusting.

By providing absolute values for the temperatures and depths at which particular rocks developed, the metamorphic petrologist can help constrain and test regional structural and tectonic models quite apart from contributing to the understanding of petrological processes themselves. This is one of the major themes of present research into metamorphism.

The combination of textural and phase equilibrium studies, together with geochronology, is at last permitting us to unravel the metamorphic development of the complex metamorphic rocks of ancient metamorphic belts, where the original tectonic setting may no longer be apparent. In metamorphic belts such as the Caledonides and the Alps, a long metamorphic history can often be broken down into a series of discrete events, which may follow one another continuously or be separated by long intervals, but each of which is characteristic of metamorphism in a different type of tectonic setting.

Lastly, however, it is worth making what should be an obvious point. No geological interpretation can be better than the data on which it is based, and there is still a shortage of good experimental and thermodynamic data on key aspects of many important metamorphic systems. As a result most geological interpretations based on the evidence of metamorphic conditions are only imperfectly constrained, once the uncertainties in the data are taken into account. If the uncertainties are ignored, the interpretations may be downright wrong.

Appendix: Schreinemakers Methods for the Construction of Phase Diagrams

Schreinemakers methods for constructing phase diagrams comprise an essentially geometric approach to establishing the number and relative location of univariant curves, invariant points and divariant fields for equilibrium between particular phases in a system, requiring only a knowledge of phase compositions. They were originally published by F.A.H. Schreinemakers in a series of 29 articles between 1915 and 1925, but although applied to geological problems by some earlier workers, they were only introduced to a wider geological audience by the publication of detailed accounts by Niggli (1954), Korzhinsky (1959) and Zen (1966).

Although originally used primarily to predict the relationship between univariant curves on a P–T diagram, the same methods can be applied equally to any phase diagram whose axes are intensive variables which control the Gibbs free energy of the mineral assemblages plotted. All such diagrams are divided into divariant fields, i.e. segments within which a specific divariant assemblage is stable, by univariant curves along which the divariant assemblages of both adjacent segments can coexist in equilibrium. Commonly used axes, in addition to P and T, include X_{CO_2} in a fluid phase (as in T–X_{CO_2} diagrams, Ch. 5), chemical potentials of fluid components (e.g. μ_{CO_2}, μ_{H_2O}, μ_{SiO_2}) and $\log f_{O_2}$. The complete set of univariant curves intersecting at an invariant point is often referred to as a '**Schreinemakers bundle**'.

1 THE BASICS: HOW MANY REACTIONS IN A SYSTEM?

Schreinemakers methods are typically applied to a system defined as a group of phases that can all be made from a set of C components. For a system of C + 2 phases it follows from the phase rule that all the phases can coexist at an invariant point. This lies at the intersection of a series of univariant curves along each of which C + 1 phases can coexist at equilibrium, i.e. along each univariant curve one phase must be missing and so there will be a univariant curve for each of the C + 2 phases. By convention, univariant curves can be conveniently labelled by the name of the phase that is *absent*, placed in brackets. For complex reactions this is obviously simpler than writing out the reaction in full.

A system containing more than C + 2 phases will have more than one univariant point, and there will be a cluster of univariant curves about each. Such systems are known as **multisystems** and will not be considered further here, because they are both more complex and less unambiguous to interpret than systems with only a single invariant point (see Day, 1972, for an approach to multisystems).

DEGENERACY

In some systems of $C + 2$ phases there are nevertheless less than $C + 2$ possible univariant reactions. For example all the possible reactions between the phases corundum (Al_2O_3), quartz (SiO_2), andalusite (Al_2SiO_5) and sillimanite (Al_2SiO_5) in the binary system $Al_2O_3 - SiO_2$ are:

corundum + quartz = sillimanite (AND)
corundum + quartz = andalusite (SIL)
sillimanite = andalusite (COR, QZ)

(Note the use of the phase-absent convention to identify each reaction.) Only three reactions are possible instead of four because one of them involves only two phases and therefore serves simultaneously as both reaction (COR) and reaction (QZ). This situation arises because this reaction effectively requires only one chemical component Al_2SiO_5. Such reactions are termed **degenerate**, because they require fewer than C components.

Degeneracy also arises in systems with more than two components, where it is no longer restricted to polymorphic transitions. In the system $MgO-SiO_2-H_2O$, illustrated in Fig. 2.5(a), several of the possible assemblages of $C + 2$ phases include degenerate reactions. For example forsterite, enstatite and quartz lie on a straight line along the $MgO-SiO_2$ join, and it is therefore possible to write a reaction in which the central phase, enstatite, breaks down to the two outer phases, forsterite and quartz. Again this will be a degenerate reaction since less than $C + 1$ phases are involved and it can be written in terms of only two components. Similarly enstatite, anthophyllite and talc plot on a straight line so that reaction between them is also degenerate.

---

2   THE FUNDAMENTAL AXIOM

The Gibbs free energy of any phase or assemblage of phases varies continuously with change in intensive parameters, and as a result any univariant curve, representing conditions where two assemblages are equally stable, must be continuous with each of the two assemblages unambiguously more stable on one side and less stable on the other. This remains true even if a third equivalent assemblage is actually more stable than either of the two under consideration. For example, it is still possible to define $P-T$ conditions within the stability of sillimanite (Fig. 2.1(a)) along which kyanite and andalusite are equally stable, and these define a univariant curve of metastable equilibrium which is simply an extension of the boundary between the andalusite and kyanite stability fields. Such a curve is known as the **metastable extension** of the stable univariant curve.

---

3   MOREY–SCHREINEMAKERS RULE

If we define a system of, say, four components and six phases, A–F, then all phases can coexist only at a single invariant point from which six stable univariant curves radiate, along each of which five of the six phases can coexist. These curves divide the diagram up into segments within each of which four phases can occur. Although various assemblages are possible, according to bulk composition, one will be unique to the segment. Consider the segment in which A, B, C, D coexist. This must be bounded by stable univariant curves

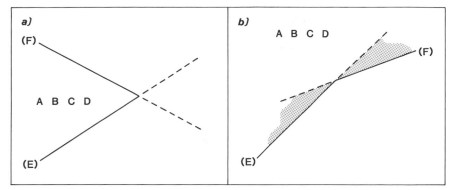

Fig. A.1 Graphical demonstration of the Morey–Schreinemakers rule, illustrated for a four-component system containing the phases A, B, C, D, E, F. *a*) Drawn correctly, shows a divariant field within which the assemblage A + B + C + D is stable, bounded by univariant curves (F) and (E) which intersect at an angle less than 180° at the invariant point. *b*) Drawn incorrectly, shows the angle between (E) and (F) greater than 180°. As a result the metastable extension of each of these curves (shown dashed) lies within the stability field of A + B + C + D. However, on crossing each metastable extension into the shaded area, an alternative assemblage becomes more stable than A + B + C + D, which is logically inconsistent with being already within the stability field of that assemblage.

along each of which A, B, C, D remain as stable phase. In other words, they must be curves (F) and (E). On the other side of each of these curves the assemblage A + B + C + D is no longer stable. Morey–Schreinemakers rule states that the angle of intersection of curves (F) and (E), or in general any two univariant curves defining the limits of a single divariant field, where they meet at the invariant point, must not exceed 180°. This rule (also known as the 180° rule) is illustrated in Fig. A.1. Because each curve has a metastable extension continuing as a smooth curve beyond the invariant point, the case in which the divariant field occupies an angle of greater than 180° (Fig. A.1(*b*)) is a logical absurdity. Within the field of stability of the assemblage A + B + C + D, lie the metastable extensions of both bounding univariant curves (F) and (E). However, on one side of each of these curves A + B + C + D is, by definition, *not* the most stable assemblage.

4   RELATIVE POSITIONING OF UNIVARIANT CURVES ABOUT THE INVARIANT POINT

Continuing with the four-component system we have just discussed, it is apparent from Fig. A.1(*a*) that, even though reaction (F) is not specified, since phase E is stable along this univariant curve (only F is absent) it must be a stable member of the assemblage in the sector to the right of this curve. By the same argument, phase F must be stable in the divariant field below curve (E). This leads us to a general rule: the stable portion of the univariant curve from which a phase P is absent (i.e. reaction (P)) must lie on the opposite side of each of the other univariant curves in the system from that on which P appears as a reaction product. Expressing it another way, the metastable portion of curve (P) lies between P-producing reactions.

In combination with Morey–Scheinemakers rule this allows us to produce only two possible solutions for the sequence in which univariant curves (and their metastable extensions) are encountered around the invariant point. However, the sequence of curves encountered in a clockwise direction around the invariant point in one solution is simply that encountered in a counter-clockwise direction in the other solution.

As an example, Fig. A.2(*a*) shows the composition of five phases; J, K, L, M, N, made up of three components. The five univariant reactions that can be written between them are:

$$K + L = M + N \quad (J)$$
$$J + M = N + L \quad (K)$$
$$N = J + K + M \quad (L)$$
$$N = J + K + L \quad (M)$$
$$K + L = J + M \quad (N)$$

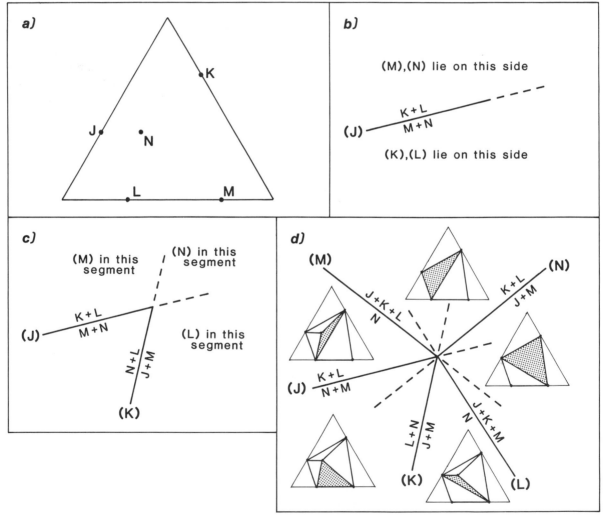

Fig. A.2   Construction of a phase diagram following the method of Schreinemakers. *a*) Compositions of the phases J to N in the three-component system. *b*) Step 1: curve (J) and the invariant point have been arbitrarily located. Constraints on the locations of the stable portions of the remaining curves are indicated. *c*) Step 2: curve (K) added to curve (J), showing additional constraints generated. *d*) The completed diagram, including compositional triangles to show the compatible assemblages in each segment; the unique assemblages are stippled.

In Fig. A.2(*b*) the univariant curve (J) has been drawn in quite arbitrarily (no axes are indicated), the invariant point in the system has been located at one end of it and the metastable extension is shown beyond the invariant point. It can be seen that the stable portions of univariant curves (K) and (L) must lie to one side, while (M) and (N) lie on the other, according to the way in which (J) is labelled. Adding curve (K) (Fig. A.2(*c*)) provides further constraints on the location of (L), (M) and (N). The completed diagram is shown in Fig. A.2(*d*). Remember that this solution is only unique in so far as it predicts the sequence in which univariant curves and their metastable extensions are encountered in going around the invariant point, the angles between them and whether they are encountered in a clockwise or counter-clockwise sense have been decided arbitrarily.

POSITIONING DEGENERATE REACTIONS

The way in which degenerate reactions are positioned follows logically in the same way as for normal reactions, except that for some the univariant curve is stable on both sides of the invariant point. In the example used above of the system corundum–quartz–andalusite–sillimanite, positioning curves (AND) and (SIL) requires curves (COR) and (QZ) to appear in the same sector. These are of course both the same reaction, which is therefore plotted within that sector in the normal manner (Fig. A.3(*a*)).

In contrast, consider the system K–feldspar–muscovite–quartz–sillimanite–andalusite–fluid, made up of the components K_2O–SiO_2–Al_2O_3–H_2O. The possible reactions are:

$$muscovite + quartz = sillimanite + K\text{–feldspar} + fluid \quad (AND)$$
$$muscovite + quartz = andalusite + K\text{–feldspar} + fluid \quad (SIL)$$
$$andalusite = sillimanite \quad (MS, KF, QZ, FL)$$

It is clear from inspection of the first reaction that it requires the stable portions of reactions (MS) and (QZ) to lie on one side of it, while (KF) and (FL) lie on the other. However, these four reactions are all one and the same. This means that univariant curve (MS, KF, QZ, FL) is stable on both sides of the invariant point and we label one end of it (MS, QZ), the other (KF, FL). This has been done in Fig. A.3(*b*).

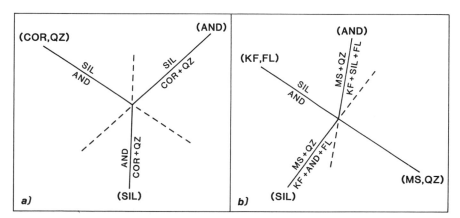

Fig. A.3 Examples of Schreinemakers bundles including degenerate reactions. *a*) System corundum–quartz–andalusite–sillimanite. Degenerate reaction (COR, QZ) stable on one side of the invariant point only. Note that most of this diagram is never actually stable. *b*) System muscovite–quartz–K–feldspar–sillimanite–andalusite–fluid. Degenerate reaction (MS, QZ, KF, FL) stable on both sides of the invariant point.

5   ORIENTATION OF THE SCHREINEMAKERS BUNDLE WITH RESPECT TO INTENSIVE VARIABLES

Although the application of Schreinemakers rules is a purely geometric exercise, it is obviously desirable to locate the curves relative to P, T or other parameters. In the case where P and T are the desired axes then, since assemblages with smaller volumes are stabilised by increased pressure, it is usually possible to decide which is the high pressure side of any curve. Molar volume data for solids are given by Robie, Hemingway & Fisher (1978) and Helgeson *et al.* (1978), and for water by Burnham, Holloway & Davis (1969). Since the molar volume of water is variable, it is obviously important to choose a value for an appropriate $P–T$ condition. Similarly higher temperatures stabilise higher entropy assemblages (data tabulated in the same sources, but be very careful to ensure consistency of units and of the particular type of entropy measurement tabulated if using more than one of them). As a rule of thumb, devolatilisation reactions normally proceed with increased temperature.

Where one or both of the axes is a parameter such as X_{CO_2}, f_{O_2} or the chemical potential of a particular component, then the fluid species or component concerned is released by reaction as its concentration, fugacity or chemical potential decreases. Thus we see in Chapter 5 that calcite + quartz break down, releasing CO_2, at low values of X_{CO_2}, if pressure and temperature are constant (Fig. 5.2).

REFERENCES

Abraham, K. & Schreyer, W., 1976. A talc–phengite assemblage in piemontite schist from Brezovica, Servia, Yugoslavia. *Journal of Petrology*, **17**, 421–39.

Aguirre, L. & Offler, R., 1985. Burial metamorphism in the western Peruvian Trough: Its relation to Andean magmatism and tectonics. *In* W.S. Pitcher, M.P. Atherton, E.J. Cobbing & R.D. Beckinsale (eds) *Magmatism at a Plate Edge: The Peruvian Andes*. Blackie, Glasgow, pp. 59–71.

Albee, A.L., 1965. Distribution of Fe, Mg and Mn between garnet and biotite in natural mineral assemblages. *Journal of Geology*, **73**, 155–64.

Anovitz, L.M. & Essene, E.J., 1987. Phase equilibria in the system $CaCO_3$ – $MgCO_3$ – $FeCO_3$. *Journal of Petrology*, **28**, 389–414.

Atherton, M.P., 1977. The metamorphism of the Dalradian rocks of Scotland. *Scottish Journal of Geology*, **13**, 331–70.

Atherton, M.P. & Edmonds, W.M., 1966. An electron microprobe study of some zoned garnets from metamorphic rocks. *Earth and Planetary Science Letters*, **1**, 185–93.

Banno, S., 1964. Petrologic studies on Sanbagawa crystalline schists in the Bessi-Ino district, central Shikoku, Japan. *Journal of Faculty of Science University Tokyo*, Sect. 2. **15**, 203–319.

Banno, S., 1986. The high pressure metamorphic belts in Japan: a review. Geological Society of America Memoir 164, pp. 365–74.

Banno, S., Sakai, C. & Higashino, T., 1986. Pressure–temperature trajectory of the Sanbagawa metamorphism deduced from garnet zoning. *Lithos*, **19**, 51–63.

Barber, J.P. & Yardley, B.W.D., 1985. Conditions of high grade metamorphism in the Dalradian of Connemara, Ireland. *Journal of the Geological Society, London*, **142**, 87–96.

Barnicoat, A., 1983. Metamorphism of the Scourian complex, N.W. Scotland. *Journal of Metamorphic Geology*, **1**, 163–82.

Barros, L.A., 1960. A ilha do Principe e a 'Linha dos Camarôes'. *Mem. Junta Investig. do Ultramar*, No. 17.

Barrow, G., 1893. On an intrusion of muscovite biotite gneiss in the S.E. Highlands of Scotland and its accompanying metamorphism. *Quarterly Journal of the Geological Society, London*, **49**, 330–58.

Barrow, G., 1912. On the geology of lower Deeside and the southern highland border. *Proceedings of the Geologists' Association*, **23**, 268–84.

Beach, A., 1973. The mineralogy of high temperature shear zones at Scourie, N.W. Scotland. *Journal of Petrology*, **14**, 231–48.

Bearth, P., 1959. Über eclogite, glaucophanschiefer, und metamorphen pillowlaven. *Schweizerische Mineralogische und Petrographische Mitteilungen*, **39**, 267–86.

Bell, T.H., 1985. Deformation partitioning and porphyroblast rotation in metamorphic rocks: a radical reinterpretation. *Journal of Metamorphic Geology*, **3**, 109–18.

Binns, R.A., 1962. Metamorphic pyroxenes from the Broken Hill district, New South Wales. *Mineralogical Magazine*, **33**, 320–38.

Binns, R.A., 1965a, b. The mineralogy of metamorphosed basic rocks from the Willyama complex, Broken Hill district, New South Wales. *Mineralogical Magazine*, **35**, 306–26; 561–87.

Black, P.M., 1974. Oxygen isotopic study of metamorphic rocks from the Ouegoa district. *Contributions to Mineralogy and Petrology*, **49**, 269–84.

Black, P.M., 1977. Regional high pressure metamorphism in New Caledonia: phase equilibria in the Ouegoa district. *Tectonophysics*, **43**, 89–107.

Bohlen, S.R., Wall, V.J. & Boettcher, A.L., 1983a. Experimental investigations and geological applications of equilibria in the system $FeO–TiO_2–Al_2O_3–SiO_2–H_2O$. *American Mineralogist*, **68**, 1049–58.

Bohlen, S.R., Wall, V.J. & Boettcher, A.L., 1983b. Geobarometry in granulites. *In* S.K. Saxena (ed) *Advances in Physical Geochemistry*, Vol. 3, Springer-Verlag, New York, pp. 141–71.

Boles, J.R. & Coombs, D.S., 1975. Mineral reactions in zeolitic Triassic tuff, Hokonui Hills, New Zealand. *Geological Society of America Bulletin*, **86**, 163–73.

Bouchez, J.L. & Pecher, A., 1981. The Himalayan main central thrust pile and its quartz-rich tectonites in central Nepal. *Tectonophysics*, **78**, 23–50.

Bowen, N.L., 1940. Progressive metamorphism of siliceous limestones and dolomite. *Journal of Geology*, **48**, 225–74.

Bowers, T.S. & Helgeson, H.C., 1983. Calculation of the thermodynamic and geochemical consequences of nonideal mixing in the system $H_2O–CO_2–NaCl$ on phase relations in geologic systems: metamorphic equilibria at high pressure and temperature. *American Mineralogist*, **68**, 1059–75.

Brodie, K.H. & Rutter, E.H., 1985. On the relationship between deformation and metamorphism with special reference to the behaviour of basic rocks. *In* A.B. Thompson & D.C. Rubie (eds) *Advances in Physical Geochemistry*, Vol. 4. Springer-Verlag, New York, pp. 138–79.

Brothers, R.N. & Yokoyama, K., 1982. Comparison of high pressure schist belts of New Caledonia and Sanbagawa, Japan. *Contributions to Mineralogy and Petrology*, **79**, 219–29.

Brown, E.H., 1974. Comparison of the mineralogy and phase relations of blueschists from the North Cascades, Washington and greenschists from Otago, New Zealand. *Geological Society of America Bulletin*, **85**, 333–44.

Brown, E.H., 1977. The crossite content of Ca-amphibole as a guide to pressure of metamorphism. *Journal of Petrology*, **18**, 53–72.

Brown, E.H., 1986. Geology of the Shuksan suite, North Cascades, Washington, USA. Geological Society of America Memoir 164, pp. 143–54.

Brown, G.C. & Mussett, A.E., 1981. *The Inaccessible Earth*. George Allen & Unwin, London.

Brown, W.H., Fyfe, W.S. & Turner, F.J., 1962. Aragonite in California glaucophane schists, and the kinetics of the aragonite–calcite transition. *Journal of Petrology*, **3**, 566–82.

Bryhni, I., Green, D.H., Heier, K.S. & Fyfe, W.S., 1970. On the occurrence of eclogite in western Norway. *Contributions to Mineralogy and Petrology*, **26**, 12–19.

Bryhni, I., Krogh, E.J. & Griffin, W.L., 1977. Crustal derivation of Norwegian eclogites: a review. *Nueus Jahrbuch fur Mineralogie Abh.*, **130**, 49–68.

Buddington, A.F., 1952. Chemical petrology of some metamorphosed Adirondack gabbroic, syenitic and quartz–syenitic rocks. *American Journal of Science*, Bowen volume, pp. 37–84.

Burg, J.P. & Chen, G.M., 1984. Tectonics and structural zonation of southern Tibet, China. *Nature*, **311**, 219–23.

Burnham, C.W., Holloway, J.R. & Davis, N.F., 1969. Thermodynamic properties of water to 1000 °C and 10,000 bars. Geological Society of America Special Publication No. 132.

Burns, R.G., 1970. *Mineralogical Applications of Crystal Field Theory*. Cambridge University Press, Cambridge.

Cann, J.R., 1969. Spilites from the Carlsberg Ridge, Indian Ocean. *Journal of Petrology*, **10**, 1–19.

Carlson, W.D. & Rosenfeld, J.L., 1981. Optical determination of topotactic aragonite–calcite growth kinetics: metamorphic implications. *Journal of Geology*, **89**, 615–38.

Carmichael, D.M., 1969. On the mechanism of prograde metamorphic reactions in quartz-bearing pelitic rocks. *Contributions to Mineralogy and Petrology*, **20**, 244–67.

Carmichael, D.M., 1970. Intersecting isograds in the Whetstone Lake area, Ontario. *Journal of Petrology*, **11**, 147–81.

Carmichael, D.M., 1978. Metamorphic bathozones and bathograds: a measure of the depth of post-metamorphic uplift and erosion on a regional scale. *American Journal of Science*, **278**, 769–97.

Carter, N.L., 1976. Steady state flow of rocks. *Reviews of Geophysics and Space Physics*, **14**, 301–60.

Chatterjee, N.D., 1972. The upper stability limit of the assemblage paragonite + quartz and its natural occurrences. *Contributions to Mineralogy and Petrology*, **34**, 288–303.

Chatterjee N.D. & Johannes, W., 1974. Thermal stability and standard thermodynamic properties of synthetic $2M_1$–muscovite $KAl_2(AlSi_3O_{10}(OH)_2)$. *Contributions to Mineralogy and Petrology*, **48**, 89–114.

Chinner, G.A., 1961. The origin of sillimanite in Glen Clova, Angus. *Journal of Petrology*, **2**, 312–23.

Chinner, G.A., 1965. The kyanite isograd in Glen Clova, Angus. *Mineralogical Magazine*, **34**, 132–43.

Chinner, G.A., 1967. Chloritoid and the isochemical character of Barrow's zone. *Journal of Petrology*, **8**, 268–82.

Chopin, C., 1981. Talc–phengite: a widespread assemblage in high-grade pelitic blueschists of the western Alps. *Journal of Petrology*, **22**, 628–50.

Chopin, C., 1984. Coesite and pure pyrope in high-grade blueschists of the western Alps: a first record and some consequences. *Contributions to Mineralogy and Petrology*, **86**, 107–18.

Chopin, C., 1987. Very-high-pressure metamorphism in the western Alps: implications for subduction of continental crust. *Philosophical Transactions of the Royal Society, London*, **A321**, 183–95.

Chopin, C. & Schreyer, W., 1983. Magnesiocarpholite and magnesiochloritoid: two index minerals of pelitic blueschists and their preliminary phase relations in the model system $MgO–Al_2O_3–SiO_2–H_2O$. *American Journal of Science*, **283A**, 72–96.

Clark, S.P. & Ringwood, A.E., 1964. Density distribution and constitution of the mantle. *Reviews of Geophysics and Space Physics*, **2**, 35–88.

Clayton, R.N., 1981. Isotopic thermometry. *In* R.C. Newton, A. Navrotsky & B.J. Wood (eds) *Advances in Physical Geochemistry*, Vol. 1. Springer-Verlag, New York, pp. 85–110.

Clemens, J.D., 1984. Water contents of silicic to intermediate magmas. *Lithos*, **17**, 273–88.

Cliff, R.A., 1985. Isotopic dating in metamorphic belts. *Journal of the Geological Society, London*, **142**, 97–110.

Cloos, M., 1982. Flow mélanges: numerical modelling and geologic constraints on their origin in the Franciscan subduction complex, California. *Geological Society of America Bulletin*, **93**, 330–45.

Clowes, R.M., Brandon, M.T., Green, A.G., Yorath, C.J., Sutherland-Brown, A., Kanasewich, E.R. & Spencer, C., 1987. LITHOPROBE-southern Vancouver Island: Cenozoic subduction complex imaged by deep seismic reflections. *Canadian Journal of Earth Sciences*, **24**, 31–51.

Coe, R.S., 1970. The thermodynamic effect of shear stress on the ortho–clino inversion in enstatite and other phase transitions characterised by a finite simple shear. *Contributions to Mineralogy and Petrology*, **26**, 247–64.

Coleman, R.G., Lee, D.E., Beatty, L.B. & Brannock, W.W., 1965. Eclogites and eclogites: their differences and similarities. *Geological Society of America Bulletin*, **76**, 483–508.

Compagnoni, R., 1977. The Sesia–Lanzo zone: high pressure, low temperature metamorphism in the Austroalpine continental margin. *Rendiconti Societa Italiana di Mineralogia e Petrologia*, **33**, 335–74.

Compagnoni, R., Dal Piaz, G.V., Hunziker, J.C., Gosso, G., Lombardo, B. & Williams, P.F., 1977. The Sesia–Lanzo zone, a slice of continental crust with alpine high pressure–low temperature assemblages in the western Italian Alps. *Rendiconti Societa Italiana di Mineralogiae Petrologia*, **33**, 123–76.

Compagnoni, R. & Maffeo, B., 1973. Jadeite-bearing metagranites and related rocks in the Mount Mucrone area (Sesia–Lanzo zone, western Italian Alps). *Schweizerische Mineralogisches und Petrographisches Mitteilungen*, **53**, 355–82.

Coombs, D.S., 1954. The nature and alteration of some Triassic sediments from Southland, New Zealand. *Transactions of the Royal Society of New Zealand*, **82**, 65–109.

Coombs, D.S., Ellis, A.J., Fyfe, W.S. & Taylor, A.M., 1959. The zeolite facies, with comments on the interpretation of hydrothermal syntheses. *Geochimica et Cosmochimica Acta*, **17**, 53–107.

Cooper, A.F., 1972. Progressive metamorphism of metabasic rocks from the Haast Schist group of southern New Zealand. *Journal of Petrology*, **13**, 457–92.

Corliss, J.B., Dymond, J., Gordon, L.I., Edmond, J.M., von Herzen, R.D., Ballard, R.D., Green, K., Williams, D., Bainbridge, A., Crane, K. & von Andel, Gj.H., 1979. Exploration of submarine thermal springs on the Galapagos rift. *Science*, **203**, 1073–83.

Correns, C.W., 1949. Growth and dissolution of crystals under linear pressure. *Discussions of the Faraday Society*, No. 5, pp. 267–71.

Cowan, D.S. & Silling, R.M., 1978. A dynamic, scaled model of accretion at trenches and its implications for the tectonic evolution of subduction complexes. *Journal of Geophysical Research*, **83**, p. 5389–96.

Crawford, M.L., 1966. Composition of plagioclase and associated minerals in some schists from Vermont, USA and South Westland, New Zealand, with inferences about the peristerite solvus. *Contributions to Mineralogy and Petrology*, **13**, pp. 269–94.

Curtis, C.D., 1985. Clay mineral precipitation and transformation during burial diagenesis. *Philosophical Transactions of the Royal Society, London*, **A315**, 91–105.

Dal Piaz, G.V. & Lombardo, B., 1986. Early alpine eclogite metamorphism in the Peninic Monte Rosa – Gran Paradiso Basement nappes of the northwestern Alps. Geological Society of America Memoir 164, 249–65.

Davis, B.T.C. & Boyd, F.R., 1966. The join $Mg_2Si_2O_6 - CaMgSi_2O_6$ at 30 kilobars and its application to pyroxenes from kimberlites. *Journal of Geophysical Research*, **71**, 3567–76.

Day, H.W., 1972. Geometrical analysis of phase equilibria in ternary system of six phases. *American Journal of Science*, **272**, 711–34.

Droop, G.T.R. & Bucher-Nurminen, K., 1984. Reaction textures and metamorphic evolution of sapphirine-bearing granulites from the Gruf complex, Italian central Alps. *Journal of Petrology*, **25**, 766–803.

Dunoyer de Segonzac, G., 1970. The transformation of clay minerals during diagenesis and low grade metamorphism: a review. *Sedimentology*, **15**, 281–346.

Durney, D.W., 1972. Solution transfer, an important geological deformation mechanism. *Nature*, **235**, 315–17.

Elliott, D., 1973. Diffusion flow laws in metamorphic rocks. *Geological Society of America Bulletin*, **84**, 2645–64.

Ellis, D.J. & Green, D.H., 1979. An experimental study of the effect of Ca upon garnet–clinopyroxene Fe–Mg exchange equilibria. *Contributions to Mineralogy and Petrology*, **71**, 13–22.

Enami, M., 1983. Petrology of pelitic schists in the oligoclase–biotite zone of the Sanbagawa metamorphic terrain, Japan: phase equilibria in the highest grade zone of a high-pressure intermediate type of metamorphic belt. *Journal of Metamorphic Geology*, **1**, 141–61.

Engel, A.E.J. & Engel, C.E., 1958. Progressive metamorphism and granitization of the major paragneiss, northwest Adirondack mountains, New York. Part 1. *Geological Society of America Bulletin*, **69**, 1369–414.

England, P.C. & Richardson, S.W., 1977. The influence of erosion upon the mineral facies of rocks from different metamorphic environments. *Journal of the Geological Society, London*, **134**, 201–13.

England, P.C. & Thompson, A.B., 1984. Pressure–temperature–time paths of regional metamorphism. I. Heat transfer during the evolution of regions of thickened continental crust. *Journal of Petrology*, **25**, 894–928.

Ernst, W.G., 1972. Occurrence and mineralogical evolution of blueschist belts with time. *American Journal of Science*, **272**, 657–68.

Ernst, W.G., Onuki, H. & Gilbert, M.C., 1970. Comparative study of low grade metamorphism in the California Coast Ranges and the outer metamorphic belt of Japan. Geological Society of America Memoir 124.

Eskola, P., 1914. On the petrology of the Orijarvi region in southwestern Finland. *Bulletin de la Commission Geologique de Finlande*, No. 40.

Eskola, P., 1915. On the relations between the chemical and mineralogical composition in the metamorphic rocks of the Orijarvi region. *Bulletin de la Commission Geologique de Finlande*, No. 44.

Eskola, P., 1922. On contact phenomena between gneiss and limestone in western Massachusetts. *Journal of Geology*, **30**, 265–94.

Essene, E., 1982. Geologic thermometry and barometry. *Reviews in Mineralogy*, **10**, 153–206.

Essene, E.J., Hensen, B.J. & Green, D.H., 1970. Experimental study of amphibolite and eclogite stability. *Physics of the Earth and Planetary Interiors*, **3**, 378–84.

Etheridge, M.A., 1983. Differential stress magnitudes during regional deformation and metamorphism: upper bound imposed by tensile fracturing. *Geology*, **11**, 231–4.

Eugster, H.P., Albee, A.L., Bence, A.E., Thompson, J.B. & Waldbaum, D.R., 1972. The two-phase region and excess mixing properties of paragonite – muscovite crystalline solutions. *Journal of Petrology*, **13**, 147–79.

Evans, B.W. & Guidotti, C.V., 1966. The sillimanite–potash feldspar isograd in western Maine, USA. *Contributions to Mineralogy and Petrology*, **12**, 25–62.

Ferguson, C.C. & Harte, B., 1975. Textural patterns at porphyroblast margins and their use in determining the time relations of deformation and crystallisation. *Geological Magazine*, **112**, 467–80.

Ferry, J.M., 1976. P, T, f_{CO_2} and f_{H_2O} during metamorphism of calcareous sediments in the Waterville–Vassalboro area, south-central Maine. *Contributions to Mineralogy and Petrology*, **57**, 119–43.

Ferry, J.M., 1983a. Mineral reactions and element migration during metamorphism of calcareous sediments from the Vassalboro Formation, south-central Maine. *American Mineralogist*, **68**, 334–54.

Ferry, J.M., 1983b. Regional metamorphism of the Vassalboro Formation, south-central Maine, USA: A case study of the role of fluid in metamorphic petrogenesis. *Journal of the Geological Society, London*, **140**, 551–76.

Ferry, J.M. & Spear, F.S., 1978. Experimental calibration of the partitioning of Fe and Mg between biotite and garnet. *Contributions to Mineralogy and Petrology*, **66**, 113–17.

Fisher, G.W., 1977. Nonequilibrium thermodynamics in metamorphism. *In* D.G. Fraser (ed) *Thermodynamics in Geology*. Reidel, Dordrecht, pp. 381–403.

Fisher, G.W., 1978. Rate laws in metamorphism. *Geochimica et Cosmochimica Acta*, **42**, 1035–50.

Forbes, R.B., Evans, B.W. & Thurston, S.P., 1984. Regional progressive high-pressure metamorphism, Seward peninsula, Alaska. *Journal of Metamorphic Geology*, **2**, 43–54.

Frey, M., 1987. Very low-grade metamorphism of clastic sedimentary rocks. *In* M. Frey (ed) *Low Temperature Metamorphism*. Blackie, Glasgow, pp. 9–58.

Frey, M., Hunziker, J.C., Frank, W., Boquet, J., Dal Piaz, G.V., Jager, E. & Niggli, E., 1974. Alpine metamorphism of the Alps. *Schweizerische Mineralogisches und Petrographisches Mitteilungen*, **54**, 248–90.

Frey, M., Teichmuller, M., Teichmuller, R., Mullis, J., Kunzi, B., Breitscmid, A., Gruner, U. & Schwizer, B., 1980. Very low-grade metamorphism in external parts of the Central Alps: Illite crystallinity, coal rank and fluid inclusion data. *Eclogae Geologicae Helvetiae*, **73**, 173–203.

Frost, B.R., 1985. On the stability of sulfides, oxides, and native metals in serpentinite. *Journal of Petrology*, **26**, 31–63.

Fry, N. & Barnicoat, A.C., 1987. The tectonic implications of high-pressure metamorphism in the western Alps. *Journal of the Geological Society, London*, **144**, 653–60.

Fry, N. & Fyfe, W.S., 1969. Eclogites and water pressure. *Contributions to Mineralogy and Petrology*, **24**, 1–6.

Fyfe, W.S., 1973. The granulite facies, partial melting and the Archaean crust. *Philosophical Transactions of the Royal Society, London*, **273**, 457–61.

Fyfe, W.S., Turner, F.J. & Verhoogen, J., 1958. Metamorphic reactions and metamorphic facies. Geological Society of America Memoir 73.

Ganguly, J., 1972. Staurolite stability and related parageneses: theory, experiments and applications. *Journal of Petrology*, **13**, 335–65.

Ghent, E.D., 1965. Glaucophane–schist facies metamorphism in the Black Butte area, northern Coast Ranges, California. *American Journal of Science*, **263**, 385–400.

Glassley, W.E., 1983. Deep crustal carbonates as CO_2 fluid sources: evidence for metasomatic reaction zones. *Contributions to Mineralogy and Petrology*, **84**, 15–24.

Goffe, B. & Velde, B., 1984. Contrasted metamorphic evolutions in thrusted cover units of the Brianconnais zone (French Alps): a model for the conservation of HP–LT metamorphic mineral assemblages. *Earth & Planetary Science Letters*, **68**, 351–60.

Goldschmidt, V.M., 1911. Die Kontaktmetamorphose im Kristianiagebiet. Videnskabelig Skrifter I. *Mathematisk Naturv. Klasse*, 11.

Goldschmidt, V.M., 1912. Die gesetze der gesteinsmetamorphose, mit beispielen aus der geologie des südlichen Norwegens. Videnskabelig Skrifter I. *Mathematisk Naturv. Klasse*, 22.

Goldsmith, J.R., 1980. The melting and breakdown reactions of anorthite at high pressures and temperatures. *American Mineralogist*, **65**, 272–84.

Goldsmith, J.R. & Newton, R.C., 1969. P–T–X relations in the system $CaCO_3$–$MgCO_3$. *American Journal of Science*, **267A**, 160–90.

Graham, C.M., 1974. Metabasite amphiboles of the Scottish Dalradian. *Contributions to Mineralogy and Petrology*, **47**, 165–85.

Graham, C.M., 1976. Petrochemistry and tectonic significance of Dalradian metabasaltic rocks of the S.W. Scottish Highlands. *Journal of the Geological Society, London*, **132**, 61–84.

Graham, C.M. & Powell, R., 1984. A garnet–hornblende geothermometer: calibration, testing and application to the Pelona Schist, southern California. *Journal of Metamorphic Geology*, **2**, 13–31.

Graham, C.M., Greig, K.M., Sheppard, S.M.F. & Turi, B., 1983. Genesis and mobility of the H_2O–CO_2 fluid phase during regional greenschist and epidote amphibolite facies metamorphism: a petrological and stable isotope study in the Scottish Dalradian. *Journal of the Geological Society, London*, **140**, 577–600.

Grant, J.A., 1973. Phase equilibria in high-grade metamorphism and partial melting of pelitic rocks. *American Journal of Science*, **273**, 289–317.

Grant, J.A., 1981. Orthoamphibole and orthopyroxene relations in high-grade metamorphism of pelitic rocks. *American Journal of Science*, **281**, 1127–43.

Grant, J.A. & Weiblen, P.W., 1971. Retrograde zoning in garnet near the second sillimanite isograd. *American Journal of Science*, **270**, 281–96.

Green, D.H. & Ringwood, A.E., 1967. An experimental investigation of the gabbro to eclogite transition and its petrological applications. *Geochimica et Cosmochimica Acta*, **31**, 767–833.

Greenwood, H.J., 1962. Metamorphic reactions involving two volatile components. Annual Report Director Geophysical Laboratory, Carnegie Institution Washington Year Book, Vol. 61, pp. 82–5.

Greenwood, H.J., 1967. Wollastonite: stability in H_2O–CO_2 mixtures and occurrences in a contact-metamorphic aureole near Almo, British Columbia, Canada. *American Mineralogist*, **52**, 1669–80.

Greenwood, H.J., 1975. The buffering of pore fluids by metamorphic reactions. *American Journal of Science*, **275**, 573–93.

Greenwood, H.J., 1977. The fundamental equations of thermodynamics. *In* H.J. Greenwood (ed) *Application of Thermodynamics to Petrology and Ore Deposits*. Mineralogical Association of Canada Short Course Handbook No. 2.

Grew, E.S., 1980. Sapphirine and quartz association from Archaean rocks in Enderby Land, Antarctica. *American Mineralogist*, **65**, 821–36.

Griffin, W.L. & Heier, K.S., 1973. Petrological implications of some corona structures. *Lithos*, **6**, 315–35.

Guidotti, C.V., 1974. Transition from staurolite to sillimanite zone, Rangeley quadrangle, Maine. *Geological Society of America Bulletin*, **85**, 475–90.

Haas, J.R., 1976. Physical properties of the coexisting phases and thermochemical properties of the H_2O component in boiling NaC1 solutions. *US Geological Survey, Bulletin 1421–A*.

Harker, A., 1932. *Metamorphism. A study of the transformation of rock masses.* Methuen, London.

Harker, R.I. & Tuttle, O.F., 1956. Experimental data on the P_{CO_2}–T curve for the reaction: calcite + quartz = wollastonite + carbon dioxide. *American Journal of Science*, **254**, 239–56.

Harley, S.L., 1983. Regional geobarometry–geothermometry and metamorphic evolution of Enderby Land, Antarctica. *In* R.L. Oliver, P.R. James & J.B. Jago (eds) *Antarctic Earth Sciences*. Cambridge University Press, pp. 25–30.

Harley, S.L., 1984a. The solubility of alumina in orthopyroxene coexisting with garnet in FeO–MgO–Al$_2$O$_3$–SiO$_2$ and CaO–FeO–MgO–Al$_2$O$_3$–SiO$_2$. *Journal of Petrology*, **25**, 665–96.

Harley, S.L., 1984b. Comparison of the garnet–orthopyroxene geobarometer with recent experimental studies, and applications to natural assemblages. *Journal of Petrology*, **25**, 697–712.

Harte, B. & Graham, C.M., 1975. The graphical analysis of greenschist to amphibolite facies mineral assemblages in metabasites. *Journal of Petrology*, **16**, 347–70.

Harte, B. & Hudson, N.F.C., 1979. Pelite facies series and the temperatures and pressures of Dalradian metamorphism in E. Scotland. *In* A.L. Harris, C.H. Holland & B.E. Leake (eds) Geological Society, London, Special Publication No. 8, pp. 323–37.

Heinrich, W. & Althaus, E., 1980. Die obere stabilitats grenze von Lawsonit plus Albit bzw. Jadeit. *Fortschrifte für Mineralogie*, **58**, 49–50.

Helgeson, H.C., Delany, J.M., Nesbitt, H.W. & Bird, D.K., 1978. Summary and critique of the thermodynamic properties of rock-forming minerals. *American Journal of Science*, **278A**.

Hess, P.C., 1969. The metamorphic paragenesis of cordierite in pelitic rocks. *Contributions to Mineralogy and Petrology*, **24**, 191–207.

Holdaway, M.J., 1971. Stability of andalusite and the aluminium silicate phase diagrams. *American Journal of Science*, **271**, 97–131.

Holdaway, M.J. & Lee, S.M., 1977. Fe–Mg cordierite stability in high grade pelitic rocks based on experimental, theoretical and natural observations. *Contributions to Mineralogy and Petrology*, **63**, 175–98.

Holland, T.J.B., 1979a. Experimental determination of the reaction paragonite = jadeite + kyanite + H$_2$O, and internally consistent thermodynamic data for part of the system Na$_2$O–Al$_2$O$_3$–SiO$_2$–H$_2$O, with applications to eclogites and blueschists. *Contributions to Mineralogy and Petrology*, **68**, 293–301.

Holland, T.J.B., 1979b. High water activities in the generation of high pressure kyanite eclogites of the Tauern Window, Austria. *Journal of Geology*, **87**, 1–27.

Holland, T.J.B., 1980. The reaction albite = jadeite + quartz determined experimentally in the range 600–1200 °C. *American Mineralogist*, **65**, 129–34.

Holland, T.J.B., 1983. The experimental determination of activities in disordered and short-range ordered jadeitic pyroxene. *Contributions to Mineralogy and Petrology*, **82**, 214–20.

Holland, T.J.B. & Powell, R., 1985. An internally consistent thermodynamic dataset with uncertainties and correlations. 2 Data and results. *Journal of Metamorphic Geology*, **3**, 343–70.

Hollister, L.S., 1966. Garnet zoning: an interpretation based on the Rayleigh fractionation model. *Science*, **154**, 1647–51.

Howie, R.A., 1955. The geochemistry of the charnockite series of Madras, India. *Transactions of the Royal Society of Edinburgh*, **62**, 725–68.

Humphris, S.E. & Thompson, G., 1978. Hydrothermal alteration of oceanic basalts by seawater. *Geochimica et Cosmochimica Acta*, **42**, 107–25.

Jaeger, J.C., 1968. Cooling and solidification of igneous rocks. *In* H.H. Hess & A. Poldervaart (eds) *Basalts. The Poldervaart treatise on rocks of basaltic composition*, Vol. 2. John Wiley & Sons, New York, pp. 503–36.

Jamieson, R.A., 1981. Metamorphism during ophiolite emplacement – the petrology of the St Anthony Complex. *Journal of Petrology*, **22**, 397–441.

Jamieson, R.A. & Craw, D., 1987. Sphalerite geobarometry in metamorphic terranes: an appraisal with implications for metamorphic pressure in the Otago schists. *Journal of Metamorphic Geology*, **5**, 87–99.

Jenkins, D.M., Newton, R.C. & Goldsmith, J.R., 1984. Relative stability of Fe-free zoisite and clinozoisite. *Journal of Geology*, **93**, 663–72.

Joesten, R., 1974. Local equilibrium and metasomatic growth of zoned calcsilicate nodules from a contact aureole, Christmas Mountains, Big Bend Region, Texas. *American Journal of Science*, **274**, 876–901.

Johannes, W. & Puhan, D., 1971. The calcite–aragonite transition re-investigated. *Contributions to Mineralogy and Petrology*, **31**, 28–38.

Kano, H. & Kuroda, Y., 1968. On the occurrences of staurolite and kyanite from the Abukuma plateau, northeastern Japan. *Proceedings of the Japan Academy*, **44**, 77–82.

Karig, D.E., 1983. Deformation in the Forearc: implications for mountain belts. *In* K.J. Hsu (ed) *Mountain Building Processes*, Academic Press, London, pp. 59–72.

Kawachi, Y., 1975. Pumpellyite–actinolite and contiguous facies metamorphism in the Upper Wakatipu district, southern New Zealand. *New Zealand Journal of Geology & Geophysics*, **17**, 169–208.

Kelling, G., 1961. The stratigraphy and structure of the Ordovician rocks of the Rhins of Galloway. *Quarterly Journal of the Geological Society, London*, **117**, 37–75.

Kennedy, W.Q., 1949. Zones of progressive regional metamorphism in the Moine Schists of the Western Highlands of Scotland. *Geological Magazine*, **86**, 43–56.

Kerrick, D.M., 1968. Experiments on the upper stability limit of pyrophyllite at 1.8 kilobars and 3.9 kilobars water pressure. *American Journal of Science*, **266**, 204–14.

Kerrick, D.M., 1974. Review of metamorphic mixed volatile (H_2O–CO_2) equilibria. *American Mineralogist*, **59**, 729–62.

Kisch, H.J., 1980. Incipient metamorphism of Cambro-Silurian clastic rocks from the Jamtland Supergroup, central Scandinavian Caledonides, western Sweden: illite crystallinity and 'vitrinite' reflectance. *Journal of the Geological Society, London*, **137**, 271–88.

Knipe, R.J., 1981. The interaction of deformation and metamorphism in slates. *Tectonophysics*, **78**, 249–72.

Koons, P.O., 1982. An experimental investigation of the behaviour of amphibole in the system Na_2O–MgO–Al_2O_3–SiO_2–H_2O at high pressures. *Contributions to Mineralogy and Petrology*, **79**, 258–67.

Koons, P.O. & Thompson, A.B., 1985. Non-mafic rocks in the greenschist, blueschist and eclogite facies. *Chemical Geology*, **50**, 3–30.

Korzhinsky, D.S., 1959. Physicochemical basis of the analysis of the paragenesis of minerals. Consultants Bureau, New York.

Kretz, R., 1959. Chemical study of garnet, biotite and hornblende from gneisses of south-western Quebec, with emphasis on distribution of elements in coexisting minerals. *Journal of Geology*, **67**, 37–402.

Kretz, R., 1966. Grain-size distribution for certain metamorphic minerals in relation of nucleation and growth. *Journal of Geology*, **74**, 147–73.

Kretz, R., 1973. Kinetics of the crystallisation of garnet at two localities near Yellowknife. *Canadian Mineralogist*, **12**, 1–20.

Kridelbaugh, S.J., 1973. The kinetics of the reaction calcite + quartz = wollastonite + carbon dioxide at elevated temperatures and pressures. *American Journal of Science*, **273**, 757–77.

Krogh, E.J., 1977. Evidence for a Precambrian continent–continent collision in western Norway. *Nature*, **267**, 17–19.

Krogh, E.J., 1980. Compatible P–T conditions for eclogites and surrounding gneisses in the Kristiansund area, western Norway. *Contributions to Mineralogy and Petrology*, **75**, 387–93.

Laird, J. & Albee, A.L., 1981. Pressure, temperature, and time indicators in mafic schist: their application to reconstructing the polymetamorphic history of Vermont. *American Journal of Science*, **281**, 127–75.

Landis, C.A. & Coombs, D.S., 1967. Metamorphic belts and orogenesis in southern New Zealand. *Tectonophysics*, **4**, 501–18.

Lappin, M.A. & Smith, D.C., 1978. Mantle equilibrated orthopyroxene eclogite pods from the basal gneisses in the Selje district, western Norway. *Journal of Petrology*, **9**, 530–84.

Lasaga, A.C., 1986. Metamorphic reaction rate laws and development of isograds. *Mineralogical Magazine*, **50**, 359–73.

Leake, B.E., 1964. The chemical distinction between ortho- and para-amphibolite. *Journal of Petrology*, **5**, 238–54.

Leggett, J.K., McKerrow, W.S., Morris, J.H., Oliver, G.J.H. & Phillips, W.E.A., 1979. The north-western margin of the Iapetus Ocean. *In* A.L. Harris, C.H. Holland & B.E. Leake (eds) *The Caledonides of the British Isles – Reviewed*. The Geological Society, London, pp. 499–511.

Le Pichon, X., 1983. Land-locked ocean basins and continental collision: the eastern Mediterranean as a case example. *In* K.J. Hsu (ed) *Mountain Building Processes*. Academic Press, London, pp. 201–12.

Lindsley, D.H., 1983. Pyroxene thermometry. *American Mineralogist*, **65**, 477–93.

Liou, J.G., 1971a. Synthesis and stability relations of prehnite, $Ca_3Al_2Si_3O_{10}(OH)_2$. *American Mineralogist*, **56**, 507–31.

Liou, J.G., 1971b. P–T stabilities of laumontite, wairakite, lawsonite and related minerals in the system $CaAl_2Si_2O_8$–SiO_2–H_2O. *Journal of Petrology*, **12**, 379–411.

Liou, J.G., Maruyama, S. & Cho, M., 1987. Very low-grade metamorphism of volcanic and volcaniclastic rocks – mineral assemblages and mineral facies. *In* M. Frey (ed) *Low Temperature Metamorphism*. Blackie, Glasgow, pp. 59–113.

Loomis, T.P., 1983. Compositional zoning of crystals: a record of growth and reaction history. *In* S.K. Saxena (ed) *Advances in Physical Geochemistry*, Vol. 3. Springer-Verlag, New York, pp. 1–60.

Luth, W.D., Jahns, R.H. & Tuttle, O.F., 1964. The granite system at pressures of 4 to 10 kilobars. *Journal of Geophysical Research*, **69**, 659–773.

Maresch, W.V., 1977. Experimental studies of glaucophane: an analysis of present knowledge. *Tectonophysics*, **43**, 109–25.

Martignole, J. & Sisi, J.–C., 1981. Cordierite – garnet – H_2O equilibrium: a geological thermometer, barometer and water fugacity indicator. *Contributions to Mineralogy and Petrology*, **77**, 38–46.

Massone, H.J., Mirwald, P.W. & Schreyer, W., 1981. Experimentalle uberprufung der reacktionscurve chlorit + quarz = talk + disthen im system MgO–Al_2O_3–SiO_2–H_2O. *Fortschrifte für Mineralogie*, **59**, 122–3.

Mather, J.D., 1970. The biotite isograd and the lower greenschist facies in the Dalradian rocks of Scotland. *Journal of Petrology*, **11**, 253–75.

Matthews, A. & Schliestedt, M., 1984. Evolution of the blueschist and greenschist facies rocks of Sifnos, Cyclades, Greece. A stable isotope study of subduction-related metamorphism. *Contributions to Mineralogy and Petrology*, **88**, 150–68.

McKenzie, D.P., 1978. Some remarks on the development of sedimentary basins. *Earth and Planetary Science Letters*, **40**, 25–32.

Mehnert, K.R., 1968. *Migmatites and the Origin of Granitic Rocks*. Amsterdam: Elsevier.

Miller, C., 1974. On the metamorphism of the eclogites and high grade blueschists from the Penine terrane of the Tauern Window. *Schweizerische Mineralogisches und Petrographisches Mitteilungen*, **54**, 371–84.

Misch, P.M., 1964. Stable association wollastonite–anorthite and other calc–silicate assemblages in amphibolite-facies crystalline schists of Nanga Parbat, northwest Himalayas. *Beiträge zur Mineralogie und Petrographie* (later *Contributions to Mineralogy and Petrology*) **10**, 315–56.

Misch, P.M. & Rice, J.M., 1975. Miscibility of tremolite and hornblende in progressive Skagit metamorphic suite, North Cascades, Washington. *Journal of Petrology*, **16**, 1–21.

Miyashiro, A., 1961. Evolution of metamorphic belts. *Journal of Petrology*, **2**, 277–311.

Miyashiro, A., 1964. Oxidation and reduction in the earth's crust, with special reference to the role of graphite. *Geochimica et Cosmichimica Acta*, **28**, 717–29.

Miyashiro, A., 1972. Pressure and temperature conditions and tectonic significance of regional and ocean floor metamorphism. *Tectonophysics*, **13**, 141–59.

Miyashiro, A., 1973. *Metamorphism and Metamorphic Belts*. George Allen & Unwin, London.

Miyashiro, A., Aki, K. & Celas Sengor, A.M., 1979. *Orogeny*. John Wiley, New York.

Mørk, M.B., 1986. Coronite and eclogite formation in olivine gabbro (western Norway): reaction paths and garnet zoning. *Mineralogical Magazine*, **50**, 417–26.

Mottl, M.J., 1983. Metabasalts, axial hot springs and the structure of hydrothermal systems at mid-ocean ridges. *Geological Society of America Bulletin*, **94**, 161–80.

Muffler, L.J.P. & White, D.E., 1969. Active metamorphism of Upper Cenozoic sediments in the Salton Sea geothermal field and the Salton trough, southeastern California. *Geological Society of America Bulletin*, **80**, 157–82.

Myers, J. & Eugster, H.P., 1983. The system Fe–Si–O: Oxygen buffer calibrations to 1500 K. *Contributions to Mineralogy and Petrology*, **82**, 75–90.

Naylor, R.S., 1971. Acadian orogeny: an abrupt and brief event. *Science*, **172**, 558–60.

Newton, R.C., 1983. Geobarometry of high grade metamorphic rocks. *American Journal of Science*, **283–A**, 1–28.

Newton, R.C., 1986. Metamorphic temperatures and pressures of Group B and C eclogites. Geological Society of America Memoir 164, pp. 17–30.

Newton, R.C. & Hasleton, H.T., 1981. Thermodynamics of the garnet–plagioclase–Al$_2$SiO$_5$–quartz geobarometer. *In* R.C. Newton, A. Navrotsky & B.J. Wood (eds) *Thermodynamics of Minerals and Melts*. Springer Verlag, New York, pp. 131–48.

Newton, R.C. & Kennedy, G.C., 1963. Some equilibrium reactions in the join CaAl$_2$Si$_2$O$_8$–H$_2$O. *Journal of Geophysical Research*, **68**, 2967–83.

New Zealand Geological Survey, 1972. *South Island Geological Map of New Zealand 1:10 000*. 1st edn. DSIR, Wellington.

Niggli, P., 1954. *Rocks and mineral deposits* (translated by R.L. Parker). W.H. Freeman & Sons, San Francisco.

Nitsch, K–H.. 1971. Stabilitätsbeziehungen von Prehnit-und Pumpellyit-haltiger Paragenesen. *Contributions to Mineralogy and Petrology*, **30**, 240–60.

Norris, R.J. & Henley, R.W., 1976. Dewatering of a metamorphic pile. *Geology* **4**, 333–6.

Norton, D. & Knight, J., 1977. Transport phenomena in hydrothermal systems: cooling plutons. *American Journal of Science*, **277**, 937–81.

O'Hara, M.J., 1961. Zoned ultrabasic and basic gneiss masses in the early Lewisian metamorphic complex at Scourie, Sutherland. *Journal of Petrology* **2**, 248–76.

O'Hara, M.J. & Yarwood, G., 1978. High pressure–temperature point on an Archaean geotherm. *Philosophical Transactions Royal Society, London*, **288A**, 441–56.

Ohta, Y., Hirajima, T. & Hiroi, Y., 1986. Caledonian high-pressure metamorphism in central western Spitzbergen. Geological Society of America Memoir 164, pp. 205–16.

Okrusch, M., Seidel, E. & Davis, E., 1978. The assemblage jadeite–quartz in the glaucophane rocks of Sifnos (Cyclades Archipeligo, Greece). *Nueus Jahrbuch für Mineralogie Abh.*, **132**, 284–308.

Oliver, G.J.H., 1978. Prehnite–pumpellyite facies metamorphism in County Cavan, Ireland. *Nature*, **274**, 242–3.

Olsen, S.N., 1984. Mass-balance and mass-transfer in migmatites from the Colorado Front Range. *Contributions to Mineralogy and Petrology*, **85**, 30–44.

O'Neil, J.R. & Ghent, E.D., 1975. Stable isotope study of coexisting metamorphic minerals from the Esplanade Range, British Columbia. *Geological Society of America Bulletin*, **86**, 1708–12.

Orville, P.M., 1969. A model for metamorphic differentiation origin of thin-layered amphibolites. *American Journal of Science*, **267**, 64–8.

Oxburgh, E.R., 1974. The plain man's guide to Plate Tectonics. *Proceedings of the Geologists' Association*, **85**, 299–358.

Oxburgh, E.R. & England, P.C., 1980. Heat flow and the metamorphic evolution of the Eastern Alps. *Eclogae geologicae Helvetiae*, **73**, 379–98.

Oxburgh, E.R. & Turcotte, D.L., 1971. Origin of paired metamorphic belts and crustal dilation in island arc regions. *Journal of Geophysical Research*, **76**, 1315–27.

Oxburgh, E.R. & Turcotte, D.L., 1974. Thermal gradients and regional metamorphism in overthrust terrains, with special reference to the eastern Alps. *Schweizerische Mineralogisches und Petrographisches Mitteilungen*, **54**, 641–62.

Perchuk, L.L., Podlesskii, K.K. & Aranovich, L.Ya., 1981. Calculation of thermodynamic properties of endmember minerals from natural parageneses. *In* R.C. Newton, A. Navrotsky & B.J. Wood (eds) *Advances in Physical Geochemistry*, Vol. 1, Springer, New York, pp. 111–29.

Phillips, G.N., 1980. Water activity changes across an amphibolite–granulite facies transition, Broken Hill, Australia. *Contributions to Mineralogy and Petrology*, **75**, 377–86.

Pitcher, W.S. & Berger, A.R., 1972. *The Geology of Donegal*. John Wiley & Sons, New York.

Platt, J.P., 1986. Dynamics of orogenic wedges and the uplift of high-pressure metamorphic rocks. *Geological Society of America Bulletin*, **97**, 1037–53.

Platt, J.P., 1987. The uplift of high-pressure–low-temperature metamorphic rocks. *Philosophical Transactions of the Royal Society, London*, **A.321**, 87–102.

Pognante, U. & Kienast, J.–R., 1987. Blueschist and eclogite transformations in Fe–Ti gabbros: a case from the western Alps ophiolites. *Journal of Petrology*, **28**, 271–92.

Poty, B.P., Stalder, H.A. & Weisbrod, A.M., 1974. Fluid inclusion studies in quartz from fissures of western and central Alps. *Schweizerische Mineralogisches und Petrographisches Mitteilungen*, **50**, 131–9.

Powell, D. & Treagus, J.E., 1967. On the geometry of S-shaped inclusion trails in garnet porphyroblasts. *Mineralogical Magazine*, **36**, 453–6.

Powell, R., 1983. Fluids and melting under upper amphibolite facies conditions. *Journal of the Geological Society, London*, **140**, 629–33.

Powell, R., Condliffe, D.M. & Condliffe, E., 1984. Calcite–dolomite geothermometry in the system $CaCO_3$–$MgCO_3$–$FeCO_3$: an experimental study. *Journal of Metamorphic Geology*, **2**, 33–41.

Pownceby, M.I., Wall, V.J. & O'Neill, H.St.C., 1987. Fe–Mn partitioning between garnet and ilmenite: experimental calibration and applications. *Contributions to Mineralogy and Petrology*, **97**, 116–26, and 539.

Putnis, A. & McConnell, J.D.C., 1980. *Principles of Mineral Behaviour*. Elsevier, Amsterdam.

Ramsay, J.G., 1962. The geometry and mechanics of formation of 'similar' type folds. *Journal of Geology*, **70**, 309–27.

Rao, B.B. & Johannes, W., 1979. Further data on the stability of staurolite. *Nueus Jahrbuch für Mineralogie, Mh.*, **1979**, 437–47.

Rast, N., 1958. The metamorphic history of the Schichallion Complex. *Transactions Royal Society of Edinburgh*, **63**, 413–31.

Rice, J.M., 1977a. Progressive metamorphism of impure dolomitic limestone in the Marysville aureole, Montana. *American Journal of Science*, **277**, 1–24.

Rice, J.M., 1977b. Contact metamorphism of impure dolomitic limestone in the Boulder Aureole, Montana. *Contributions to Mineralogy and Petrology*, **59**, 237–59.

Richardson, C.J., Cann, J.R., Richards, H.G. & Cowan, J.G., 1987. Metal-depleted root zones of the Troodos ore-forming hydrothermal systems, Cyprus. *Earth and Planetary Science Letters*, **84**, 243–53.

Richardson, S.W., 1968. Staurolite stability in a part of the system Fe–Al–Si–O–H. *Journal of Petrology*, **9**, 467–88.

Richardson, S.W., Gilbert, M.C. & Bell, P.M., 1969. Experimental determination of kyanite–andalusite and andalusite–sillimanite equilibrium; the aluminium silicate triple point. *American Journal of Science*, **267**, 259–72.

Richardson, S.W. & Powell, R., 1976. Thermal causes of the Dalradian metamorphism in the central Highlands of Scotland. *Scottish Journal of Geology*, **12**, 237–68.

Ridley, J., 1984. Evidence of a temperature-dependent 'blueschist' to 'eclogite' transformation in high-pressure metamorphism of metabasic rocks. *Journal of Petrology*, **25**, 852–70.

Ridley, J., 1985. The effect of reaction enthalpy on the progress of a metamorphic reaction. *In* A.B. Thompson & D.C. Rubie (eds) *Advances in Physical Geochemistry*, Vol. 4. Springer-Verlag, New York, pp. 80–97.

Robie, R.A., Hemingway, B.S. & Fisher, J.R., 1978. Thermodynamic properties of minerals and related substances at 298.15 K and 1 bar (10^5 pascals) pressure and at higher temperatures. *US Geological Survey Bulletin 1452*.

Robinson, D., 1971. On the inhibiting effect of organic carbon on contact metamorphic recrystallisation of limestones. *Contributions to Mineralogy and Petrology*, **32**, 245–50.

Robinson, D., 1987. Transition from diagenesis to metamorphism in extensional and collision settings. *Geology*, **15**, 866–9.

Robinson, P. & Jaffe, H.W., 1969. Chemographic exploration of amphibole assemblages from central Massachusetts and southwestern New Hampshire. Mineralogical Society of America Special Paper 2, pp. 251–74.

Robinson, P., Spear, F.S., Schumacher, J.C., Laird, J., Klein, C., Evans, B.W. & Doolan, B.L., 1982. Phase relations of metamorphic amphiboles: natural occurrence and theory. *Reviews in Mineralogy*, **9B**, 1–227.

Roedder, E., 1972. Composition of fluid inclusions. *In* M. Fleischer (ed) *Data of Geochemistry*, 6th edn, Ch. JJ. US Geological Survey Professional Paper 440–JJ.

Rosenfeld, J.L., 1970. Rotated garnets in metamorphic rocks. Geological Society of America Special Paper 129.

Rubie, D.C., 1983. Reaction-enhanced ductility: the role of solid–solid univariant reactions in deformation of the crust and mantle. *Tectonophysics*, **96**, 331–52.

Rubie, D.C., 1984. A thermal–tectonic model for high-pressure metamorphism and deformation in the Sesia zone, western Alps. *Journal of Geology*, **92**, 21–36.

Rumble, D., Ferry, J.M., Hoering, R.C. & Boucot, A.J., 1982. Fluid flow during metamorphism at the Beaver Brook fossil locality. *American Journal of Science*, **282**, 866–919.

Rutland, R.W.R., 1965. Tectonic overpressures. *In* W.S. Pitcher & G.S. Flinn (eds) *Controls of Metamorphism*. Oliver & Boyd, Edinburgh, pp. 119–39.

Salje, E., 1986. Heat capacities and entropies of andalusite and sillimanite: the influence of fibrolitisation on the phase diagram of the Al_2SiO_5 polymorphs. *American Mineralogist*, **71**, 1366–71.

Schiffman, P. & Liou, J.G., 1980. Synthesis and stability relations of Mg–Al pumpellyite, $Ca_4Al_5MgSi_6O_{21}(OH)_7$. *Journal of Petrology*, **21** 441–74.

Schiffman, P. & Liou, J.G., 1983. Synthesis of Fe-pumpellyite and its stability relations with epidote. *Journal of Metamorphic Geology*, **1**, 91–101.

Schreyer, W., 1973. Whiteschist: a high pressure rock and its geological significance. *Journal of Geology*, **81**, 735–9.

Schreyer, W. & Seifert, F., 1969. Compatibility relations of the aluminium silicates in the system $MgO–Al_2O_3–SiO_2–H_2O$ and $K_2O–MgO–Al_2O_3–H_2O$ at high pressures. *American Journal of Science*, **267**, 371–88.

Sclater, J.G., Jaupart, C. & Galson, D., 1980. The heat flow through oceanic and continental crust and the heat loss of the Earth. *Reviews of Geophysics and Space Physics*, **18**, 269–311.

Seyfried, W.E., 1987. Experimental and theoretical constraints on hydrothermal alteration processes at mid-ocean ridges. *Annual Reviews of Earth and Planetary Science*, **15**, 317–35.

Shaw, D.M., 1956. Geochemistry of pelitic rocks. Part III: major elements and general geochemistry. *Geological Society of America Bulletin*, **67**, 919–34.

Sisson, V.B., Crawford, M.L. & Thompson, P.H., 1981. CO_2– brine immiscibility at high temperatures, evidence from calcareous metasedimentary rocks. *Contributions to Mineralogy and Petrology*, **78**, 371–78.

Skippen, G.B., 1971. Experimental data for reactions in siliceous marbles. *Journal of Geology*, **70**, 451–81.

Skippen, G.B., 1974. An experimental model for low pressure metamorphism of siliceous dolomitic marbles. *American Journal of Science*, **274**, 487–509.

Skippen, G.B. & Trommsdorf, V., 1975. Invariant phase relations among minerals on T–X $_{fluid}$ sections. *American Journal of Science*, **275**, 561–72.

Skippen, G.B. & Trommsdorf, V., 1986. The influence of NaCl and KCl on phase relations in metamorphosed carbonate rocks. *American Journal of Science*, **286**, 81–104.

Slaughter, J., Kerrick, D.M. & Wall, V.J., 1975. Experimental and thermodynamic study of equilibria in the system $CaO–MgO–SiO_2–H_2O–CO_2$. *American Journal of Science*, **275**, 143–62.

Sleep, N.H., 1979. A thermal constraint on the duration of folding with reference to Acadian geology, New England, USA. *Journal of Geology*, **87**, 583–9.

Smith, B.K., 1985. The influence of defect crystallography on some properties of orthosilicates. *In* A.B. Thompson & D.C. Rubie (eds) *Advances in Physical Geochemistry*, Vol. 4. Springer Verlag, New York, pp. 98–117.

Smith, D.C., 1984. Coesite in clinopyroxene in the Caledonides and its implications for geodynamics. *Nature*, **310**, 641–44.

Smith, P. & Parsons, I., 1974. The alkali-feldspar solvus at 1 kilobar water-vapour pressure. *Mineralogical Magazine*, **39**, 747–67.

Spear, F.S., 1980. NaSi $\rightleftharpoons$ CaAl exchange equilibrium between plagioclase and amphibole. An empirical model. *Contributions to Mineralogy and Petrology*, **72**, 33–41.

Spear, F.S., 1981. An experimental study of hornblende stability and compositional variability in amphibolite. *American Journal of Science*, **281**, 697–734.

Spear, F.S., Selverstone, J., Hickmont, D., Crowley, P. & Hodges, K.V., 1984. P–T paths from garnet zoning: a new technique for deciphering tectonic processes in crystalline terranes. *Geology*, **12**, 87–90.

Speiss, F.N., MacDonald, K.C., Atwater, T., Ballard, R., Carvanza, A., Cordoba, D., Cox, C., Diaz Garcia, V.M., Francheteau, J., Guerrero, J., Hawkins, J., Haymon, R., Hessler, R., Juteau, T., Kastner, M., Larson, R., Luyendyke, B., MacDougall, J.D., Miller, S., Normark, W., Orcutt, J. & Rangin, C., 1980. East Pacific Rise; hot springs and geophysical experiments. *Science*, **207**, 1421–33.

Spooner, E.T.C. & Fyfe, W.S., 1973. Sub-sea-floor metamorphism, heat and mass transfer. *Contributions to Mineralogy and Petrology*, **42**, 287–304.

Spray, J.G. & Roddick, J.C., 1980. Petrology and ^{40}Ar/^{39}Ar geochronology of some Hellenic sub-ophiolite metamorphic rocks. *Contributions to Mineralogy and Petrology*, **72**, 43–55.

Spry, A., 1963a. The occurrence of eclogite on the Lyell Highway, Tasmania. *Mineralogical Magazine*, **33**, 589–94.

Spry, A., 1963b. Origin and significance of snowball structure in garnet. *Journal of Petrology*, **4**, 211–22.

Spry, A., 1969. *Metamorphic Textures*, Pergamon, Oxford.

Sturt, B.A. & Harris, A.L., 1961. The metamorphic history of the Loch Tummel area, central Perthshire, Scotland. *Liverpool and Manchester Geological Journal*, **2**, 289–711.

Tarney, J. & Windley, B.F., 1977. Chemistry, thermal gradients and evolution of the lower continental crust. *Journal of the Geological Society, London*, **134**, 153–72.

Taylor, H.P. & Coleman, R.G., 1968. O^{18}/O^{16} ratios of coexisting minerals in glaucophane-bearing metamorphic rocks. *Geological Society of America Bulletin*, **79**, 1727–56.

Thompson, A.B., 1970. Laumontite equilibria and the zeolite facies. *American Journal of Science*, **269**, 267–75.

Thompson, A.B., 1971a. P_{CO_2} in low grade metamorphism: zeolite, carbonate, clay mineral, prehnite relations in the system $CaO-Al_2O_3-CO_2-H_2O$. *Contributions to Mineralogy and Petrology*, **33**, 145–61.

Thompson, A.B., 1971b. Analcite–albite equilibria at low temperatures. *American Journal of Science*, **271**, 79–92.

Thompson, A.B., 1976. Mineral reactions in pelitic rocks: Parts I and II. *American Journal of Science*, **276**, 401–54.

Thompson, A.B., 1982. Dehydration melting of pelitic rocks and the generation of H_2O-undersaturated granitic liquids. *American Journal of Science*, **282**, 1567–95.

Thompson, A.B., 1983. Fluid-absent metamorphism. *Journal of the Geological Society, London*, **140**, 533–47.

Thompson, A.B. & England, P.C., 1984. Pressure–temperature–time paths of regional metamorphism. II. Their inference and interpretation using mineral assemblages in metamorphic rocks. *Journal of Petrology*, **25**, 929–55.

Thompson, A.B., Tracy, R.J., Lyttle, P.T. & Thompson, J.B., 1977. Prograde reaction histories deduced from compositional zonation and mineral inclusions in garnet from the Gassetts schist, Vermont. *American Journal of Science*, **277**, 1152–67.

Thompson, J.B., 1957. The graphical analysis of mineral assemblages in pelitic schists. *American Mineralogist*, **42**, 842–58.

Tilley, C.E., 1925. Metamorphic zones in the southern Highlands of Scotland. *Quarterly Journal of the Geological Society*, **81**, 100–12.

Tilley, C.E., 1951. A note on the progressive metamorphism of siliceous limestones and dolomites. *Geological Magazine*, **88**, 175–8.

Tomasson, J. & Kristmansdottir, H., 1972. High temperature alteration minerals and thermal brines, Reykjanes, Iceland. *Contributions to Mineralogy and Petrology*, **36**, 123–34.

Touret, J., 1971a. Le facies granulite en Norvège méridionale I: les associations mineralogiques. *Lithos*, **4**, 239–49.

Touret, J., 1971b. Le facies granulite en Norvège méridionale II: les inclusions fluides. *Lithos*, **4**, 423–36.

Touret, J., 1977. The significance of fluid inclusions in metamorphic rocks. *In* D.G. Fraser (ed) *Thermodynamics in Geology*. Reidel, Dordrecht, pp. 203–27.

Tracy, R.J., 1978. High grade metamorphic reactions and partial melting in pelitic schist, west-central Massachusetts. *American Journal of Science*, **278**, 150–78.

Tracy, R.J. & Robinson, P., 1983. Acadian migmatite types in pelitic rocks of Central Massachusetts. *In* M.P. Atherton & C.D. Gribble (eds) *Migmatites, Melting and Metamorphism*. Shiva, Nantwich, pp. 163–73.

Tracy, R.J., Rye, D.M., Hewitt, D.A. & Schiffries, C.M., 1983. Petrologic and stable-isotopic studies of fluid-rock interactions south-central Connecticut: I. The role of infiltrations in producing reaction assemblages in impure marbles. *American Journal of Science*, **283A**, 589–616.

Trommsdorf, V., 1966. Progressive metamorphose kieseliger karbonatgesteine in den Zentralalpen zwischen Bernina und Simplon. *Schweizerische Mineralogisches und Petrographisches Mitteilungen*, **46**, 431–60.

Trommsdorf, V., 1972. Change in *T–X* during metamorphism of siliceous dolomitic rocks of the central Alps. *Schweizerische Mineralogisches und Petrographisches Mitteilungen*, **52**, 567–71.

Trommsdorf, V. & Evans, B.W., 1972. Progressive metamorphism of antigorite schist in the Bergell tonalite aureole (Italy). *American Journal of Science*, **272**, 423–37.

Trommsdorf, V., Skippen, G.B. & Ulmer, P., 1985. Halite and sylvite as solid inclusions in high-grade metamorphic rocks. *Contributions to Mineralogy and Petrology*, **89**, 24–9.

Turner, F.J., 1981. *Metamorphic Petrology – Mineralogical, Field and Tectonic Aspects*. 2nd edn. McGraw-Hill, New York.

Vallance, T.G., 1965. On the chemistry of pillow lavas and the origin of spilites. *Mineralogical Magazine*, **34**, 471–82.

Vallance, T.G., 1967. Mafic rock alteration and the isochemical development of some cordierite–anthophyllite rocks. *Journal of Petrology*, **8**, 84–96.

Valley, J.W. & O'Neill, J.R., 1978. Fluid heterogeneity during granulite facies metamorphism in the Adirondacks: Stable isotope evidence. *Contributions to Mineralogy and Petrology*, **85**, 158–73.

Vielzeuf, D., 1980. Orthopyroxene and cordierite secondary assemblages in the granulitic paragneisses from Lherz and Saleix (French Pyrénées). *Bulletin de Mineralogie*, **103**, 66–78.

Voll, G., 1960. New work on petrofabrics. *Liverpool and Manchester Geological Journal*, **2**, 503–67.

Walcott, R.I., 1987. Geodetic strain and the deformational history of the North Island of New Zealand during the late Cainozoic. *Philosophical Transactions of the Royal Society, London*, **A321**, 163–82.

Walker, K.R., 1969. A mineralogical, petrological and geochemical investigation of the Palisades Sill, New Jersey. Geological Society of America Memoir 115, pp. 175–87.

Walther, J.V. & Orville, P.M., 1982. Rates of metamorphism and volatile production and transport in regional metamorphism. *Contributions to Mineralogy and Petrology*, **79**, 252–7.

Warburton, J., 1986. The ophiolite-bearing schistes lustres nappe in alpine Corsica: a model for the emplacement of ophiolites that have suffered HP/LT metamorphism. Geological Society of America Memoir 164, pp. 313–31.

Weaver, B.L. & Tarney, J., 1983. Elemental depletion in Archaean granulite facies rocks. *In* M.P. Atherton & C.D. Gribble (eds) *Migmatites, Melting and Metamorphism*. Shiva, Nantwich, pp. 250–63.

Weaver, C.E., 1984. Shale-slate metamorphism in southern Appalachians. *Developments in Petrology*, **10**, Elsevier, Amsterdam.

Weber, K. & Behr, H.–J., 1983. Geodynamic interpretation of the mid-European Variscides. *In* H. Martin & F.W. Eder (eds) *Intracontinental Fold Belts*. Springer-Verlag, Berlin, pp. 427–69.

Wells, P.R.A., 1977. Pyroxene thermometry in simple and complex systems. *Contributions to Mineralogy and Petrology*, **62**, 129–39.

White, J.C. & White, S.H., 1981. On the structure of grain boundaries in tectonites. *Tectonophysics*, **78**, 613–28.

White, S.H. & Knipe, R.J., 1978. Microstructure and cleavage development in selected slates. *Contributions to Mineralogy and Petrology*, **66**, 165–74.

Wickham, S.M. & Oxburgh, E.R., 1985. Continental rifts as a setting for regional metamorphism. *Nature*, **318**, 330–3.

Wickham, S.M. & Oxburgh, E.R., 1987. Low-pressure regional metamorphism in the Pyrénées and its implications for the thermal evolution of rifted continental crust. *Philosophical Transactions of the Royal Society, London*, **A321**, 219–41.

Williams, H. & Smyth, W.R., 1973. Metamorphic aureoles beneath ophiolite suites and alpine peridotites: tectonic implications with west Newfoundland examples. *American Journal of Science*, **273**, 594–621.

Winkler, H.G.F., 1976. *Petrogenesis of Metamorphic Rocks* 4th edn. Springer-Verlag, New York.

Wiseman, J.D.H., 1934. The central and south-west Highland epidiorites: a study in progressive metamorphism. *Quarterly Journal of the Geological Society, London*, **90**, 354–417.

Wood, B.J. & Banno, S., 1973. Garnet–orthopyroxene and orthopyroxene–clinopyroxene relationships in simple and complex systems. *Contributions to Mineralogy and Petrology*, **42**, 109–24.

Wood, B.J. & Walther, J.V., 1984. Rates of hydrothermal reactions. *Science*, **222**, 413–15.

Wright, T.O. & Platt, L.B., 1982. Pressure dissolution and cleavage in the Martinsburg shale. *American Journal of Science*, **282**, 122–35.

Wyllie, P.J., 1962. The effect of 'impure' pore fluids on metamorphic dissociation reactions. *Mineralogical Magazine*, **33**, 9–25.

Yardley, B.W.D., 1977a. An empirical study of diffusion in garnet. *American Mineralogist*, **62**, 793–800.

Yardley, B.W.D., 1977b. The nature and significance of the mechanism of sillimanite growth in the Connemara Schists, Ireland. *Contributions to Mineralogy and Petrology*, **65**, 53–8.

Yardley, B.W.D., 1981a. Effect of cooling on the water content and mechanical behaviour of metamorphosed rocks. *Geology*, **9**, 405–8.

Yardley, B.W.D., 1981b. A note on the composition and stability of Fe-staurolite. *Nueus Jahrbuch für Mineralogie Monatschafte Jg 1981*, 127–32.

Yardley, B.W.D., 1982. The early metamorphic history of the Haast Schists and related rocks of New Zealand. *Contributions to Mineralogy and Petrology*, **81**, 317–27.

Yardley, B.W.D., 1986. Fluid migration and veining in the Connemara Schists, Ireland. *In* J.V. Walther & B.J. Wood (eds) *Advances in Physical Geochemistry*, Vol. 5, Springer-Verlag, New York, pp. 109–31.

Yardley, B.W.D., Barber, J.P. & Gray, J.R., 1987. The metamorphism of the Dalradian rocks of western Ireland and its relation to tectonic setting. *Philosophical Transactions of the Royal Society, London*, **A321**, 243–68.

Yardley, B.W.D., Leake, B.E. & Farrow, C.M., 1980. The metamorphism of Fe-rich pelites from Connemara, Ireland. *Journal of Petrology*, **21**, 365–99.

Yoder, H.S. & Tilley, C.E., 1962. Origin of basalt magmas: an experimental study of natural and synthetic rock systems. *Journal of Petrology*, **3**, 342–52.

Zen, E-an., 1961. The zeolite facies: an interpretation. *American Journal of Science*, **259**, 401–9.

Zen, E-an., 1966. Construction of pressure–temperature diagrams for multi-component systems after the method of Schreinemakers – a geometrical approach. *US Geological Survey Bulletin 1225*.

Zwart, H.J., 1962. On the determination of polymetamorphic mineral associations, and its application to the Bosost area (central Pyrénées). *Geologisch Rundschau*, **52**, 38–65.

Zwart, H.J., 1967. The duality of orogenic belts. *Geologie en Mijnbouw*, **46**, 283–309.

GLOSSARY OF MINERAL NAMES AND ABBREVIATIONS USED IN THE TEXT

Mineral Name	Abbreviation	Chemical Formula
actinolite	ACT	see amphibole
adularia		see feldspar
aegirine		see pyroxene
albite	AB	see feldspar
almandine	ALM	see garnet
Al-silicate	ALS	Al_2SiO_5
amphibole group:		
clinoamphiboles		
actinolite	ACT	$Ca_2(Mg,Fe^{2+})_5Si_8O_{22}(OH)_2$
barroisite		$CaNa(Mg,Fe^{2+})_3(Al,Fe^{3+})_2AlSi_7O_{22}(OH)_2$
crossite		$Na_2(Mg,Fe^{2+})_3(Al,Fe^{3+})_2Si_8O_{22}(OH)_2$
cummingtonite	CUM	$(Mg,Fe^{2+})_7Si_8O_{22}(OH)_2$
glaucophane	GL	$Na_2(Mg,Fe^{2+})_3Al_2Si_8O_{22}(OH)_2$
hornblende	HBL	$Na_{0-1}Ca_2(Mg,Fe^{2+},Fe^{3+},Al)_5Al_{2-1}Si_{6-7}O_{22}$ $(OH)_2$
tremolite	TR	$Ca_2Mg_5Si_8O_{22}(OH)_2$
orthoamphiboles		
anthophyllite	ANT	$(Mg,Fe^{2+})_7Si_8O_{22}(OH)_2$
gedrite	GED	$(Mg,Fe^{2+})_5Al_4Si_6O_{22}(OH)_2$
analcite	AC	see zeolite
anatase		TiO_2
andalusite	AND	Al_2SiO_5
andradite		see garnet
ankerite		$Ca(Mg,Fe,Mn)(CO_3)_2$
anorthite	AN	see feldspar
anthophyllite	ANT	see amphibole
antigorite	ANG	see serpentine
apatite		$Ca_5(PO_4)_3(OH,F,Cl)$
aragonite	AG	$CaCO_3$
augite		see pyroxene
barroisite		see amphibole
biotite	BIO	see mica
calcite	CTE	$CaCO_3$
carpholite	CRP	$(Mn,Mg,Fe^{2+})(Al,Fe^{3+})_2Si_2O_6(OH)_4$
celadonite		see clay minerals

Mineral Name	Abbreviation	Chemical Formula
chlorite	CHL	$(Mg,Fe^{2+},Al)_{12}(Si,Al)_8O_{20}(OH)_{16}$
chloritoid	CTD	$(Fe^{2+},Mg)_2Al_4Si_2O_{10}(OH)_4$
chrysotile		see serpentine
clay mineral group:		
celadonite		$K_2Al_2(Mg, Fe^{2+})_2Si_8O_{20}(OH)_4$
illite		$K_{1-5}Al_{5-5.5}Si_{7-6.5}O_{20}(OH)_4$
kaolinite		$Al_4Si_4O_{10}(OH)_8$
montmorillonite (smectite)		$(\frac{1}{2}Ca,Na)_{0.7}(Al,Mg,Fe)_4(Si,Al)_8O_{20}(OH)_4$ $.nH_2O$
nontronite		$(\frac{1}{2}Ca,Na)_{0.7}Fe^{3+}{}_4(Si,Al)_8O_{20}(OH)_4.nH_2O$
clinopyroxene	CPX	see pyroxene
clinozoisite		see epidote
coesite		SiO_2
cordierite	CD	$(Mg,Fe^{2+})_2Al_4Si_5O_{18}$
corundum	COR	Al_2O_3
crossite		see amphibole
cummingtonite	CUM	see amphibole
diopside	DI	see pyroxene
dolomite	DO	$CaMg(CO_3)_2$
enstatite	EN	see pyroxene
epidote group:		
clinozoisite	CZ	$Ca_2Al_3Si_3O_{12}(OH)$
epidote	EP	$Ca_2Fe^{3+}Al_2Si_3O_{12}(OH)$
fayalite		see olivine
feldspar group:		
adularia		$(Na,K)AlSi_3O_8$
albite	AB	$NaAlSi_3O_8$
anorthite	AN	$CaAl_2Si_2O_8$
K-feldspar	KF	(no specific structural state) $KAlSi_3O_8$
orthoclase	OR	$KAlSi_3O_8$
plagioclase	PL	$(Ca,Na)Al(Al,Si)Si_2O_8$
forsterite	FO	see olivine
garnet group:	GT	
almandine	ALM	$Fe_3Al_2Si_3O_{12}$
andradite		$Ca_3Fe^{3+}{}_2Si_3O_{12}$
grossular	GR	$Ca_3Al_2Si_3O_{12}$
pyrope	PP	$Mg_3Al_2Si_3O_{12}$
spessartine		$Mn_3Al_2Si_3O_{12}$
gedrite	GED	see amphibole
glaucophane	GL	see amphibole
graphite		C
grossular	GR	see garnet
hematite		Fe_2O_3
heulandite		see zeolite
hornblende	HBL	see amphibole
hypersthene		see pyroxene
illite		see clay minerals
ilmenite	ILM	$FeTiO_3$
jadeite	JD	see pyroxene
kaolinite		see clay minerals
K-feldspar	KF	see feldspar

Mineral Name	Abbreviation	Chemical Formula
kyanite	KY	Al_2SiO_5
larnite		Ca_2SiO_4
laumontite	LM	see zeolite
lawsonite	LAW	$CaAl_2Si_2O_7(OH)_2.H_2O$
magnetite		Fe_3O_4
margarite		see mica
mesolite		see zeolite
mica group:		
biotite	BIO	$K_2(Mg,Fe^{2+},Fe^{3+}Al)_6\ Al_{2-3}Si_{6-5}O_{20}(OH)_4$
margarite		$Ca_2Al_8Si_4O_{20}(OH)_4$
muscovite	MS	$K_2Al_6Si_6O_{20}(OH)_4$
paragonite	PG	$Na_2Al_6Si_6O_{20}(OH)_4$
phengite		$K_2(Mg,Fe^{2+})_{0-2}Al_{6-4}Si_6O_{20}(OH)_4$
phlogopite		$K_2Mg_6Al_2Si_6O_{20}(OH)_4$
sericite		(fine phengite $\pm$ paragonite $\pm$ other sheet silicates)
mordenite		see zeolite
montmorillonite		see clay minerals
muscovite	MS	see mica
nontronite		see clay minerals
olivine group:	OL	
fayalite		Fe_2SiO_4
forsterite	FO	Mg_2SiO_4
omphacite	OM	see pyroxene
orthopyroxene	OPX	see pyroxene
paragonite	PG	see mica
phengite		see mica
phlogopite		see mica
plagioclase	PL	see feldspar
prehnite	PR(PREH)	$Ca_2Al_2Si_3O_{10}(OH)_2$
pumpellyite	PU(PUMP)	$Ca_4(Mg,Fe^{2+})(Al,Fe^{3+})_5Si_6O_{23}(OH)_3.2H_2O$
pyrite		FeS_2
pyrope	PP	see garnet
pyrophyllite	PYP	$Al_4Si_8O_{20}(OH)_4$
pyroxene group:		
clinopyroxenes	CPX	
aegirine		$NaFeSi_2O_6$
augite		$(Na,Ca,Fe,Mg,Al)_2(Al,Si)_2O_6$
diopside	DI	$CaMgSi_2O_6$
jadeite	JD	$NaAlSi_2O_6$
omphacite	OM	$(Ca,Na)(Mg,Al)Si_2O_6$
orthopyroxenes	OPX	
enstatite	EN	$Mg_2Si_2O_6$
hypersthene		$Mg_{1-1.4}Fe^{2+}_{1-0.6}Si_2O_6$
pyrrhotine		$Fe_{(1-x)}S$
quartz	QZ	SiO_2
rutile	RT	TiO_2
sapphirine	SA	$(Mg,Fe)_2Al_4SiO_{10}$
scolecite		see zeolite
sericite		see mica
serpentine group:	SP	
antigorite	ANG	$Mg_3Si_2O_5(OH)_4$

Mineral Name	Abbreviation	Chemical Formula
chrysotile		$Mg_3Si_2O_5(OH)_4$
sillimanite	SIL	Al_2SiO_5
spessartine		see garnet
sphene		$CaTiSiO_4(O,OH)$
spinel	SL	$MgAl_2O_4$
spurrite		$Ca_4Si_2O_8.CaCO_3$
staurolite	ST	$(Fe^{2+},Mg)_4Al_{18}Si_{7.5}O_{44}(OH)_4$
stilbite		see zeolite
stilpnomelane		$(K,Ca)_{0-1.4}(Fe,Mg,Al)_{5.9-8.2}Si_8O_{20}(OH)_4$ $(O,OH,H_2O)_{3.6-8.5}$
talc	TC	$Mg_6Si_8O_{20}(OH)_4$
thomsonite		see zeolite
tourmaline		$Na(Fe^{2+},Mg)_3Al_6B_3Si_6O_{27}(OH,F)_4$
tremolite	TR	see amphibole
vesuvianite (idocrase)		$Ca_{10}(Mg,Fe^{2+})_2Al_4Si_9O_{34}(OH,F)_4$
wairakite		see zeolite
wollastonite	WO	$CaSiO_3$
zeolite group:		
analcite	AC	$NaAlSi_2O_6.H_2O$
heulandite		$(Ca,Na_2)Al_2Si_7O_{18}.6H_2O$
laumonite	LM	$CaAl_2Si_4O_{12}.4H_2O$
mesolite		$Na_2Ca_2Al_6Si_9O_{30}.8H_2O$
mordenite		$(Na_2,K_2,Ca)Al_2Si_{10}O_{24}.7H_2O$
scolecite		$CaAl_2Si_3O_{10}.3H_2O$
stilbite		$(Ca,Na_2K_2)Al_2Si_7O_{18}.7H_2O$
thomsonite		$NaCa_2Al_5Si_5O_{20}.6H_2O$
wairakite		$CaAl_2Si_4O_{12}.2H_2O$
zircon		$ZrSiO_4$
zoisite	ZO	$Ca_2Al_3Si_3O_{12}(OH)$

INDEX

In the case of commonly used terms, such as mineral names, only the more important references are given.

Page numbers in **bold** type refer to figures, those in *italics* to tables.

Abukuma plateau, **189**, 200
accessory minerals, 8, *64*, 65, 127, 143
   in relation to phase rule, 33–34
accretionary prism/wedge, 191, 194
acicular, 161
actinolite, 92, *94*, 98–101, 105, 121
activity, 55, 116
adsorbtion, 148
AFM projection, 60–63, **61**, **66**, **69**, 71, **81**, **83**
age dating, 177, 180–181, 208, 214
albite, 8, 53, 54, *94*, 100–101, 116, 120–121, 190
albite–epidote hornfels facies, **50**, 51, *94*
alkali feldspar, 52–53
alkali feldspar geothermometer, **57**, 58
Alpine metamorphism, *see* Lepontine metamorphism
Alpinotype belts, 188
Al-silicate polymorphs, 11, **33**, 35–36, 48, 78–79, 85, **86**, **87**
   see also under individual names
AMF projection, **123**, 124
amphibole, 91–93, **118**, 143–145, 163
   miscibility gaps, 92
   see also under individual names
amphibole zone, 144, **144**
amphibolite, 26, 91
amphibolite facies, **50**, 51, *94*, 100, 117, 121
anatexis, 74, 120
anchizone, 4, 63
andalusite, 10–11, 78–82, 85, **86**, **87**, *94*, 191
andalusite zone, 10, 80–81, **81**
anisotropic growth, 161
ankerite, 143
ankerite zone, 143, **144**
anorthite, 55, 89, *94*, 101, **135**

anthophyllite, 43–44, **45**, **46**, **47**, 122–124, **123**
antigorite, 43–46, **45**, **46**, **47**, 54
aragonite, 102, **127**, 182–183
armoured relics, *see* relic inclusions, inclusions
Arrhenius plot, 182, **183**
augen gneiss, **24**

back-arc spreading, **14**, 192, 197
baric type, 188, *189*
   changes with time, 200–201
barroisite, 93, 190
Barrovian zones, 9, 64–71, 79, 99–100, 143
Barrow, G.M., 4, 7
Barth, T.F.W., 58
bathozones, 88, **89**, 191
Bay of Islands, Newfoundland, 213–214
Bergell aureole, Italy, 42–46, **43**, 53, **54**, **209**
biotite, 8, 10, **61**, 62, 64–65, **66**, **69**, **70**, 78, 82, *94*, 143, 175
biotite zone, 8, *64*, 64–65, **66**, 80, 99, 143, **144**
black smokers, 121
blastomylonite, 158
blocking temperature, 56, 119
blueschist, 26, 102–106, 108, 113–116, **115**
   preservation of, 202–204, 215
   see also glaucophane schist
blueschist facies, **50**, 51, *95*, 102–106, 113–116
boiling, 18–20, **20**, 38, **39**
Bowen, N.L., 52
Broken Hill, Australia, 109, 119
buffering, 37
   external, 134, **134**, 143
   internal, 134, 137–139, **138**
burial metamorphism, 12, 95

calcite, 98–99, 127–129, **127**, 182–184
calcite–dolomite geothermometer, 58, 141
calc-silicate, 21, 126, 141–145, 176
Ca-poor amphiboles, 120, 122
　　see also under individual names
carbon dioxide, *see* CO₂
carpholite, 82–83, **83**, *95*
Cascade arc, 193, **193**
cataclasis, 13, 158
cataclastic metamorphism, *see* dynamic
　　metamorphism, mylonite
catazone, 4
cation exchange reaction, 39–40, 55–56,
　　88–90
Central Alps, 130–133, **131, 132, 138,**
　　139, 208–211, **209**
charnockite, 27, 110
chemical equilibrium, 29–30, 34–37, 40,
　　42–49, 147
chemical potential, 35, 124, 151
chlorite, 9, 64–68, **66, 69, 70,** 80–81, *94–*
　　95, 100–101, 120–121, 210
chlorite zone, 8, 64, *64,* 99
chloritoid, *64,* 67–68, **69,** 82–83, **83,** *94–*
　　95, 164
clay minerals, 63, *94,* 98
cleavage, 12, 21, 168–170
clinoenstatite, 17
clinopyroxene, 93, *94–95,* 109–110
　　see also under individual names
clinozoisite, 93, 109
　　see also epidote
closed system, 124
CO₂, 119–120
coesite, 13, 117, 210
compatible assemblages, 33, **33, 220**
component, 30, 32–35, 38–40, 124–125,
　　217–221
　　inert, 124–125
　　perfectly mobile, 124–125
Connemara, Ireland, **26,** 71, **72, 73,** 79,
　　173, 177, 205, **206**
contact metamorphism, 1, 10, 12, 42–46,
　　128, 178–180
continental collision, 207–208, 212–213
continuous reactions, 38–39, 67, **70,** 71–
　　72, 88–90, 164
convergent margin, 194–197
Coombs, D.S., 51, 95
cordierite, 76–78, **79,** 80–81, 90, *94–95,*
　　191, 205
cordierite–anthophyllite rocks, 120, 122–
　　124
cordierite–garnet geobarometer, 90

cordierite zone, 10, 80, **81**
corona texture, **111,** 111–112, 164
corundum, *94*
country rock, 12
creep, 167
crenulation cleavage, 169, **169**
critical point, 19
critical pressure, 20
crossite–epidote schist, 106, 116
crustal stretching, 15, 205, 207
crystal defects, 148–150, **149**
crystallisation, 158–161
cummingtonite, *94,* 102
Cyclades, Greece, 105, 108, **115,** 116

Dalradian supergroup, 7, 10
decussate texture, 155, **156**
deformation, 167–168
deformation lamellae, 168
deformation–metamorphism interactions,
　　175–176
deformation textures, 168–170
degenerate reactions, 218–219
degrees of freedom, 30–31, 38–40, 65,
　　125, 133
dependent variable, 31, 37
depth zones, 4
deviatoric stress, 17, **19,** 176
devolitilisation reactions, 41, 135–136, **135**
diagenesis, 63
diffusion, 40, 48, 150–151, **153,** 160, 162–
　　163, 168, 175, 184
diopside, 130–133, **131, 132,** 145
diopside zone, **144,** 145
discontinuous reactions, 38, **70,** 71–72
disequilibrium textures, 41, 92, 161–164
dislocation flow, **167,** 167–168
dislocations, 148–149, **149,** 155
distribution coefficient, 40, 49, 56, 66, 163
divariant assemblage, 44
divariant field, 32, 217–222
divariant reactions, *see* continuous reactions
dolomite, 130–131, **131, 132**
dolomitic marble, 129–133
duration of metamorphism, 180
dynamic metamorphism, 12

Eastern Mediterranean, 212
eclogite, 27, 106–109, 116–117, 212
eclogite facies, **50,** 51, 93, *95,* 103, 105–
　　109, 116–117, 176
edenite, 101
element partitioning, 49
　　see also distribution coefficient

endothermic reactions, 37, 184
end member, *see* solid solution
enthalpy, 36–37
    see also heat of reaction
entropy, 36, 52, 222
Eoalpine metamorphism, 210–212
epidosite, 120
epidote, 93, *94*, 99–101, 121
epitaxial growth, **159**, 160
epizone, 4, 63
equilibrium, *see* chemical equilibrium
Eskola, P., 4, 49, 50, 141
European Alps, 130–133, **131**, **138**, 139,
    203, 207–212, **209**
exothermic reactions, 37
extensive property, 34

fabrics, 168–171, **173**, 175
facies classification, *see* metamorphic facies
facies series, 188
Fanad aureole, Ireland, 10, **11**
Fe–Mg exchange, 39–40
fibrolite, 85, **87**, **159**, 165–166, **173**
    see also sillimanite
Fick's first law, 151
fluid composition, *see* metamorphic fluid
fluid flow pathways, 145–146
fluid inclusions, 19, 119
fluid overpressures, 176
fluids, *see* metamorphic fluids
foliation, 21
forsterite, 130–133, **131**, **133**
Franciscan terrane, California, 102, 107,
    115, **115**
fugacity, 37, 55

Ganly, P., 4
garnet, 8, 10, **24**, *64*, 65–68, **66**, **69**, 70,
    76–80, 93, *94–95*, 100, 103, 106–
    107, 142, 163–164, **173**, 174, 175,
    210
    chemical zoning, 151, 161–163, **162**
garnet–biotite geothermometer, 56, **57**
garnet–clinopyroxene geothermometer,
    117
garnet–hornblende geothermometer, 117
garnet–orthopyroxene geobarometer, 58,
    117
garnet zone, 8, *64*, 65–67, **66**, **69**, 99, *141*
geobarometry, 52–55, 58, 89–90, 113–
    119
geotherm, 15, **17**
geothermal field metamorphism, **97**, 98
geothermal gradient, 14, **17**, **20**, 188

geothermometry, 51–58, 88, 113–119
Gibbs free energy, 34–36, **36**, 52, 151–
    153, **152**, 217–218
Gibbs phase rule, *see* phase rule
glaucophane, 92, *95*, 102–106, 190
glaucophane schist, 4
    see also blueschist
gneiss, 21
Goldschmidt, V.M., 4, 141
grade, 8
grain boundary, 48, 150, 153, **154**, 175,
    184
grain boundary sliding, 167
grain growth, 155, 158, 182
granoblastic polygonal texture, 155, **156**
granulite, 27, 109–112, 118–120
granulite facies, **50**, 51, 93, *95*, 109–110,
    117–120
graphite, 42, *84*
greenschist, 26
greenschist facies, **50**, 51, *94*, 99–100,
    105–106, 113, 117, 121
Greenwood, H.J., 128, 136
growth zoning, **162**, 162–163

H_2O, 18–20
Haast schist, New Zealand, **96**, 100, 106,
    180, 200–201, **201**
Hall, Sir J., 4
Harker, A., 4, 17
heat, 13–15, **16**
heat capacity, 14
heat flow, 13–15, **16**, 184, 191–192, 197
heat of reaction, 37, 181, 185
heat supply (rate of), 181, 184–185
helicitic texture, 171
Helvetic zone, 208
Hercynian belt, *see* Variscan belt
Hercynotype belts, 188
high-*P* metamorphism, 82–85, 102–109,
    113–117, *189*, 190, 192, 200, 202–
    204, 211–212
Himalayas, 207, 212
hornblende, 92, *94–95*, 99–102, 109, 142,
    175, 190
hornblende hornfels facies, **50**, 51, *94*
hornfels, 12, **23**, 26, 161
Hutton, J., 1, 4
hydraulic fracturing, 18, 41, 176
hydrothermal fluid, 10
    see also metamorphic fluid
hydrothermal metamorphism, 3, 12, 98, 120–
    125
hypersthene, 93, *95*

idioblastic texture, 27, 155
illite crystallinity, 63
ilmenite, 87–88, 90, 109
ilmenite–garnet geothermometer, 89
impact metamorphism, 13, 28
inclusion fabrics, 171–175, **172**, **173**
inclusions, 8, 27
    see also relic inclusions
independent variable, 31, 37
index mineral, 8
infiltration, 13, 142–145
intensive property, 34
invariant point, 32, **33**, 217–221
ionic reactions, 165–166, **166**
isobaric invariant point, 137, 138
isobaric T–X_{CO_2} diagram, 129, **129**
isobaric univariant curve, 129
isochemical metamorphism, 5, *6*
isograd, 8, **9**, 10, 137, 142–143, 186

jadeite, **53**, 54, 102, 105, 116, 176

K_D, *see* distribution coefficient
K-feldspar, 10, 64, 75–78, *94–95*, 142–143, 145, 191
kinetics, *see* reaction rates
Korzhinsky, D.S., 124
kyanite, 8, 10–11, *64*, 68–69, **69**, 78, 83–84, **83**, 87–88, *94–95*, 108, 165–166
kyanite zone, 8, 68–69, **69**, 99, **141**

larnite, 129
laumontite, *94*, 99, 113
lawsonite, 93, *95*, 102–106, 113, 190
lawsonite–albite facies, 93, 113
Lepontine metamorphism, 130–133, **131**, **132**, 211–212
Lesser Antilles arc, 193, **193**
leucosome, 27, **72**, **73**, 74–75
Lewisian terrane, Scotland, 109, 119
Ligurian Appenines, Italy, 9
limestone, 5, 126
lineation, 21, 168
line defects, *see* dislocations
low grade metamorphism, 63, 194, 205
low-P metamorphism, 78–82, 101–102, *189*, 191, 200, 204–207

magnetite, 13, *94*
Maine, 79
Makran trench, 193, **193**
mantle, 15, 107
marble, 4–5, 21, **26**, 126–141
margarite, 142

Massachusetts, 76, **77**
melanosome, 74
melt, 76, 78, **79**
melting, 26, 79, 119–120
    see also anatexis
mesozone, 4
metabasite, *6*, 9, 10, 21, 91–124
metamorphic aureole, 10, 42–46, 79
metamorphic belt, 187
metamorphic facies, 4, 49–51, 92–120, *94–95*
metamorphic field gradient, 199
metamorphic fluid, 5, 18–21, 28, 41, 99, 127–129, 133–139, 143, 145–146, 182, 184–185
    composition, 18, 41, 128–129, 133–135
    eclogite formation, 108–109
    immiscibility, 146
    granulite formation, 119–120
metamorphic geotherm, 199–201, **199**, **201**
metamorphic reactions, 5, 30, 38–40, 43–46, 135–136
metamorphic segregation, 170
metamorphic sole, 213–214
metamorphic terrane, 192
metamorphic textures, 21–27, 147, 154–175, 205
    deformation, 155–158
    reaction mechanism, 164–166
    timing, 170–175
metamorphic zones, 7–11, 43, *141*, 200
metasomatism, 5, 10, 12, 121, 124, 145, 175
    see also open systems
metastable assemblage, 38
metastable extension, 218–221
Meteor Crater, Arizona, 13
migmatite, **25**, 27, **72**, **73**, 74–75, 78, 205, **206**
mineral assemblage, 5, 8
mineral paragenesis, 8
mobile belt, 187
moderate-P metamorphism, 188, **189**
    see also Barrovian zones
molar volume, 18–19, **20**, 35–36, 52
Morey–Schreinemakers rule, 218–219
mortar texture, 158
multisystem, 217
muscovite, 8, 52–53, **53**, **61**, 62–70, 75–76, *94*, 165–166, 175
muscovite–paragonite geothermometer, 89
mylonite, 12, 22, **25**, 158

Na-amphibole, 102
    see also glaucophane

nebulite, **73**, 74
New Zealand, recent plate tectonics, 194–197, **196**
nucleus, 152–153, 158
nucleation, 151–153, 158, 160, 166, 175, 181
nucleation–growth theory, 151–154

olivine, *94–95*, 110
  see also forsterite
oligoclase, 67, 100
oligoclase zone, 100
omphacite, *95*, 103, 190
open systems, 124–125
  see also metasomatism
ophiolite, 9, 213–214
  emplacement, 213
orogenic belts, 13
  see also metamorphic belts
orogeny, 187
orthoamphibole, 78, 93
  see also under individual names
orthopyroxene, 78, 93, *94–95*, 109–110, 175
  see also under individual names
Oslo region, Norway, 4
Ouegoa district, New Caledonia, 84–85, **84**, 103, **104**
overgrowths, 158, **159**
overprinting, 211, 214
  see also superimposed metamorphism
overstepping, 38, 133, **134**, 151–153, **152**, 184–185
oxidation, 80, 105
oxygen fugacity, 37, 42
oxygen isotope geothermometry, 56–57, 113, 115

paired metamorphic belts, 188–192, 207
palaeosome, 74
paragonite, 52–53, **53**, 63, *94*, 103–105, 107
P-domain, **169**, 170
pelite, 6, 7–11, 21, 60–90
Penninic zone, 208–211
Perchuk, L.L., 56
peristerite gap, 67, 100
permeability, 146, 176
petrogenetic grid, 52, **54**, 85, **86**, **89**, **114**, 139–141
phase, 29, 32–35, 38–40, 217–219
phase change, *see* metamorphic reaction
phase diagrams, 32–33, 43–46, **45**, **46**, **47**, 60–63, **61**

phase rule, 30–32, 38–40, 48, 124–125, 133, 217–219
phengite, 63–65, 82–84, *95*
phlogopite, 142
phyllite, 21
phyllonite, 158
pillow lavas, 2, 9, 107
plagioclase, *94–95*, 109–110, 142, 175
  composition, 67, 93, 100
plagioclase–garnet geobarometer, 55, 89
plagioclase–hornblende geothermometer, 117
plagioclase projection, *see* AMF projection
planar fabric, 21
plate tectonic setting, 191–197, 204–208
poikiloblast, 27, 160–161, 170–175
polymorphic transition, 69–70, 165, 182–183, 185
point defect, 148
pore fluid, *see* metamorphic fluid
porosity, 41, 146, 176
porphyroblast, 27, 154, 160, 170–175, **172**, **173**, **174**
porphyroclast, **157**, 158
preferred orientation, 21, 168–169
prehnite, 93, *94*, 96–99
prehnite–pumpellyite facies, **50**, 51, 93, *94*, 97–98, 113, 121, 205
pressure, 15–18, 35–36
  confining, 15
  effective, 18, **19**
  fluid, 18, **19**, 41, 128, 176
  lithostatic, 15, 17, **19**, 41, 128
pressure shadow, 171, **172**
pressure solution, **167**, 168
progressive metamorphism, 8, 164, 199
projection, 44–46, **46**, **47**, 60–63, **61**, **123**, 124
protomylonite, 158
psammite, 21
pseudomorph, 27, 48, 105
pseudotachylite, 26
P-T diagram, 15, **20**, 32, **33**, 217
P-T-t path, 198–201, **199**, **201**, 203, 205
pumpellyite, **2**, 93, *94*, 96–99, 102, 113, 190
pyrophyllite, 63, 83, **83**
pyroxene, 93, 109
  see also under individual names
pyroxene hornfels facies, **50**, 51, *94*

Q-domain, **169**, 170
quartz, 8, 13, 62, 78, *95*, 107, 110, 121, 127–129, **127**, **129**

quartz veins, *see* veins
quartzite, 21

Ramberg, H., 128
rate constant, 182
rate limiting step, 181, 184
rate of metamorphism, 177–185
    see also reaction rate
reaction mechanism, 164–166, **166**
reaction rate, 40–41, 181–186
reaction rims, 164, 184–185
    see also corona texture, overgrowths
reaction textures, 49, 111–112
recovery, 155
recrystallisation, 5, 154–158, **154**, **156**, **157**
redox reactions, 42
reduction, 42
reflectance (of organic matter), 63
regional metamorphism, 1, 9, 12, 180
relic igneous minerals, 96, **97**, 107
relic inclusions, 27, 48, 107, **156**, 163–164, 200, 210
    see also inclusions
restite, 74
retrograde metamorphism, 10, 47–48
retrograde minerals, 48
retrograde zoning, 163
Reykjanes geothermal field, Iceland, **97**, 98
Robinson, P., 122
rock strength, 15
rutile, 87–88, 90, 107
Ryoke belt, Japan, 188, **189**, 191, 200

Sanbagawa belt, Japan, 84, **84**, 102, 106, 188, **189**, 190, 200
sanidinite facies, *50*, **94**
sapphirine, 78, *95*
schist, 8, 21, **24**
schistosity, 12, 21, 168, 171
Schreinemakers bundle, 217
Scottish Highlands, 1, 7, **7**, 11, *64*, 64–70, 80–82, 99–100, 141
sea floor metamorphism, 10, 28, 120–122
semi-pelite, 21
sericite, 48, 63, 102
serpentinisation, 42
serpentinite, 21, 27
Sesia–Lanzo zone, Italy, 105, **115**, **209**
sillimanite, 8, 10, 11, *64*, 69–70, 75–78, 82, 85, **86**, **87**, *94–95*, 165–166, 191
sillimanite zone, 8, 10, *64*, 69–71, **69**, 79, 82, 99, 141
    upper, 75–76, **77**, 82, 186

skarn, 126
slate, 5, 7, 21, **23**, 64, 170
slip plane, 148, **149**
snowball texture, 171, **172**, **173**, 174
solid solution, 29, 31–32, 34, 39–40, **53**, 52–55, 91–93, 125, 161–163
Sorby, H.C., 168
spilite, 120
spotted slate, 80
spurrite, 129
stacking fault, 150
standard state, 55
staurolite, **2**, 8, *64*, 68–72, **94**, **156**, 163–164
staurolite zone, *64*, 67–68, **69**, 99, *141*
stilpnomelane, 64, **94**, 100, 102
stress differences, *see* deviatoric stress
stress minerals, 4, 17, 175
strain, 149, 155
strain energy, 155, 158, 175
stromatic migmatite, **72**, 74, 77
subcretion, *see* underplating
subduction, **14**, 191–195
sub-grains, 150, 155
sulphidation, 80
supercritical fluid, 19, 127
superimposed metamorphism, 13, 201, 212
    see also overprinting
superplastic flow, 176
surface energy, 148, 152, 155, 158
surface imperfections, 149–150
surface steps (kinetic sense), 181, 184–186
symplectite, 164
syntectonic recrystallisation, 155, **157**, 158, 168
system, 29
    CaO–MgO–SiO$_2$–H$_2$O, 46, **47**
    CaO–MgO–SiO$_2$–CO$_2$–H$_2$O, 130–133, **132**
    CaO–Na$_2$O–MgO–FeO–Al$_2$O$_3$–SiO$_2$–H$_2$O, 122–124, **123**
    H$_2$O–NaCl, 38, 39
    K$_2$O–MgO–FeO–Al$_2$O$_3$–SiO$_2$–H$_2$O, 60–72
    MgO–SiO$_2$–H$_2$O, 42–44, **45**, **46**, 218
    Na$_2$O–K$_2$O–Al$_2$O$_3$–SiO$_2$–H$_2$O, 31, 52, **53**

talc, 43–44, **45**, **46**, **47**, 83, **83**, *95*, 108, 130–131, **131**, **132**, 139–140, 210
Taringatura Hills, New Zealand, 95–96, **96**, **97**
Tauern Window, Austria, 108, **209**, 211

tectonic overpressures, 17, 194
tectonic setting, *see* plate tectonic setting
temperature, 13–15, 36, **36**, 183
terminology, 21–27, 155, 166–169
textures, *see* metamorphic textures
thermal aureole, *see* metamorphic aureole
thermal diffusivity, 178
thermal metamorphism, *see* contact
     metamorphism
thermal modelling, 178–180, 197–200,
     211
thermodynamic equilibrium, *see* chemical
     equilibrium
Thompson, J.B. Jr., 122
tie-line, 32, 44, 65–67, 130–133
tie-line flip, 44
Tilley, C.E., 129, 141
topotaxy, 158, 182
transport steps, 181, 184
tremolite, 43, **47**, 130–133, **131**, **132**
trench, **14**, 193
tschermakite, 101–102
two-pyroxene geothermometer, 58
T–X diagrams, **39**, **70**, 71–72, 129

ultrabasic rocks, 42–46
ultramylonite, 158
underplating, 193–195, **196**, **204**
underthrusting, 191
     see also subduction
univariant curve, 32, **33**, 38, 52–53, 217–
     221
univariant reactions, *see* discontinuous
     reactions
uplift, 15, 182, 202–204

Vancouver Island, Canada, 194
Variscan belt, 207–208, 210
Vassalboro Formation, Maine, 143–145,
     144
veins, 176, **177**
vein-type migmatite, 74, **77**
vesuvianite, 142
volcanic arc, **14**, 204–205
volcanogenic sediments, 91, 95, 205
volume changes, 222
von Rosenbusch, 4

water, *see* H_2O
water–rock ratio, 121–122
Waterville Formation, Maine, 143–145,
     144
White Hills Peridotite, Newfoundland,
     213–214, *214*
whiteschist, 108
Willyama Complex, Australia, 109–110,
     118–119
Winkler, H.G.F., 139
wollastonite, 127, **127**, 133–137, **134**, **135**,
     183–184

X_{CO_2}, 128–129, **129**, 134–136, 143, 217
X_{Mg}, 39, 40, 65, 68, **70**, 71, 80, 92

zeolite facies, *94*, 95–97, **96**, **97**, *121*
zeolite minerals, 93, *94*, 95–98, **97**, *121*
Zermatt–Saas zone, 105, 107, **203**, **209**
zoisite, 93, 103, 105, 126, **135**, 136, 142,
     145, *189*
zoisite zone, 144–145, **144**
zones, *see* metamorphic zones